SATELLITE MONITORING OF THE EARTH

WILEY SERIES IN REMOTE SENSING

Jin Au Kong, Editor

Tsang, Kong, Shin • THEORY OF MICROWAVE REMOTE SENSING

Hord • REMOTE SENSING: METHODS AND APPLICATIONS

Elachi • INTRODUCTION TO THE PHYSICS AND TECHNIQUES OF REMOTE SENSING

Szekielda • SATELLITE MONITORING OF THE EARTH

Maffett • TOPICS FOR A STATISTICAL DESCRIPTION OF RADAR CROSS SECTION

SATELLITE MONITORING OF THE EARTH

Karl-Heinz Szekielda
Adjunct Professor
Department of Geology and Geography
Hunter College
City University of New York

A WILEY-INTERSCIENCE PUBLICATION
JOHN WILEY & SONS
New York • Chichester • Brisbane • Toronto • Singapore

Library of Congress Cataloging in Publication Data:

Szekielda, Karl-Heinz.
Satellite monitoring of the earth/Karl-Heinz Szekielda.
p. cm.—(Wiley series in remote sensing)
"A Wiley-Interscience publication."
Bibliography: p.
Includes index.
ISBN 0-471-61330-4
1. Remote sensing—Equipment and supplies. I. Title.
II. Series.
G70.6.S93 1988
621.36′78—dc19 88-21978
CIP

Printed in the United States of America

10 9 8 7 6 5 4 3 2 1

To Aurora and Kristina

CONTENTS

PREFACE

The original outline for this book dates back to 1977 and reflects the course work taught while I was Visiting Full Professor at the University of Hamburg. After extensive revisions, this version includes research results from my own studies and results from many other investigators.

The book is a synopsis on state-of-the-art use of satellite technology for monitoring the earth. It is also an introduction to satellite remote sensing which covers, in a multidisciplinary approach, the different areas in which remote sensing can be applied. Chapter 2 satellite presents, in a chronology, platforms and selected sensors which have orbited the earth; and in this context, Chapter 3 describes atmospheric considerations involved in interactions between energy and atmospheric constituents. A detailed description of the spectral signatures of objects is presented in Chapter 4 as a base for the reader to better understand the interpretation of data referred to in the different chapters. Concepts in data interpretation outlined in Chapter 5 basically lead to the understanding of data as discussed in subsequent chapters. Chapter 6 "Observations Over The Oceans," takes into account phenomena which can be observed with recent satellite technology, and selected case studies are cited to enhance principal data products. Chapter 7 "Observations Over The Continents," includes the identification and recognition of geological features in connection with lithology and structures, magnetic anomalies, and thermal inertia. Observations over water and on vegetation are included in the same chapter.

Several individuals to whom I am grateful made material available and credits are given in appropriate places in the book. I would like to give special thanks to the many personnel who have assisted me at the different NASA research centers as well as those from NASA Headquarters in Washington, D.C.

Needless to say, I wrote this book in my personal capacity and the views expressed here do not necessarily reflect those of the United Nations or any other organization.

KARL-HEINZ SZEKIELDA

New York, New York
November 1988

ACRONYMS

ALT	Radar Altimeter
APT	Automatic Picture Transmission
ATS	Application Technology Satellite
AVCS	Advanced Vidicon Camera System
AVHRR	Advanced Very High Resolution Radiometer
BUV	Backscatter Ultraviolet
CCD	Charged-Couple Devices
CCT	Computer Compatible Tapes
CNES	Centre National d'Etudes Spatiales
CZCS	Coastal Zone Color Scanner
DFVLR	Deutsche Forschungs- und Versuchsanstalt für Luft- und Raumfahrt e.v.
EMS	Electromagnetic Spectrum
ERDAS	Earth Resources Data Analysis Systems
EREP	Earth Resources Experiment Package (Skylab)
ERTS	Earth Resources Technology Satellite
ESMR	Electrically Scanning Microwave Radiometer
ESSA	Environmental Science Service Administration
GAC	Global Area Coverage
GIS	Geographic Information System
GOES	Geostationary Operational Environmental Satellite
GOSSTCOMP	Global Ocean Sea Surface Temperature Computer Program
GSFC	Goddard Space Flight Center
HCMM	Heat Capacity Mapping Mission
HCMR	Heat Capacity Mapping Radiometer
HIRS	High Resolution Infrared Radiation Sounder
HRIR	High Resolution Infrared Radiometer
HRPT	High Resolution Picture Transmission
HRV	Haute Résolution Visible (or High Resolution Visible Range Instruments)

IFOV	Instantaneous Field of View
ITC	Intertropical Convergence
ITOS	Improved Tiros Operational Satellite
LAC	Local Area Coverage
LACIE	Large Area Crop Investigation Experiment
LAGEOS	Laser Geodynamics Satellite
LFC	Large Format Camera
MCSST	Multichannel Sea Surface Temperature
MEH	Mean Estimate Histogram
MOMS	Modular Optoelectronic Multispectral Scanner
MRIR	Medium Resolution Infrared Radiometer
MRSE	Microwave Remote Sensing Experiment
MSS	Multispectral Scanner
MSU	Microwave Sounding Unit
MTF	Modular Transfer Function
NASA	National Aeronautics and Space Administration
NDVI	Normalized Difference Vegetation Indes
NETD	Noise Equivalent Temperature Difference
NOAA	National Oceanic and Atmospheric Administration
NVI	Normalized Vegetation Index
RBV	Return Beam Vidicon
SAR	Synthetic Aperture Radar
SCAMS	Scanning Microwave Spectrometer
SBUV	Solar Backscatter Ultraviolet
SCMR	Surface Composition Mapping Radiometer
SIR	Shuttle Imaging Radar
SMMR	Scanning Multichannel Microwave Radiometer
SMS	Synchronous Meteorological Satellite
SPAS	Shuttle Pallet Satellite
SPOT	Système Probatoire d'Observation de la Terre
SRIR	Scanning Radiometer Infrared
SST	Sea Surface Temperature
SSU	Stratospheric Sound Unit
STZ	Subtropical Convergence Zone
TDRSS	Tracking and Data Relay Satellite System
THIR	Temperature Humidity Infrared Radiometer
TIPS	TM Image Processing System
Tiros	Television Infrared Observation Satellite
TM	Thematic Mapper
TOMS	Total Ozone Mapping Spectrometer
TOVS	Tiros Operational Vertical Sounder
UTM	Universal Transverse Mercator
VHRR	Very High Resolution Radiometer
VI	Vegetation Index
VIRGS	VISSR Image Registration and Gridding System
VISSR	Visible-Infrared Spin-Scan Radiometer
VTPR	Vertical Temperature Profile Radiometer
XBT	Expandable Bathythermograph

1

INTRODUCTION

The detection of objects by remote sensing is based on the processing of energy that is either reflected or emitted due to thermal agitation. The wavelengths of reflected energy used by remote sensing techniques range generally from about 0.4 to 3.0 μm while emitted energy generally predominates above 3 μm. Energy received by a remote sensing device, such as photographic cameras and multispectral scanners, is modified by the atmosphere through which the energy is transmitted into the sensor. Therefore, remote sensing from satellite or aircraft altitudes needs to take into account simultaneously the atmospheric conditions.

For the visible part of the electromagnetic spectrum (EMS), the nature of reflected light is a product of the incident light as well as absorption and reflectance properties of the object. Most natural scenes undergo temporal changes due to the physiology of vegetation and its coverage on the ground. Therefore, it is essential to study the specific spectral reflectance, transmission, and absorption characteristics of targets to be investigated under laboratory and field conditions.

Water poses a complex situation because it is partly transparent in the blue and green wavelengths, which means that reflection occurs not only from the surface, but the reflected energy is reinforced by light reemerging at the surface after scattering throughout the volume of water. Because light penetrates clear water, there may also be a reflection from the bottom in shallow water.

The units of measurement in the EMS are wavelength and frequency: Wavelength is the distance from wave peak, and frequency is the number of waves or cycles passing a fixed point in a given time. The EMS extends from very short-wave, high-frequency gamma rays up to long-wave low-frequency radio communications. In this range, the waves normally travel at the speed of light, and the relationship between frequency and wavelength is given by the general expression $f = c/\lambda$, where f is the frequency, c is the speed of light, and λ is the wavelength.

The electromagnetic energy considered in remote sensing spans a wide spectrum ranging from 10^{-10} to 10^{10} μm, the most used portion of the EMS being in the range of optical wavelengths from 0.3 to 15.0 μm. Radiation originating from the sun and reflected by objects is referred to as reflective energy and covers the spectrum roughly between 0.38 and 3.0 μm, which is subdivided into the visible wavelengths of about 0.38 to 0.72 μm, according to the response of the human eye. The region between 0.72 and 3.0 μm, referred to as the reflective infrared (IR) is subdivided into the near IR, 0.72 to 1.3 μm, and the middle IR, 1.3 to 3.0 μm. The spectral region between 7.0 to 15.0 μm is termed the far IR, or emissive or thermal IR, although reflection of solar radiation still occurs in this wavelength region.

Electromagnetic radiation in empty space propagates without limit, but within an atmospheric environment radiation occurs through a changing medium, especially from high altitudes. The atmospheric transmission in connection with the EMS is shown in Figure 1-1. The interaction of radiation from a target with the medium must be taken into account if the physical nature of a target is to be determined quantitatively. This demonstrates that the sequence of the remote sensing process has to be studied in detail, because some energy or radiation emitted by a source may be lost through interaction with the atmosphere. The interactions between radiation and matter in both medium and target are of the same types. If the ground is the target, atmospheric effects can hinder remote sensing in certain parts of the EMS.

For remote sensing in the visible part of the EMS, the sun is usually the illuminating source for monitoring the earth's surface. That the illumination angles change as a function of latitude, time of day, and time of year should be taken into account. For instance, if we consider an elevated surface facing the sun at right angles to the ground track of the sun, the side facing the sun will be brightly illuminated while the side facing away will be in the shade. In the early morning and late afternoon the sunrays make a small angle with the earth's surface. In consequence, topographic trends, for example, not oriented along the sun's ground track will be subjected to varying degrees of differential solar illumination. Areas located at high latitudes in the north or south, where the incoming sunrays strike the earth perpendicularly at noon, are illuminated at progressively lower sun angles for a particular time of day.

Objects reflect, absorb, and transmit radiation, as summarized in the equation

$$\rho + \alpha + \tau = 1$$

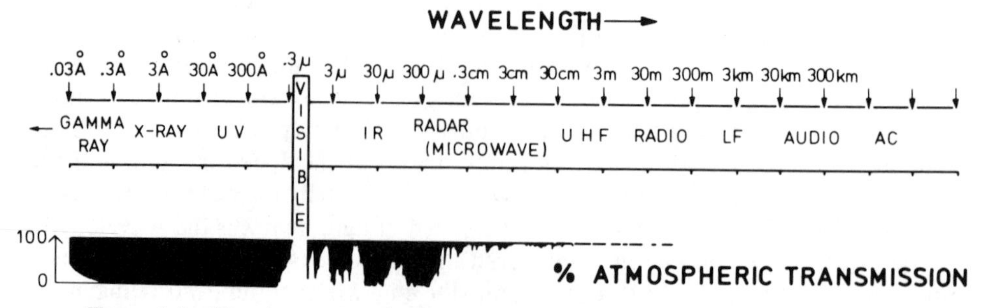

Figure 1-1. The electromagnetic spectrum and generalized atmospheric transmission.

$$\text{where } \rho = \frac{\text{reflected radiation}}{\text{incident radiation}}$$

$$\alpha = \frac{\text{absorbed radiation}}{\text{incident radiation}}$$

$$\tau = \frac{\text{transmitted radiation}}{\text{incident radiation}}$$

If object is opaque, τ will equal 0, and the equation becomes

$$\rho + \alpha = 1 \qquad \text{or} \qquad \alpha = 1 - \rho$$

Assuming that an ideal absorber is an ideal radiator, we obtain

$$\varepsilon = 1 - \rho$$

This equation expresses the Kirchhoff radiation law, which can be applied only in that portion of the spectrum where the materials in question are opaque. Thus, by detecting spectral absorption, reflectance, emission, and/or scattering properties of a surface, we can identify the material constituting the observed target.

All matter above absolute zero is a source of electromagnetic energy, and when matter is in thermal equilibrium with its surroundings, it absorbs radiated energy or emits an equal amount. In this case, the object is said to be a blackbody radiator, whereas those that do not absorb all incident radiation are called graybodies. Planck's distribution law predicts that radiation from a blackbody is distributed according to wavelength. It describes the energy radiated per unit of wavelength per unit volume of the radiator:

$$M_\lambda = \frac{\varepsilon c_1}{\lambda^3(e^{c_2/\lambda T} - 1)}$$

where M_λ = spectral radiant exitance, W/m^2 μm
ε = emittance (emissivity), dimensionless
c_1 = first radiation constant, 3.7413×10^8 W, $(\mu\text{m})^4/\text{m}^2$
λ = radiation wavelength, μm
c_2 = second radiation constant, 1.4388×10^4 μm K
T = absolute radiant temperature, K

If M_λ is differentiated with respect to wavelength, the derivative set equal to 0, and the resulting equation solved for λ_{max}, an expression for the wavelength at which M_λ is a maximum is obtained:

$$\lambda_{max} = \frac{2898}{T}$$

where T is in K.

Planck's law for several temperatures is plotted in Figure 1-2, which shows that if temperature increases, the peak of M_λ moves to shorter wavelengths and the area under the curve increases.

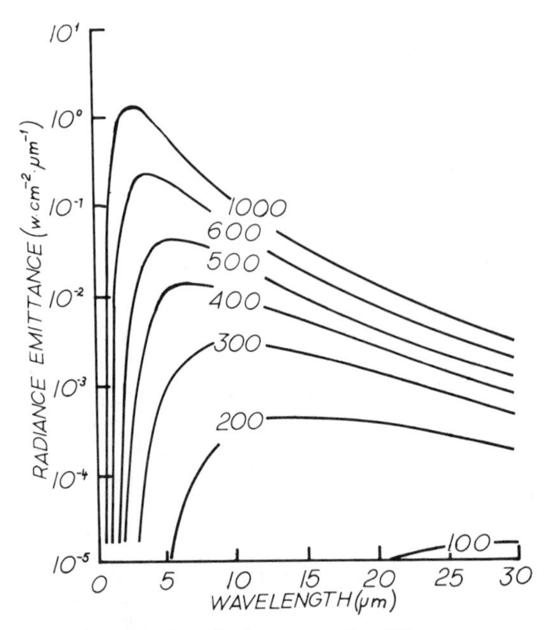

Figure 1-2. Blackbody radiation curves for different temperatures.

Another basic law of importance in remote sensing concepts is the Stefan-Boltzmann radiation law

$$M = \varepsilon \sigma T^4$$

where $\sigma = 5.6693 \times 10^{-8}$ W/m^2 K^4 is the Stefan-Boltzmann radiation constant.

With respect to illuminating surface features on earth, it is important to note that, based on the temperature of the sun, the solar peak is at about 0.5 μm, as shown in Figure 1-3, and is an excellent energy source for remote sensing by use

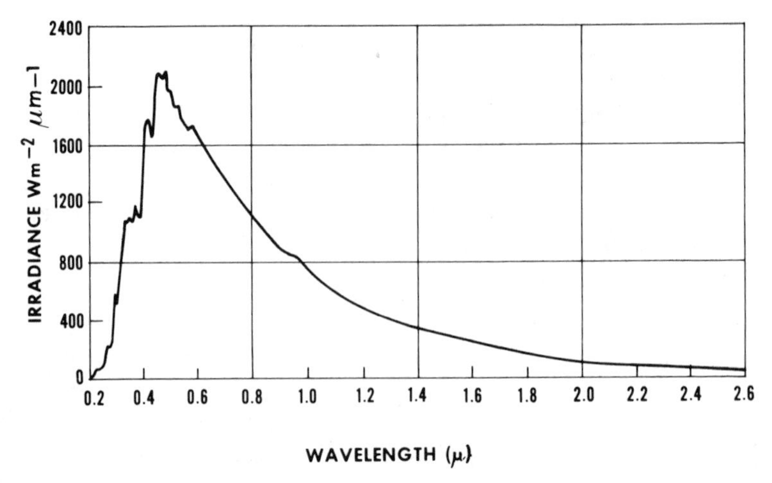

Figure 1-3. The NASA/ASTM standard curve of extraterrestrial solar spectral irradiance (solar constant: 1353 W m^{-2}). Courtesy of NASA.

of reflected light in the 0.4–0.75 μm range. The earth (average temperature of about 300 K) has its main radiant power peak at about 9.6 μm and has a broad distribution curve. This gives access to the use of a band from about 5 to 25 μm for monitoring the earth by emitted energy, which means that most of the remote sensing systems used are accomplished with a passive system in contrast to active systems that supply the required radiant energy for sensing.

There are no true blackbodies in nature, although many objects, such as water, which is opaque in the IR (3–15 μm), are similar to graybodies. They obey Kirchhoff's law, which can be written as,

$$\rho(\lambda, \theta') + \varepsilon(\lambda, \theta') = 1$$

where ρ is the specular reflectivity (given by Fresnel's equation for a flat surface) and ε is the specular emissivity, defined as the ratio of the emittance of a body to that of a real blackbody at the same temperature and wavelength. Emissivity also has angular (θ) dependence. Emissivities of water, for instance, as summarized by Bramson et al. (1965), are listed in the table:

$\lambda(\mu m)$	$\theta = 90°$	$\theta = 40°$
10.5	$\varepsilon = 0.9916$	$\varepsilon = 0.9833$
3.8	$\varepsilon = 0.9752$	$\varepsilon = 0.9608$

These facts show that observations should be constrained to lower nadir angles (θ) and that longer-wavelength observations are preferable.

Microwave radiation that is thermally emitted by an object is called brightness temperature, T_B. It is expressed in units of temperature, because, for radiation wavelengths in the microwave range, the radiation emitted from a perfect emitter is proportional to its physical temperature T. However, most real objects emit only a fraction of the radiation that a perfect emitter would emit at the same physical temperature. This fraction defines the emissivity ε of the object. In the microwave region, $\varepsilon = T_B/T$, which is a basic equation of passive microwave radiometry.

At microwave frequencies (1–500 GHz) and at temperatures typical of the earth and its atmosphere (200–300 K), the Rayleigh-Jeans approximation for the intensity of thermal radiation from a blackbody can be applied:

$$I = e\frac{2\pi ck}{\lambda^4}T$$

where e = emissivity
c = speed of light
λ = wavelength
k = Boltzmann's constant

It is common practice to define an equivalent blackbody temperature T_B such that

$$T_B = eT = \frac{I\lambda^4}{2\pi ck}$$

so the equivalent blackbody temperature emitted from the earth's surface depends on the emissivity and the surface temperature (Beer, 1980).

The radiative transfer for an isotropic, nonscattering medium is given as

$$\frac{dT_B(\theta, \phi)}{dX} = \gamma[T(X) - T_B(\theta, \phi)]$$

where $T_B(\theta, \phi)$ = radiance in direction (θ, ϕ) expressed as equivalent blackbody temperature
$T(X)$ = thermodynamic temperature of absorbing medium
γ = absorption coefficient in units of X^{-1}

The derivative on the left side is along a ray path. The equation becomes meaningful if one examines the effect on microwave radiation incident at an angle θ on an absorbing slab of uniform temperature T_0, absorptivity γ_0, and thickness δ. The radiation coming out the other side consists of the incident radiation attenuated by the slab $T_B e^{-\gamma_0 \delta \sec \theta}$ + a reradiation component $(1 - T_B e^{-\gamma_0 \delta \sec \theta})T_0$. If the expression $\gamma_0 \delta \sec \theta$ is small, or if $T_0 - T_B$ is small, the slab has no effect on the intensity of the radiation; however, if the slab is opaque ($\gamma_0 \delta \sec \theta = \infty$), the radiation emerging is characterized by the thermodynamic temperature of the slab independent of the intensity of the incident radiation.

Since few natural surfaces are specular reflectors of microwaves, T_s must be averaged over the appropriate distribution angles:

$$T_s = \frac{\int T_z(\theta, \phi) P(\theta, \phi)\, d\Omega}{\int P(\theta, \phi)\, d\Omega}$$

where $T_z(\theta, \phi)$ is the downwelling radiation intensity from the direction (θ, ϕ) and $P(\theta, \phi)$ is the probability of radiation from the direction being reflected from the surface in the direction of the radiometer. The integral in the denominator makes the normalization of P arbitrary. There is little data or theory available to determine this distribution, which requires at least an approximation which assumes a specular distribution. It can also be assumed that the surface is infinitely rough, and the Lambertian approximation can be applied:

$$P(\theta, \phi) = \cos \theta$$

In order to understand features recorded in the microwave region of the EMS, one must know the approximate depths from which the radiation emanates. For most emitters the amount of external emission from a given depth below the surface tends to decrease approximately exponentially with depth. The optical or skin depth is then the thickness of the top layer, from which approximately 63% of the radiation emanates. For a 1.55-cm wavelength the optical depth is of the order of millimeters for water, soils, and first-year sea ice; of the order of centimeters for multiyear sea ice; and of the order of meters for dry snow.

Final considerations are the transmission and emission properties of the earth's atmosphere through which the surface is viewed by satellite sensors. Although the dry polar atmosphere is rather transparent and nonemissive at a 1.55-cm wavelength, atmospheric effects are apparent in the satellite recordings, mainly at the lower latitudes.

Radar techniques are applied to monitor especially in those regions with frequent or almost permanent, cloud coverage. Radio detection and ranging (radar) [i.e., the use of radio frequencies (rf)—electromagnetic energy in the radio portion of the spectrum] detects objects by reflecting radio energy from them and measures the distance (ranging) by recording the time it takes for the signal to go out and back. The transmitter generates short pulses of rf energy, which radiate from a directional antenna as a block of energy traveling at the speed of light. Some of this energy is reflected back to the antenna from the ground. Reflections from near objects arrive first, and subsequent signals come in from features farther and farther away.

2

PLATFORMS AND SENSORS

2-1 GENERAL CONSIDERATIONS

This section on platforms is based on data from selected spacecraft that brought new instruments into orbit and delivered data that are still valid and available in data centers. Valuable photographs were made possible by manned spacecraft (Mercury, Gemini, and some other programs), leading later to the Space Shuttle with the large format camera (LFC). The section does not, however, go into detailed descriptions of manned spacecraft, since they are well-covered or have remote sensing only as a minor part in the overall objectives of the missions.

The applications of data from meteorological and resources satellites are overlapping in many resources areas. Although many satellites were considered experimental and delivered a high data flow for a broad user community, only recently have resources satellites become truly operational; the U.S. *Landsat* system was transferred to the private sector in 1985, and the French SPOT program started in 1986.

At an early stage of the development of satellite remote sensors, useful information was retrieved from the early satellite systems, with the 1970s bringing about a broader user community. In the sections on sensors, we focus on scanners and describe cameras, for which some samples of data products will be shown in later chapters. An overview on the principles of the most important radiometers to measure quantitatively selective parts of the EMS will be given. However, it should be kept in mind that some sensors simultaneously measure different parts of the EMS and the priority for application of the data might be in different disciplines.

2-2 SATELLITE PLATFORMS

A historical overview of the weather satellites and their prototypes, from *Tiros* (television infrared observation satellite) in the 1960s to *ITOS* (improved Tiros

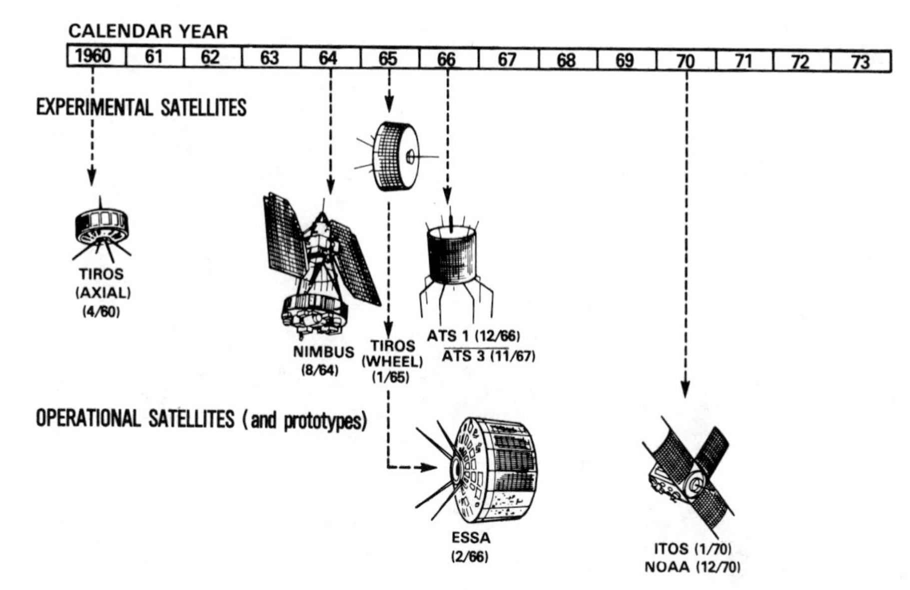

Figure 2-2-1. Meteorological satellite program (1960–1973). Courtesy of NASA.

operational satellite) in the 1970s, is given in Figure 2-2-1. An overview of 15 years of operations with meteorological satellites has been given by Smith et al. (1986).

The program of bringing meteorological satellites into orbit evolved rapidly to the creation of the world's first daily global operational system in 1966. The *ITOS-1* system, under the National Oceanographic Atmospheric Administration (NOAA), was later developed to routinely monitor night and day weather conditions on a global scale with a single spacecraft. This was accomplished with a dual-channel scanning radiometer with associated single processor and recorders for day and night observations and for recording surface temperatures of the earth, sea and, cloud tops.

Figure 2-2-2 shows an *ITOS* satellite. To stabilize the *ITOS* altitude, the control system was configured principally around a dual-spin stabilization technique coupled with magnetic torquers. The main body of the *ITOS* satellite was despun so that it rotated at one revolution per orbit while the momentum wheel spun at 150 rpm. In this mode of operation, the sensors were continuously oriented toward the earth.

In the early 1960s, the *Nimbus* satellite program was initiated by NASA to develop an observational system capable of meeting the research and development needs of atmospheric and earth scientists. The general objectives of the program were (1) to develop advanced passive radiometric and spectrometric sensors for daily global surveillance of the earth's atmosphere and thereby provide a data base for long-range weather forecasting; (2) to develop and evaluate new active and passive sensors for sounding the earth's atmosphere and mapping surface characteristics; (3) to develop advanced space technology and ground techniques for meteorological and other spacecraft for observing the earth; (4) to develop new techniques and knowledge useful for the exploration of other plane-

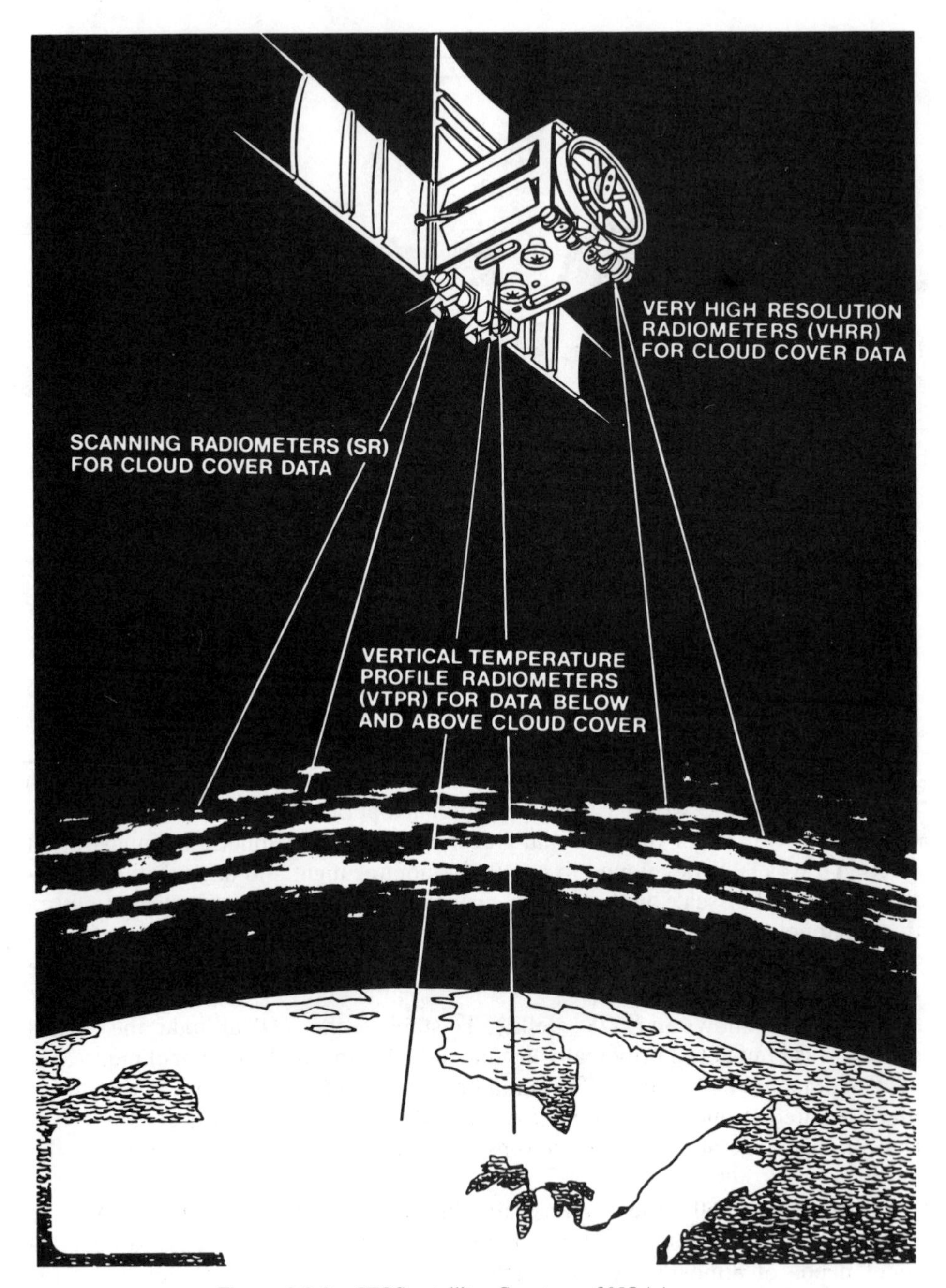

Figure 2-2-2. ITOS satellite. Courtesy of NOAA.

tary atmospheres; and (5) to participate in global observation programs (World Weather Watch) by expanding daily global weather observations.

The *Nimbus* orbit was selected to meet the diverse experiments and data retrieval requirements. *Nimbus* was placed in near circular orbit, sun-synchronous, having a local high noon equator crossing, and an 81° retrograde inclination.

Successive orbits crossed the equator at a longitude separation of 26°. The period for this orbit was about 107 minutes.

The *Nimbus* series started with the launching of *Nimbus 1* in August 1964, carrying three meteorological sensors: (1) the advanced vidicon camera system (AVCS), (2) the automatic picture transmission (APT), and (3) the high-resolution infrared radiometer (HRIR). Over the years, the payload was modified with each flight.

The *Nimbus* spacecraft system contained an improved active three-axis stabilization system, designed to maintain the spacecraft body axes, with the yaw axis pointing normal to the earth and the roll axis aligned to the spacecraft velocity vector. The improved attitude control subsystem of *Nimbus 4* permitted fine control to $\pm 1°$ in all axes. In view of the high pointing accuracy and the lack of more precise orientation data, attitude corrections were not utilized in geographic location procedures. All the data obtained were geographically located by using the orbit ephemeris data only. Figure 2-2-3 shows the *Nimbus 4* spacecraft with its solar panels unfolded.

Launched in October 1978, the *Nimbus 7* satellite was the last in a series of research and development satellites that collected experimental data related to the earth's atmospheric and oceanic processes.

The *Nimbus* satellite series proved successful, and data were later used and incorporated into the planning and flying of the earth resources technology satellite (ERTS), later renamed *Landsat*.

A cornerstone in sensor development and testing was the earth resources experiment package (EREP) on *Skylab,* which was designed to be a logical step in the application of equipment and techniques used in aerial surveys and ERTS. During three missions from 1973 to 1974, EREP sensors were operated individually or as a group, depending on the scientific requirements of the investigation.

Launched with a 50° inclination, EREP's coverage was limited to an area 50° north and south of the equator. Because it completed an orbit every 93 min, it repeated its ground track every five days. However, in spite of this, EREP's coverage along *Skylab*'s ground track was not complete.

The operational polar-orbiting environmental satellite system, designated *Tiros-N,* was placed into operation in 1978. Eight spacecraft in this series provided global observational service from 1978 through 1984. Figure 2-2-4 shows the general design of the *Tiros-N*.

Tiros-N was designed to operate in near polar sun-synchronous orbit. The orbital period was about 102 min, which produced 1.2 orbits per day over the same area. Because the number or orbits per day was not an integer, the suborbital tracks did not repeat on a daily basis, although the local solar time of the satellite's passage was essentially unchanged for any latitude. For this reason, the orbital equator crossing occurred at varying longitudes. Tiros-N operated with a southbound equator crossing (descending node) at approximately 0300 local solar time (LST) and a northbound equator crossing (ascending node) at approximately 1500.

In the mid-1970s, geosynchronous satellites became operational for meteorological monitoring, especially the geostationary operational environmental satellites (GOES) and the synchronous meteorological satellite (SMS).

The GOES/SMS spacecraft were placed in a circular orbit at about 35,800 km and travel at about 11,000 km/hr. At this altitude and speed, a satellite remains continuously above the same point on the earth, and thus is termed geostationary,

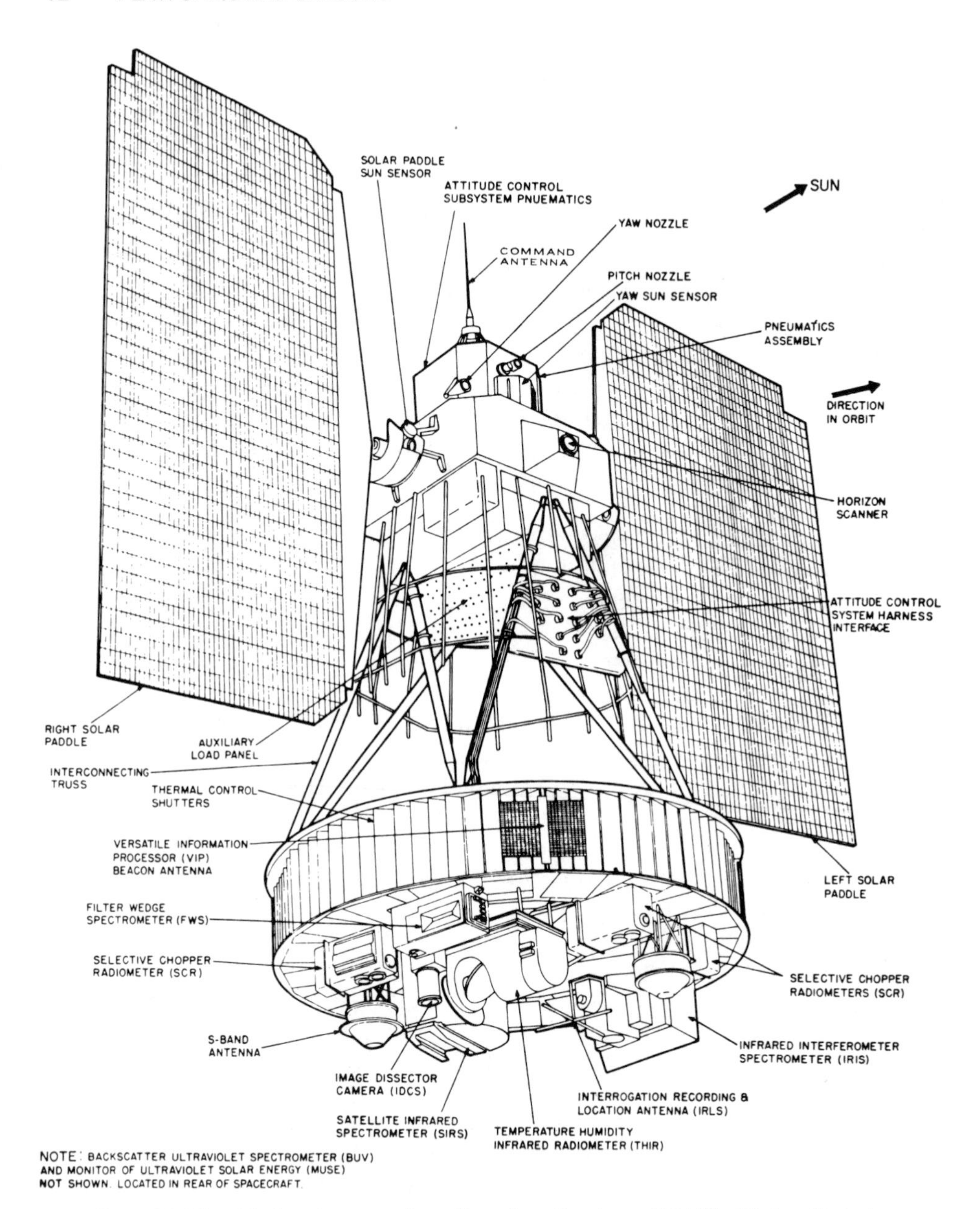

Figure 2-2-3. *Nimbus 4* spacecraft configuration. Courtesy of NASA, Nimbus Project.

geosynchronous, earth synchronous, or, merely, synchronous. The spacecraft is controlled for proper earth imaging by an attitude control subsystem that maintains the spin rate at 100 rpm and aligns the spacecraft spin axis parallel to the earth's polar axis and, thus, perpendicular to the earth's orbital plane.

The GOES operational system is shown schematically in Figure 2-2-5 and consists of the spacecraft ground stations, central data distribution system, and data collection platforms. The principal ground station is the command and data acquisition station.

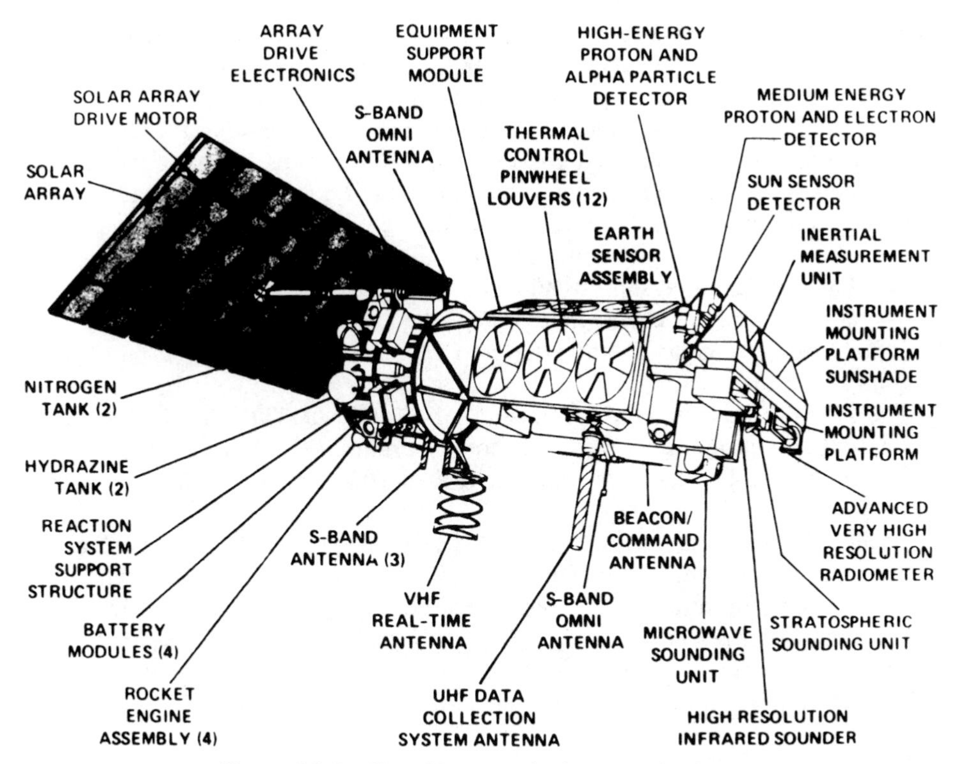

Figure 2-2-4. *Tiros-N* spacecraft. Courtesy of NOAA.

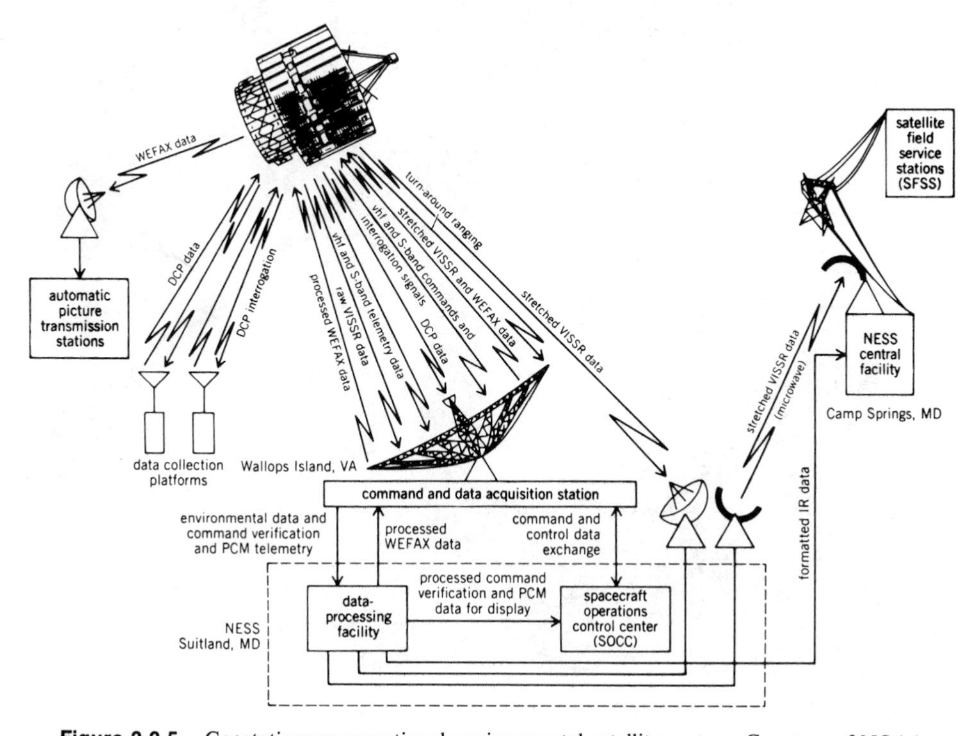

Figure 2-2-5. Geostationary operational environmental satellite system. Courtesy of NOAA.

A small, versatile, low-cost spacecraft, which used three-axis stabilization for the earth-viewing instruments carried by it, was used for the heat capacity mapping mission (HCMM). The spacecraft structure primarily consisted of two major components: (1) a base module, which contained the necessary attitude control, data handling, communications, command and power subsystems for the instrument module, and (2) an instrument module, which contained the heat capacity mapping radiometer (HCMR) and its supporting equipment. Two solar paddles were mounted on the spacecraft structure along the velocity vector (see Fig. 2-2-6). A yo-yo despin system was located on the base module cone section and below the solar array. Despin was accomplished by the transfer of some or all of the spacecraft's angular momentum into increased momentum of the yo-yo weights and cables.

The HCMM was targeted for an orbital altitude of 620 km (circular) at an inclination appropriate to sun-synchronous operation (97.9°). The 2:00 P.M. local time ascending node placed the satellite over northern mid-latitudes at 1:30 P.M. and 2:30 A.M. local time.

As already mentioned, the first ERTS (ERTS-1) represented a modified version of the *Nimbus* satellite. Launched in 1972, its objectives were to obtain information on agricultural and forestry resources, geology and mineral resources, hydrology and water resources, geography, cartography, oceanography, and meteorology. It, and later *Landsat,* was placed in a near polar orbit with an apogee of 907 km and a perigee of 900 km. The satellites were earth oriented had stabilized

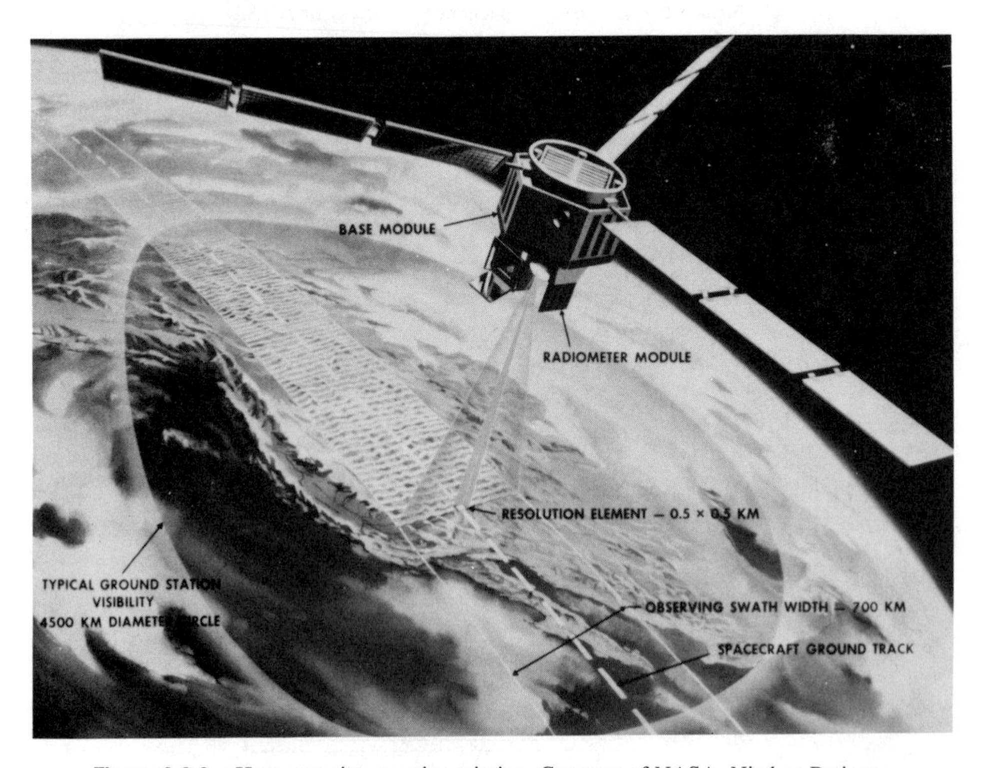

Figure 2-2-6. Heat capacity mapping mission. Courtesy of NASA, Nimbus Project.

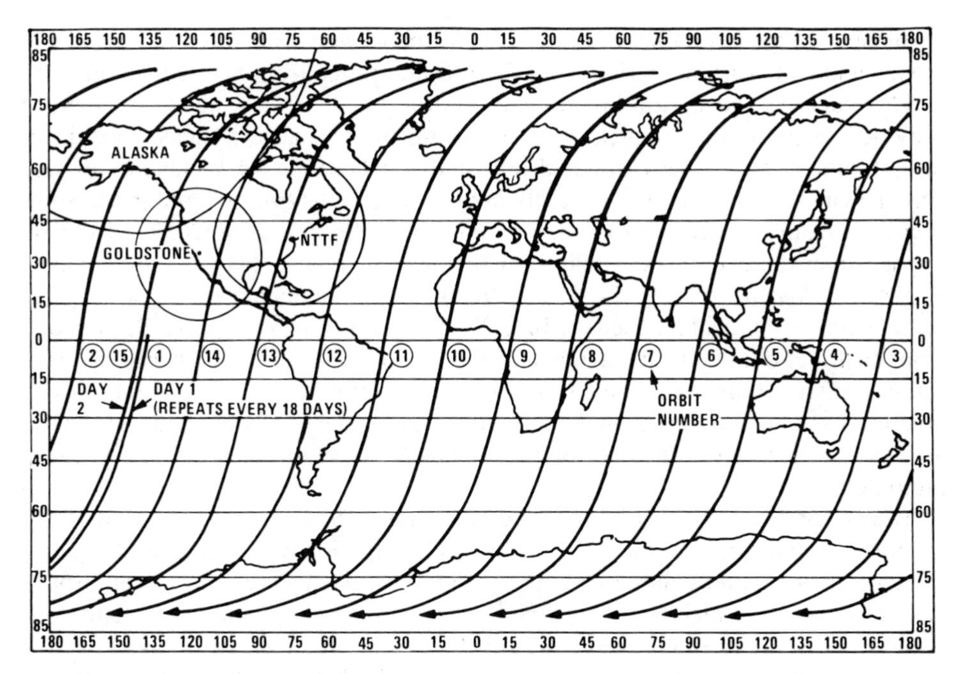

Figure 2-2-7. Typical *Landsat* daily ground tracks for daylight passes. Courtesy of NASA.

platforms and carried a four-channel multispectral scanner (MSS) and a three-camera return beam vidicon to monitor outgoing radiation in both the visible and near IR during the first missions. In addition, ERTS and *Landsat* were equipped with data collection platforms to obtain data from remotely located but individually equipped ground stations and to relay them to central acquisition stations. A nearly global coverage capability for storage of about 30 min of camera or scanner data was provided by two wideband video tape recorders. The spacecraft's orientation was controlled to within $\pm 0.7°$ in all three axes. The orbital parameters were selected in a way that the daily coverage swath was shifted 1.43° in longitude, corresponding to 159 km, with a revolution progress in a westwardly direction until all the area between orbit N and orbit $N + 1$ was covered (see Fig. 2-2-7). The complete coverage cycle, for the first series of *Landsat,* consisted of 251 revolutions, which required 18 days, whereby the area between 81°N and 81°S latitude was included. *Landsat 1* and *Landsat 2* operated for some time, simultaneously, whereby the relative phasing of their orbits was established in such a way as to overfly a given area every nine days.

Starting with *Landsat 4,* the earth resources satellites varied from the original *Nimbus* platform concept. In 1984 *Landsat 5* was placed in an orbit with an equatorial node crossing that was displaced with respect to the orbit of the *Landsat 4* spacecraft.

Based on the sun-synchronous orbit of *Landsat,* the geometric relationship between the orbit's descending mode and the mean sun's projection into the equatorial plane remains nearly constant throughout the whole mission.

The circular orbits of *Landsat* (as well as *Nimbus*) satellites do not allow a constant attitude with respect to earth. This is due to polar flattening and to perturbing forces upon the satellites, such as the gravitational effects of the earth, moon, and sun, varying periodically the attitude range from 880 to 930 km.

The tracking and data relay satellite system (TDRSS) has been deployed to provide a means of communicating with earth-orbiting spacecraft on a global basis in real time. *TDRSS-East* (located above the Atlantic at 41°W longitude) was launched in April 1983 and is currently in orbit. Problems encountered in transmitting data from the ground to orbiting spacecraft through *TDRSS-East* have no impact upon the use of the satellite for thematic mapper (TM) data collection.

NASA's *Seasat* (see Fig. 2-2-8) was the first satellite dedicated to studying the ocean surface. It was launched on June 27, 1978 into an orbit of about 800 km and stopped functioning due to a power failure after 100 days in operation. This satellite carried four microwave sensors: a radar altimeter (ALT), a wind scatterometer, a scanning multichannel microwave radiometer (SMMR), and a synthetic aperture radar (SAR). This microwave payload was supported by visible and IR radiometers.

The système probatoire d'observation de la terre (SPOT) program was decided by the French government in 1978, with the participation of Sweden and Belgium, to be operational and commercial. The program is managed by the French Space Agency (CNES), which is responsible for program development and satellite operations. *SPOT 1* was launched in March 1986, and *SPOT 2* is to be available for launch possibly in 1988 as a backup for *SPOT 1*. Plans are being made for the launch of *SPOT 3* and *SPOT 4* in 1990 onward in order to ensure the necessary service continuity expected from an operational spaceborne remote sensing program.

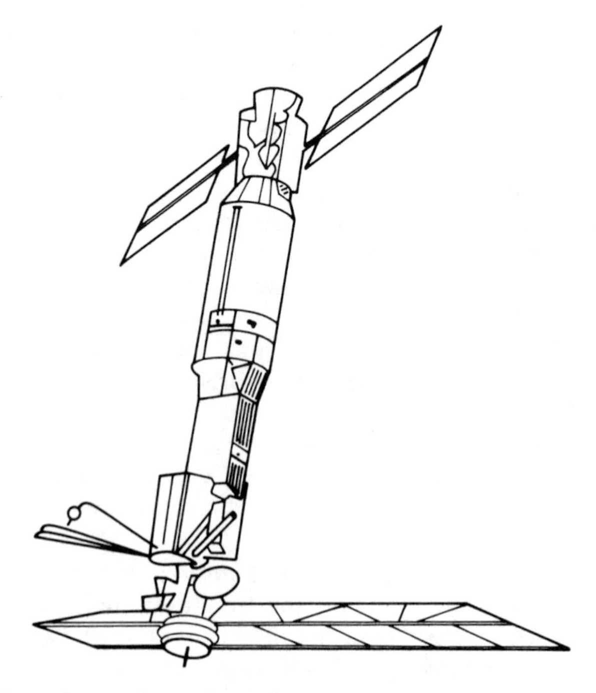

Figure 2-2-8. Generalized sketch of *Seasat*. Courtesy of NASA.

Figure 2-2-9. SPOT satellite. Courtesy of CNES.

The organizational structure adopted for the management of the SPOT program distinguishes between the functions of technical system management, executed by CNES, and the responsibility assigned to SPOT IMAGE, a commercial corporation, for relations with the user community and data distribution.

The SPOT satellite was launched by Europe's Arianespace from the launch center located at Kourou in French Guyana in March 1986. The SPOT system comprises a satellite made of the SPOT multimission platform and the mission payload and a ground segment composed of one image-receiving station and control stations. The satellite was designed with a maximum payload mass of 800 kg, is in circular sun-synchronous orbit at altitudes between 570 and 1200 km, at local time of ascending or descending node of 8.00 A.M. to 4.00 P.M., and is expected to have a lifetime of two years or more. Figure 2-2-9 shows the *SPOT* satellite.

2-3 SATELLITE REMOTE SENSORS

2-3-1 Sensors for the Visible and Near Reflected Infrared

Skylab Multispectral Scanner

One of the most important scanners for signature research and response of targets to specific wavelengths from space was the *Skylab* MSS, shown in Figure 2-3-1. This optical electromechanical scanner collected incoming radiant energy using a rotating mirror in the image plane to conically scan the scene viewed. A spherical mirror was the major element of a folded reflecting telescope that had a 43.2-cm entrance pupil. The energy scanned in the image plane passed through a reflective corrector mirror and through a field stop that was the entrance slit of a prism spectrometer. A dichroic mirror then separated the short wavelengths from the long-thermal-wavelength band.

The spectrally dispersed electromagnetic energy received from the scene simultaneously irradiated 13 detectors, which responded to the following wavelengths:

Band No.	Wavelength (μm)
1	0.41– 0.45
2	0.44– 0.52
3	0.49– 0.56
4	0.53– 0.61
5	0.59– 0.67
6	0.64– 0.76
7	0.75– 0.90
8	0.90– 1.08
9	1.00– 1.24
10	1.10– 1.35
11	1.48– 1.85
12	2.00– 2.43
13	10.20–12.50

Each detector produced an electronic signal that corresponded to the average value of the radiance received in its spectral band from the spot on the surface in the instrument's 0.182-mrad field of view, which had an instantaneous ground coverage of 79 m, a corresponding sweep angle viewed from the sensor at 10.4°, and provided a ground swath width of about 74 km.

Data from the MSS consisted of computer-compatible tapes of raw or calibrated data with ancillary information, such as tape format descriptions, calibration data, data acquisition characteristics like time, spacecraft attitude, field of view, and so on.

Landsat Multispectral Scanner System

The *Landsat* multispectral scanner system (MSS) consists of a double-reflector-type telescope, scanning mirror, filter detectors, and the necessary electronics.

The spectral response as shown in Figure 2-3-2 is mainly between 0.5–0.6 μm, 0.6–0.7 μm, 0.7–0.8 μm, and 0.8–1.1 μm. The scan mirror oscillates 2.89° to

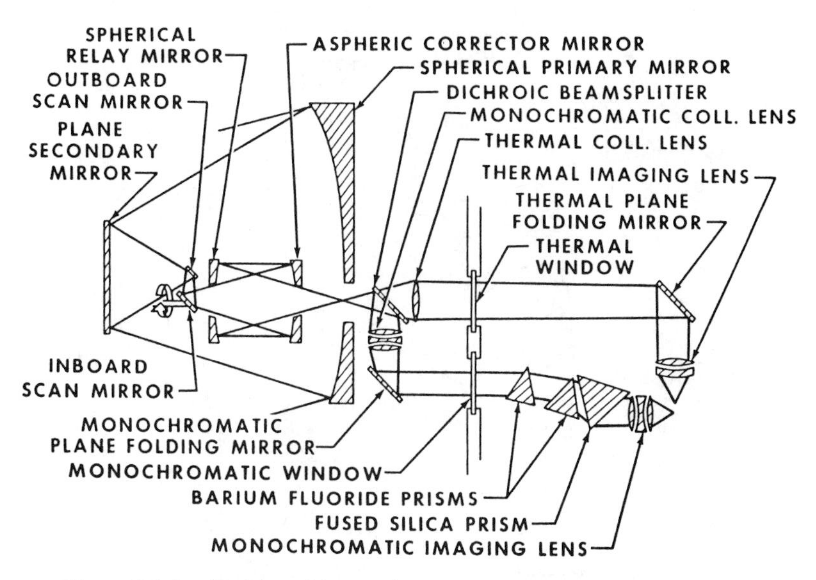

Figure 2-3-1. Skylab multispectral scanner system. Courtesy of NASA.

either side and has a swath of 185 km on the ground. The value usually cited for the ground-projected instantaneous field of view at nadir for the *Landsat 1, Landsat 2,* and *Landsat 3* MSS in the visible and near IR bands is 79 m × 79 m. Actually, the values are 76.05 ± 0.39 m, 76.27 ± 0.36 m, and 76.23 ± 0.73 m for the *Landsat 1, Landsat 2,* and *Landsat 3* systems, respectively.

The orbit and monitoring system were designed so that imagery sidelapped 14% at the equator. At higher latitudes the sidelap of adjacent *Landsat* coverage swaths increased with greater than 50% sidelap, and duplicate coverage in high

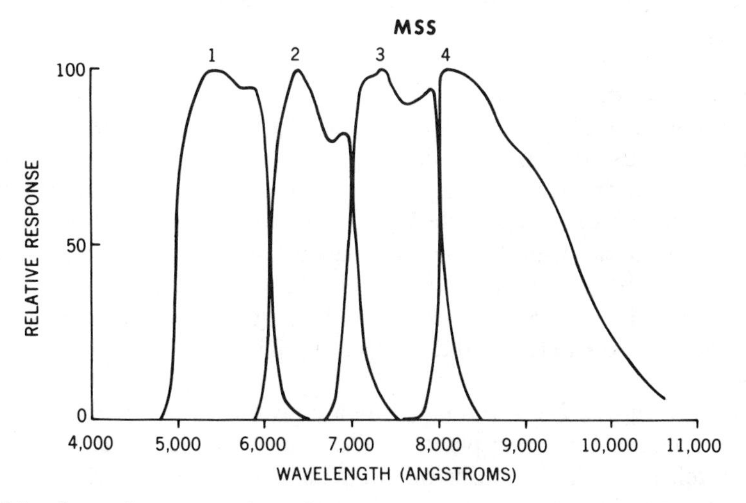

Figure 2-3-2. Spectral response of *Landsat* multispectral scanner system (MSS-1). Courtesy of NASA.

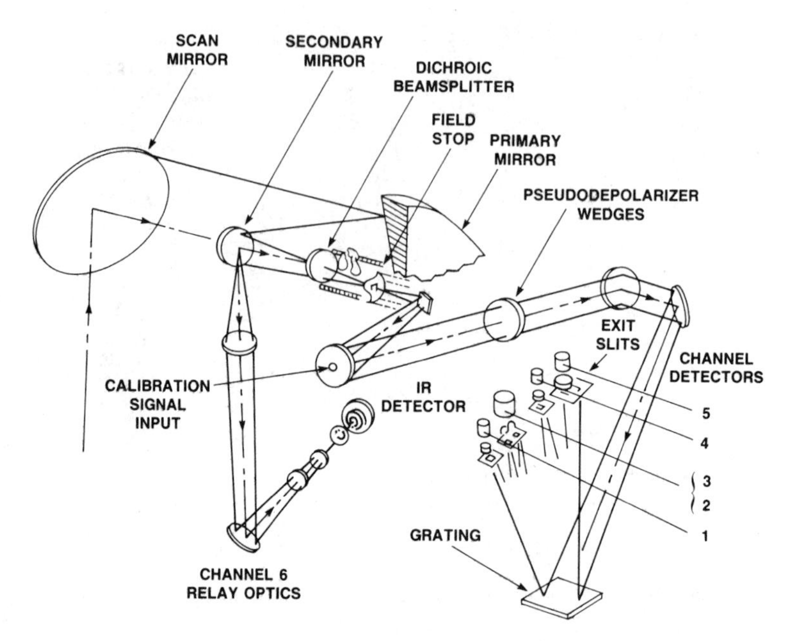

Figure 2-3-3. Coastal zone color scanner (CZCS) optical arrangement. Courtesy of Dr. J. A. Yoder, NASA HQ.

latitudes was achieved on sequential days. Also, the design was such that even though an image was not obtained on day M, it could be composed from portions of images taken on days $M - 1$ and $M + 1$. The *Landsat* orbit has also been designed so that the swaths viewed during one 18-day coverage cycle repeat.

The data as recorded by *Landsat* MSS can either be electronically transmitted to receiving stations or stored aboard the satellite on a tape recorder. The data are computer processed and are available in the form of computer-compatible tapes, black-and-white images, or color composites, based on the recordings in the different bands.

Coastal Zone Color Scanner

Instruments on several satellites have the ability to sense ocean color although their spectral bands, spatial resolution, and dynamic range have been optimized for land and meteorological use. The coastal zone color scanner (CZCS), flown on *Nimbus,* was the first instrument developed for the measurement of ocean color. The signal-to-noise ratios in the spectral channels sensing reflected solar radiance were higher than those required in the past because of the ocean's low reflectance and the fact that the majority of the reflected energy is backscattered solar radiation from the atmosphere rather than reflected solar energy from the ocean (Hovis et al., 1980).

As shown in Figure 2-3-3, the CZCS is a conventional multichannel scanning radiometer utilizing a rotating plane mirror at a 45° angle to the optic axis of a Cassegrian telescope. The rotating mirror scans 360°; however, only 140° of data centered on each side of the spacecraft nadir is collected. During the rest of the scan, the instrument acquires a view of internal instrument sources for calibration of the various channels. The radiation collected by the telescope is divided into two portions by a dichroic beam splitter. One portion is transmitted to a field stop, which is also the entrance aperture of a small polychromator. The radiant energy entering the polychromator is disbursed and reimaged in five wavelengths on five silicon detectors in the focal place of the polychromator. The portion of the beam reflected off the dichroic mirror is directed to a cooled mercury-cadmium-telluride detector sensing in the 10.5–12.5 μm region.

The spectral parameters of the CZCS are summarized as follows:

Channel No.	Spectral Range (μm)	Indicator
1	0.433– 0.453	Chlorophyll absorption
2	0.510– 0.530	Chlorophyll correlation
3	0.540– 0.560	Gelbstoffe
4	0.660– 0.680	Chlorophyll absorption
5	0.700– 0.800	Surface vegetation
6	10.5 –12.5	Surface temperature

The scanning takes place at 0.7 rad about the nadir in a plane normal to spacecraft orbit track. This plane of view can be moved 0.35 rad in 0.035-rad steps (about the spacecraft pitch axis) with respect to nadir. This means that the CZCS scan mirror can be tilted from nadir to look either forward or behind the spacecraft line of flight. This feature was built into the instrument to avoid the glint caused by ocean waves that would obscure any scattering from below the surface. The angle of tilt of the scan mirror is determined by the solar elevation angle. The spectral response of channels 1 through 5 is illustrated in Figure 2-3-4. The 10.5–12.5 μm

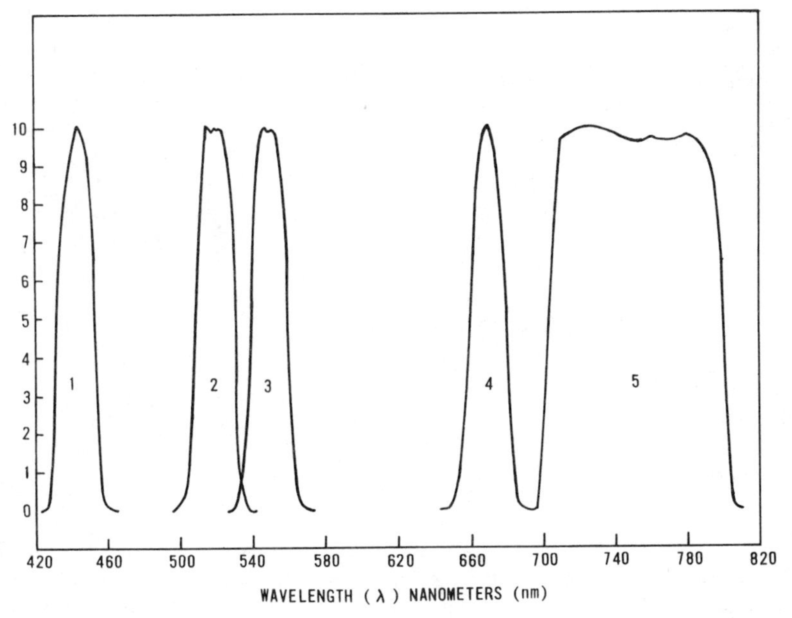

Figure 2-3-4. CZCS spectral response for channels 1 through 5. Courtesy of NASA, Nimbus Project.

channel measures equivalent blackbody temperature with a noise equivalent temperature difference of less than 0.35 K at 270 K. Atmospheric interference with this channel, principally from water vapor absorption can produce measurement errors of several degrees. Temperature gradients, however, can be seen quite well because of the extremely low noise equivalent temperature difference (NETD) of this sensor.

In-flight calibration of the CZCS is accomplished for the first five bands by using a built-in incandescent light source. This in-flight calibration source has been calibrated with the instrument itself as a transfer against the referenced sphere output. The light source is redundant in the instrument so that if one light fails another one can be commanded to operate. After launch, the first calibration source is used routinely, with the second light source tested occasionally to verify its stability. In-flight calibration for channel 6 has been done by viewing the blackened housing of the instrument whose temperature is monitored and viewing deep space during the rotation of the scan mirror.

Nimbus 7 has its orbit from south to north in daylight; the scan mirror is positioned to look behind the satellite when the spacecraft is south of the subsolar point and ahead of the spacecraft when it is north of the subsolar point.

Thematic Mapper

The thematic mapper (TM) is significantly more complicated and more sophisticated than the MSS that served originally as the prime sensor on *Landsat*, as shown in Table 2-3-1. The TM possesses approximately twice the number of spectral bands that previously existed on the MSS. Furthermore, the spatial reso-

TABLE 2-3-1. Comparison of Landsat TM and MSS Measurements Capabilities

	Thematic Mapper			Multispectral Scanner		
Spectral resolution	Band Number	Bandwidth (μm)	Radiometric Sensitivity $(NE\Delta\rho)^a$	Band Number	Bandwidth (μm)	Radiometric Sensitivity $(NE\Delta\rho)^a$
	1	0.45– 0.52	0.8	1	0.5–0.6	0.57
	2	0.53– 0.61	0.5	2	0.6–0.7	0.57
	3	0.62– 0.69	0.5	3	0.7–0.8	0.65
	4	0.78– 0.91	0.5	4	0.8–1.1	0.70
	5	1.57– 1.78	1.0			
	6	10.42–11.66	0.5 K[b]			
	7	2.08– 2.35	2.4			
Ground instantaneous field of view	30 m (bands 1–5 & 7) 120 m (band 6)			83 m (bands 1–4)		
Signal quantization level	256 (8-bit word)			64 (6-bit word)		
Data rate	85 Mbit/sec			15 Mbit/sec		

[a] Radiometric sensitivity specified as the noise equivalent difference in surface reflectance.
[b] Thermal band sensitivity specified as the noise equivalent difference in apparent temperature.
Courtesy of NASA.

lution of the TM bands situated in the visible and reflective IR spectrum is $2\frac{1}{2}$ times greater than the spatial resolution of comparable MSS bands. The TM and MSS were originally designed to obtain measurements over a comparable range of radiation intensity. The output signal of the TM detectors is quantized into 256 different gray levels, whereas the scale range employed in the digital quantization of raw MSS data has only 64 gray levels. In addition, individual detectors within the TM are more sensitive to changes in incident radiation than are the detectors employed in the MSS. The inherent sensitivity of the TM detector elements has been increased and the range over which detector signals are quantized has been broadened, resulting in a major improvement in the overall radiometric sensitivity of the TM.

Band 1, with a spectral range of 0.45–0.52 μm, is principally applied for coastal water mapping, differentiating between soil and vegetation, and differentiating between coniferous and deciduous trees. Band 2, with a spectral range of 0.52–0.60 μm, detects green reflectance by healthy vegetation (bands 1 and 2 are in the visible light). Band 3, with a spectral range from 0.63 to 0.69 μm, is in the red and detects the chlorophyll absorption of different plant species. Band 4, with a spectral range of 0.76–0.90 μm, is in the near IR and is applied to biomass surveys and water body delineation. Band 5, with a spectral range of 1.55–1.75 μm, measures the moisture of vegetation and differentiates between snow and clouds. Band 6, with a spectral range of 10.4–12.5 μm, operates in the far IR or thermal region, measures plant heat stress, and is used for other thermal mapping. Band 7, with a spectral range of 2.08–2.35 μm, is used primarily for hydrothermal mapping.

The TM is an object-space line scanner with 100 signal channels or detectors. The scan mirror sweeps the TM line of sight back and forth seven times each second in a direction normal to the orbital ground track. Data are collected during forward and reverse scans. A telescope images the scanned scene energy onto the prime focal plane. Located at a short distance in front of the prime focal plane is a two-mirror image-plane scanner called a scan line corrector. Its function is to rotate the TM's line of sight during the scanning to compensate for the orbital motion of the spacecraft. This action generates scan swaths that are perpendicular to the ground track and parallel to each other.

The detector assemblies for spectral bands 1, 2, 3, and 4 are located at the prime focus of the telescope. Each band employs a 16-element monolithic silicon detector array to convert the sensed scene energy into low-level electrical signals, which are amplified and converted into an 85-mb data stream. A two-mirror all-reflective relay is used to reimage the energy for bands 5, 6, and 7 to a second focal plane located in the radiative cooler. The cooled focal plane contains two 16-element monolithic indium-antimonide detector arrays for sensing the shortwave IR scene energy in bands 5 and 7. The focal plane also contains one four-element monolithic mercury-cadmium-telluride detector array for sensing the thermal IR scene energy in band 6. The arrays for bands 5 and 7 are configured identically to the silicon arrays of bands 1 through 4, but are half the size (0.05 mm) because of the relay magnification of 0.5. The detectors for band 6 are 2 mm^2.

The orbital performance of the TM detectors is monitored on a continuous basis through the use of internal calibration lamps and a blackbody, which serve as reference radiation standards. The focal plane detectors are exposed to these calibration standards during each complete cycle of the oscillating scan mirror.

The response of the *Landsat 4* TM detectors to these internal calibration standards has remained highly linear during the first year and a half of orbital operation. The internal calibration standards can be used to relate the orbital response of individual detectors to prelaunch calibration tests in which the same detectors were exposed to a series of light sources of known intensity.

Modular Optoelectronic Multispectral Scanner

The modular optoelectronic multispectral scanner (MOMS) is sponsored by the West German Federal Ministry of Research and Technology and is developed by Messerschmitt-Bölkow-Blohm (MBB) together with the German Aerospace Research Agency (DFVLR). In two space missions, MOMS was mounted on the shuttle pallet satellite (SPAS) aboard Shuttle flights in June 1983 and in February 1984. The missions served the dual purpose of verifying the technical operation of the sensor in space and of conducting geoscientific and application-oriented experiments around the globe (Bodechtel, 1986).

For the first time, MOMS used charge-couple device (CCD) technology via the push-broom principle (see Fig. 2-3-5). Scan line extension beyond one CCD array (up to six arrays per focal plane are feasible) is carried out by the dual optics principle. The most useful characteristic of MOMS is its modular structure of filters, optics, sensors, and preamplifier electronics allocated to specific spectral bands (see Fig. 2-3-6).

Due to the experimental character of the missions, the recording time was

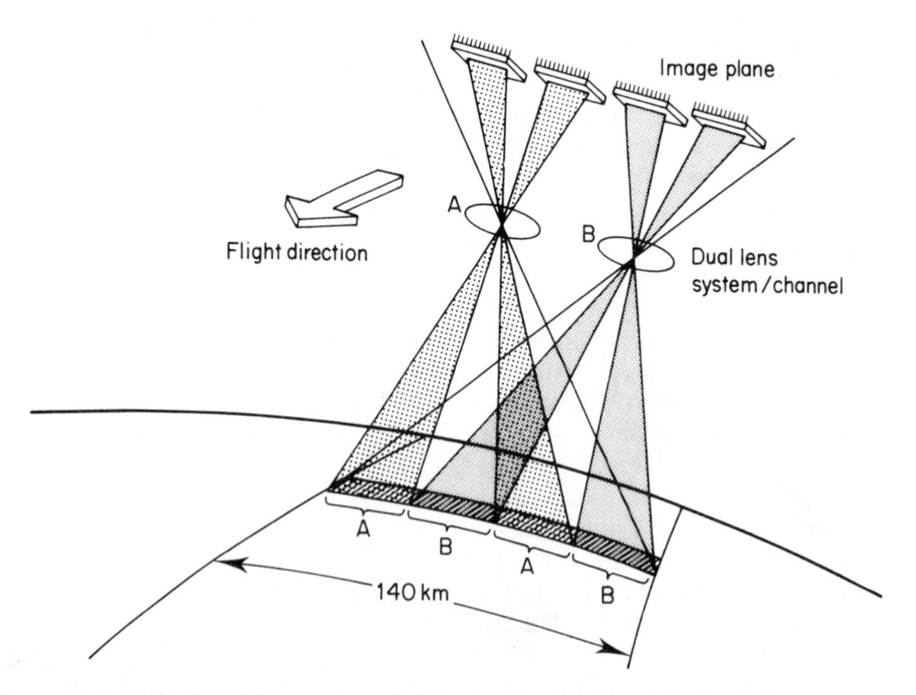

Figure 2-3-5. MOMS-01 CCD scanning principle; scan lines are extended with the dual lens system. After Bodechtel (1986). Reproduced with permission from Graham & Trotman, Ltd.: Szekielda, K.-H., 1986. *Satellite Remote Sensing for Resources Development.*

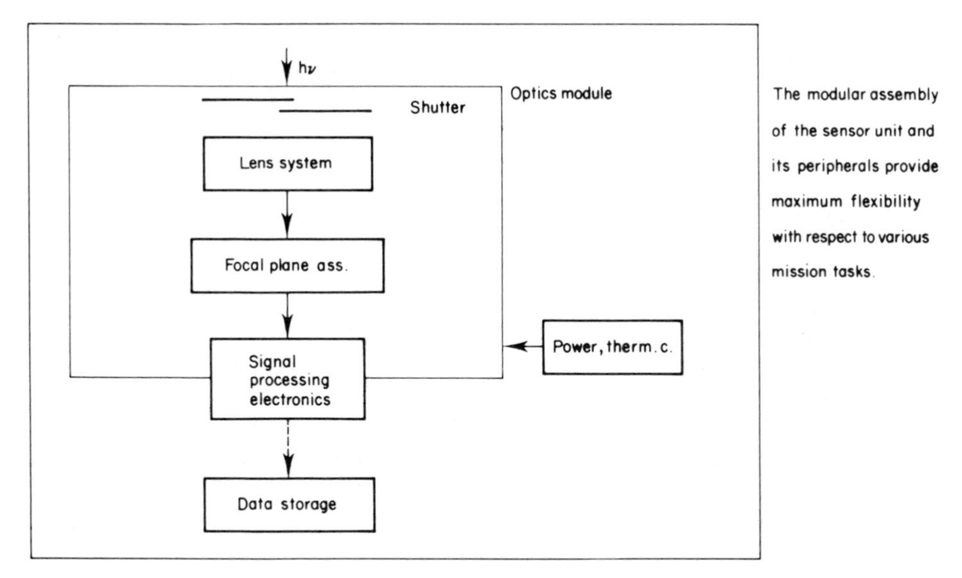

Figure 2-3-6. MOMS modular concept. After Bodechtel (1986). Reproduced with permission from Graham & Trotman, Ltd.: Szekielda, K.-H., 1986. *Satellite Remote Sensing for Resources Development.*

limited to about 30 min. For a swath width of 140 km, this is equal to an area of roughly 1,800,000 km^2 or nearly 150 image frames.

The MOMS information and commands for imaging were controlled during the missions from the mission control center at the Johnson Space Center in Houston or from aboard the Shuttle. To maximize the benefit from the MOMS missions, scientists set up an information network for real-time controlled data acquisition. Also, in order to identify areas most suitable for image sequences, researchers collected geoscientific information and weather data from geostationary weather satellites 1½ hours before the actual data take. This allowed MOMS to record a total number of 26 image frames under mostly optimal weather conditions.

SPOT

Système Probatoire d'Observation de la Terre (SPOT) spacecraft carries two identical sensors, called haute résolution visible (HRV) (or high-resolution visible range instruments) made of static solid-state arrays of detectors and operating in the visible and near IR part of the spectrum. Among the innovative features of SPOT are the relatively high ground resolution of the imagery (it produces 10 m in the panchromatic mode and 20 m in the multispectral mode) and the ability of its sensors to point up to 27° east and west of the local vertical axis. This latter feature offers increased opportunities to obtain views of an area. It also permits stereoscopic observations by combining views taken at different angles from the vertical and therefore opens up the possibility of third-dimension or altitude determination, an important requirement for cartographic applications (see Fig. 2-3-7).

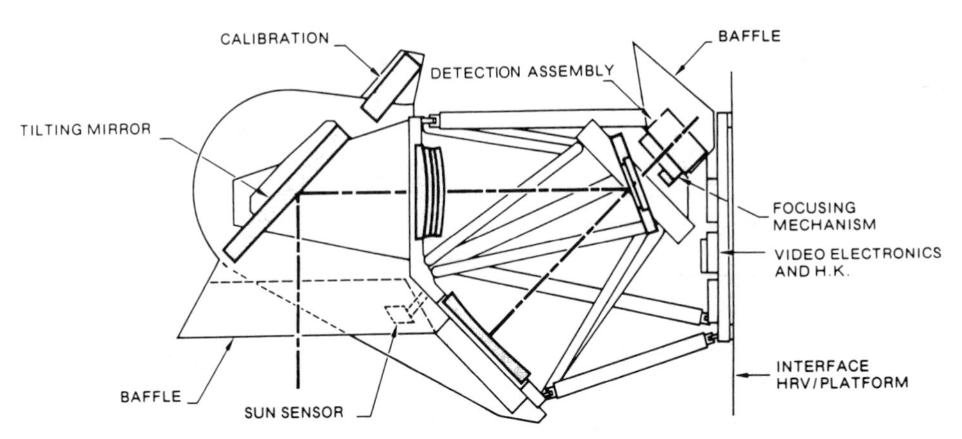

Figure 2-3-7. The principles of SPOT's haute resolution visible. Courtesy of SPOT-Image.

The principal characteristics of SPOT are summarized in Table 2-3-2 (Brachet, 1986).

The SPOT system operates in two modes: (1) the multispectral mode, by which observations are made in three spectral bands with a pixel size of 20 m—a green band from 0.50 to 0.59 μm, a red band from 0.61 to 0.68 μm, and a near IR band from 0.79 to 0.89 μm; and (2) the panchromatic mode by which observations are made in a single broad band from 0.51 to 0.73 μm with a pixel size of 10 m. The multispectral bands have been selected to take advantage of interpretation methods developed over the last 10 years. They have been designed to allow the best

TABLE 2-3-2. SPOT: Principal Characteristics

Orbit	Circular at 832 km
	Inclination: 98.7°
	Descending node at 10 h 30 min A.M.
	Orbital cycle: 26 days
Haute resolution visible (HRV)	Two identical instruments
	Pointing capability: ±27° east or west of the orbital plane
	Ground swath: 60 km each at vertical incidence
	Pixel size
	10 m in panchromatic mode
	20 m in multispectral mode
	Spectral channels
	panchromatic: 0.51–0.73 μm
	multispectral: 0.50–0.59 μm
	0.61–0.68 μm
	0.79–0.89 μm
Images	Two onboard recorders with 23 min capacity each
Transmission	Direct broadcast at 8 GHz (50 Mbit/sec)
Weight	1750 kg
Size	2 × 2 × 3.5 m plus solar panel (9 m)

After Brachet (1986). Reproduced with permission from Graham & Trotman, Ltd. Szekielda, K.-H. 1986. *Satellite Remote Sensing for Resources Deelopment.*

discrimination among crop species and among different types of vegetation using three channels only. The panchromatic band offers the best geometric resolution (10 m) and makes it possible to comply with cartographic standards for maps at a scale of 1 : 1,000,000 or to update at a scale of 1 : 50,000, and in some cases 1 : 25,000, for thematic applications.

The steerable mirror, which provides off-nadir viewing capability, can be tilted sideward (to the east or to the west) step by step from 0 to 27°, thus allowing scene centers to be targeted anywhere within a 950-km-wide strip centered on the satellite track. This technique provides a quick revisit capability on specific sites. For instance, at the equator, the same area can be targeted seven times during the 26 days of an orbital cycle (i.e., 98 times in one year, with an average revisit of 3.7 days). At 45° latitude, the same area can be targeted 11 times in a cycle (i.e., 157 times in one year, with an average of 2.4 days, a maximum time lapse of 4 days, and a minimum time lapse of 1 day).

Data transmitted by SPOT satellites are received at the Toulouse (France) and Kiruna (Sweden) main stations in direct readout mode for data taken over Europe and polar zones. These two stations also receive the worldwide data from the onboard recorders. Each station has a receiving capacity of 250,000 scenes per year.

The preprocessing centers attached to both stations have a capacity of 70 system-corrected scenes per day or 20 precision-processed scenes per day. A system-corrected scene can be preprocessed within 48 hours from its acquisition at the ground station in standard procedure, whereas a precision-processed scene requires five to seven days. Data can also be received at other receiving stations in the world that receive only real-time data over the station's visibility range.

Advanced Vidicon Camera System

The first vidicon camera, the AVCS, was flown on *Nimbus 2* and provided television pictures of the cloud distribution on a global scale and of terrestrial features over cloud-free regions. The system consisted of three vidicon cameras with a square field of view of 37°. The central camera was oriented downward along the yaw axis, and the side cameras were mounted in the yaw pitch plane at an angle of 35° with respect to the central camera. Successive frames were taken at 91-sec intervals, providing about 20% overlap of earth coverage of approximately 3520 km × 74 km.

The three cameras used 1-in vidicons with 800-line resolution. The size of the resolution element varied from 0.81 km at nadir to approximately 2.41 km at the corners, at a 1.111-km altitude. A 40-msec exposure time was used, and the image was scanned by an electron beam in 6.5 sec. The resulting signal was frequency modulated and recorded on three tracks of a tape recorder, one track for each camera. A fourth track recorded a continuous timing signal from the spacecraft clock so that picture exposure times could be identified upon relay to the ground receiving stations.

Return Beam Vidicon

The return beam vidicon (RBV) camera system on *Landsat* contained three independent cameras covering the spectral bands from blue-green (0.47–0.575 μm) through yellow-red (0.58–0.68 μm) to near IR (0.69–.83 μm), each viewing the same area.

The field of view of the RBV was 185 km × 185 km in area and was stored on the photosensitive surface of the camera tube. After shuttering, the image was scanned by an electron beam to produce a video signal output and reading from each camera sequentially every 3.5 sec.

The RBV camera system on *Landsat 3* was significantly different from the RBV systems on previous satellites. In *Landsat 3,* two panchromatic cameras produced two side-by-side images, rather than three overlapping multispectral images, of the same scene. Each RBV camera sensor is designed to cover an area of 99 km × 99 km, with a total swath width of approximately 185 km recording one broad spectral band from 0.505 to 0.750 μm. Four RBV images coincided approximately with one MSS frame. A focal length of 23.6 cm, or nearly twice that of *Landsat 1* and *Landsat 2* RBV, nearly doubled the resolution for ground area mapping with a ground resolution of 38 m × 38 m.

Earth Terrain Camera

The earth terrain camera was flown during *Skylab*'s EREP mission. The camera had an f/4 lens with a focal length of 45.7 cm and was compensated for *Skylab*'s forward motion. The field of view of 14.24° provided a ground coverage of about 109 km × 109 km. Film width was 12.7 cm, which provided a usable image that was 11.4 cm^2. Shutter speeds were 1/100, 1/140, and 1/200 of a second. Sequence photography intervals were possible from 0 to 25 frames per minute. Stereoscopic viewing was made possible by using a rate of 10.5 frames per minute to obtain overlaps of 60%.

Spacelab Metric Camera

The spacelab metric camera was flown on the Space Shuttle. The experiment was composed of a standard ZEISS RMK A 30/23 aerial survey camera (Fig. 2-3-8) modified to interface with the spacelab system. The camera was equipped with a ZEISS Topar A 305mm f/5.6 lens with two interchangeable magazines, each loaded with 150 m of 24-cm-wide film: one with a black-and-white film and the other with a false-color IR film (see Table 2-3-3 and Schroeder, 1986). In addition, the system consisted of two filters, a remote control unit, camera suspension mount, and stowage containers. The camera, film magazines, and filters were stowed in special containers in experiment racks during launch and landing. The camera interfaces to the optically flat high-quality window via a suspension mount, which is permanently mounted to the window adapter plate. The remote control unit was installed in an experiment rack and remained there during the entire mission. In orbit, the camera was assembled on the suspension mount, fitted with a magazine, and electrically connected to the remote control unit by payload specialists onboard.

During camera operation, the Space Shuttle was positioned so that the optical axis of the camera was oriented vertically to the earth, with an accuracy of 0.5°. With a relative forward velocity between the spacecraft and the earth of approximately 7.55 km/sec, exposure times must be restricted to between 1/500 and 1/1000 sec in order to minimize image motion.

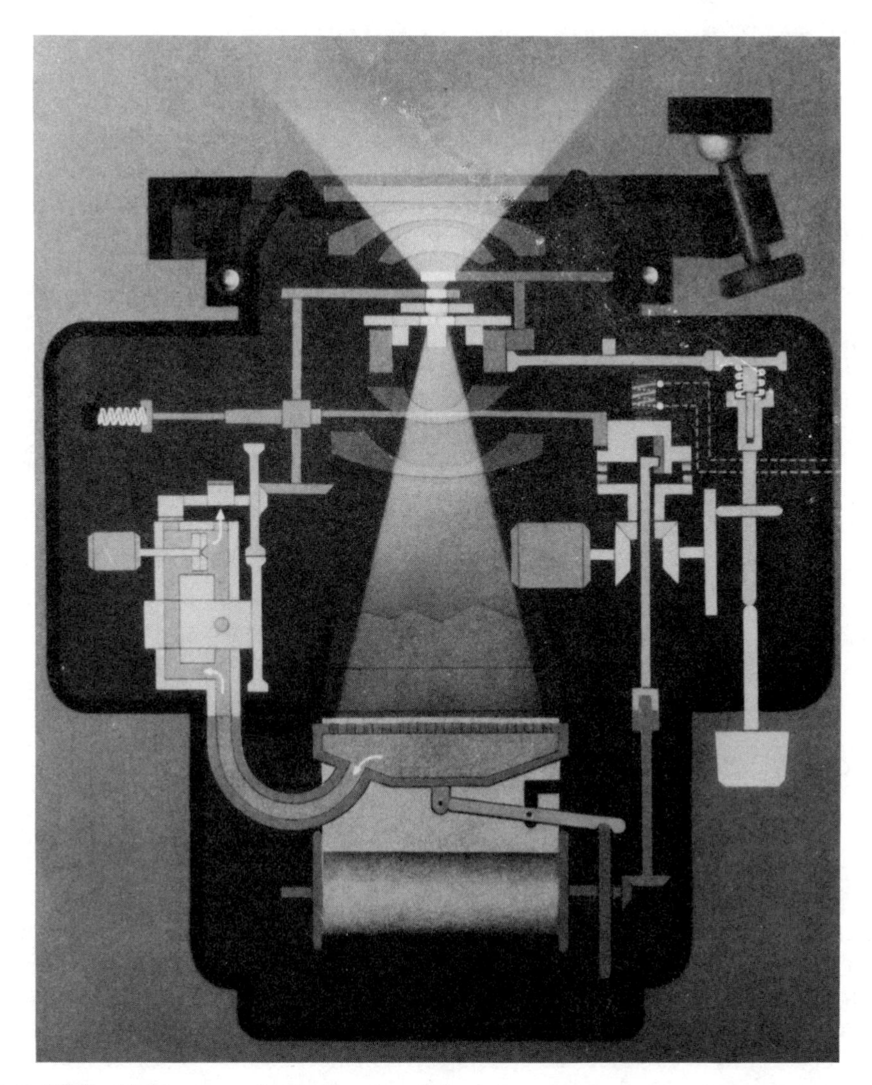

Figure 2-3-8. Metric camera for the first *Spacelab* mission. Courtesy of Zeiss, Federal Republic of Germany.

The focal length of 305 mm and the nominal orbit altitude of 250 km yield an image scale of 1 : 820,000, which results in an image format of 23 cm × 23 cm for a ground coverage of approximately 189 km × 189 km per image.

The films selected for experiment were the Kodak double-X aerographic film 2405 (black-and-white) and the Kodak aerochrome infrared film 2443 (color). Both films were flown with an orange filter with a cutoff wavelength of 0.535 μm. For the black-and-white film the main application is in topographic mapping, and for the color infrared, photo interpretation and thematic mapping.

The camera was operated in automatic and manual modes. In the nominal case, the camera operation was controlled automatically by the onboard computer system.

TABLE 2-3-3. Characteristics of *Spacelab* Metric Camera

Type	Modified ZEISS RMK A 30/23
Lens	Topar A 1 with 7 lens elements
Calibrated focal length	305.128 mm
Max. distortion	6 μm (measured)
Resolution	39 1 p/mm AWAR on Aviphot Pan 30 film
Film flattening	By blower motor incorporated in the camera body
Shutter	Aerotop rotating disk shutter (between the lens shutter)
Shutter speed	1/250 sec–1/1000 sec in 31 steps
F/STOPS	5.6–11.0 in 31 steps
Exposure frequency	4–6 sec and 8–12 sec
Image format	23 × 23 cm
Film width	24 cm
Film length	150 m = 550 image frames
Dimensions	
camera	46 × 40 × 52 cm
magazine	32 × 23 × 47 cm
Mass	
camera	54.0 kg
magazine	24.5 kg (with film)

Courtesy of Zeiss, Federal Republic of Germany.

Large Format Camera

The LFC was flown on the Space Shuttle to demonstrate an instrument capable of superior spatial resolution in the visible and, to some extent, the near IR spectra with minimal geometric distortion for precision cartographic applications. At the same time, the mission was to provide a wide-area stereoscopic capability.

During the mission, three altitudes (370 km, 272 km, 239 km) were selected for LFC operations to evaluate the optical performance and usefulness of four different film types: two black-and-white, one color, and one color infrared. For LFC parameters, see Table 2-3-4.

All commands to the camera were issued from the ground in real time or as stored program commands requiring no manual intervention by the crew. The TDRSS-A was used to relay any commands from the ground that could not be sent directly to the spacecraft.

The spacing of the exposures by the LFC permitted the acquisition of stereo imagery. Individual exposures were timed to provide a precise degree of overlap between frames, with each five-frame set providing a series of forward overlaps ranging from 80 to 20%. Associated with this exposure pattern, an increasing base-to-height ratio ranging from 0.3 to 1.2 was provided, allowing various degrees of vertical exaggeration of the terrain below to be established.

From a Space Shuttle altitude of 300 km, the LFC was able to distinguish objects the size of a single-family house within an area of 100,000 km^2 covered by each photograph. A Shuttle mission can collect over 2300 photos, each which 23

TABLE 2-3-4. Operational Parameters for the Large Format Camera

Parameter	Value
Lens	
focal length	30.5 cm (12 in)
aperture	f/6.0 (T/16)
spectral range	400 to 900 nm
distortion	10 mm (average)
Shutter	
type	3 disk rotary intralens with capping blade
speed range	4 to 32 msec
setting	External command or AEC
Forward motion type	Translation of platen
Compensation range	11 to 41 mrad/sec
Camera internal pressure	2 psig
Film flattening	Vacuum
Image format	22.9 by 45.7 cm (9 by 18 in)
Forward overlap modes	10, 60, 70, and 80%
Base-to-height ratios	0.3 to 1.2
Filters	
antivignetting	Front of lens
minus-haze	Intralens
minus-blue	Intralens
Minimum cycle time	7.5 sec
FOV	
along track	73.7°
across track	41.1°
Film	
width	24.1 cm (9.5 in)
roll length	1220 m (4000 ft)
exposures/roll	2500

cm × 46 cm, are overlapped by as much as 80% for stereoscopic three-dimensional observation of the earth.

Solar Backscattered Ultraviolet Radiometers (Ozone Measurements)

Measurements of backscattered solar ultraviolet radiation are associated with two experiments, which involved the use of backscatter ultraviolet (BUV) on *Nimbus 4* (1970–1977) and both the solar backscatter ultraviolet (SBUV) and total ozone mapping spectrometers (TOMS) on *Nimbus 7,* launched in October 1978 (Heath et al., 1983). The instruments measure ozone by the BUV method but differ slightly in the specific wavelengths observed or in resolution and coverage. The BUV measurements center, for instance, at 0.38 μm, whereas the SBUV measurements are centered at 0.343 μm.

For nadir-pointing instruments (BUV and SBUV) the horizontal resolution is about 200 km × 200 km. An optics diagram for the SBUV is given in Figure 2-3-9.

Total ozone mapping spectrometers measure only total ozone, the values of which represent columns varying in cross section from 50 km × 50 km at nadir to 130 km × 300 km at the scan extremes. The vertical resolution of the ozone

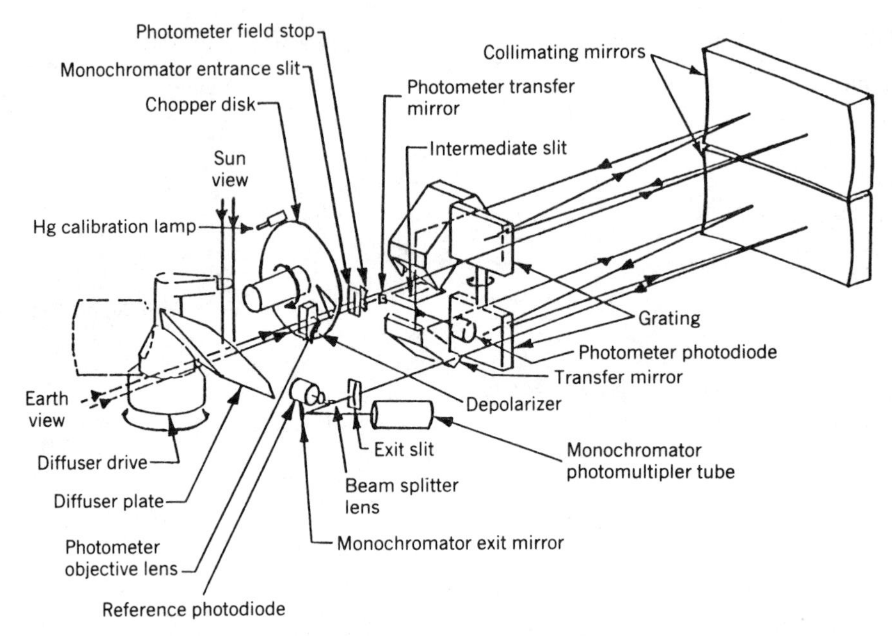

Figure 2-3-9. Solar backscatter ultraviolet optics diagram. Courtesy of Dr. A. Krueger, NASA GSFC.

profiles from BUV and SBUV is about 8 km (above 25 km) varying to about 15 km (below 25 km). Vertically, ozone profiles are produced from BUV and SBUV measurements taken at 19 levels from 100 to 0.3 mb.

Measurements of BUV began in April 1970 and have continued to the present, except for the period May 1977 to November 1978, when routine data recording and processing of *Nimbus 4* BUV ceased before successor instruments SBUV and TOMS were launched aboard *Nimbus 7*.

Ozone values are derived for each instrument scan. The interval between scans is 32 sec for the SBUV and BUV and 8 sec for the TOMS. Thirty-five samples are retrieved for each scan. Except for polar night, complete global coverage is obtained in about two weeks for the BUV and SBUV, but daily for the TOMS.

2-3-2 Infrared Sensors

Temperature Humidity Infrared Radiometer

A temperature humidity infrared radiometer (THIR) was flown on the *Nimbus* satellites. The two-channel high-resolution scanning radiometer was designed to perform two major functions: first the 10.5–12.5 μm window channel provided both day and night cloud top or surface temperatures; second, a water vapor channel at 6.5–7.0 μm (see Fig. 2-3-10) gave information on the moisture content of the upper troposphere and stratosphere and the location of jet streams and frontal systems. The ground resolution at the subpoint was 8 km for the window channel and 22 km for the water vapor channel. Both channels operated continuously to provide day and night global coverage.

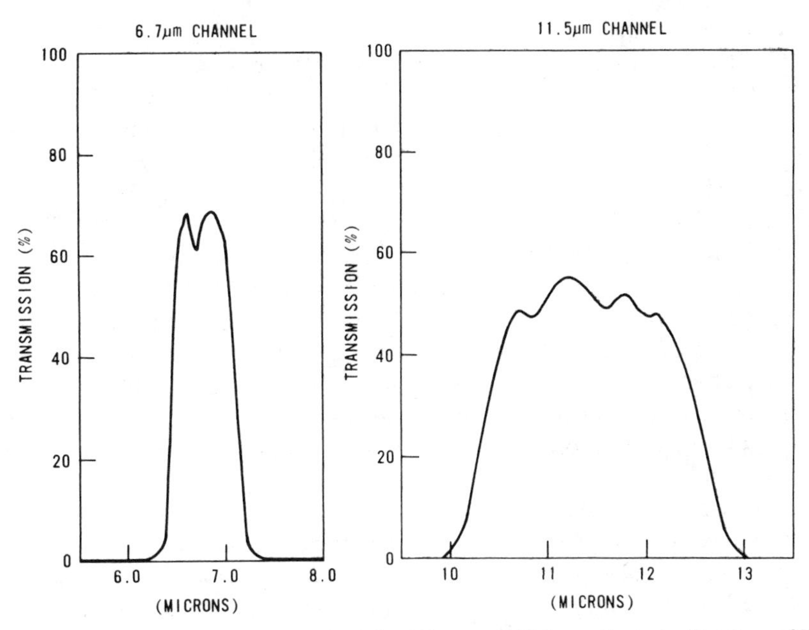

Figure 2-3-10. Relative spectral response of the 6.7-μm and 11.5-μm channels. Courtesy of NASA, Nimbus Project.

The optical system shown in Figure 2-3-11 consisted of a scan mirror, a telescope with primary and secondary mirrors, and a dichroic beam splitter. The scan mirror inclined at 45° to the optical axis, rotated at 48 rpm, and scanned in a plane perpendicular to the direction of the motion of the satellite (see Fig. 2-3-11). The scan mirror rotation was such that, when combined with the velocity vector of the satellite, a right-handed spiral resulted, so the field of view scanned across the earth from east to west during the day and west to east at night, when the satellite was traveling northward and southward, respectively (McCulloch, 1972).

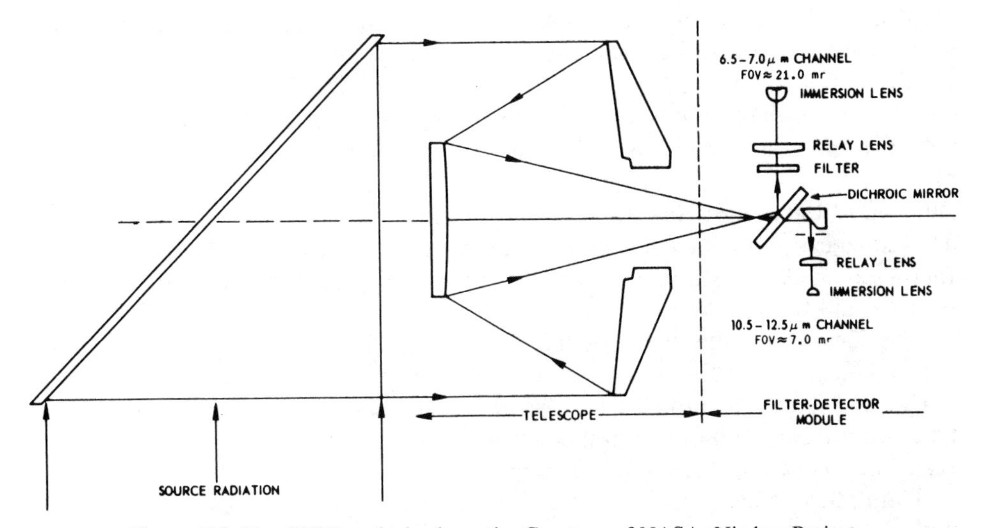

Figure 2-3-11. THIR optical schematic. Courtesy of NASA, Nimbus Project.

Heat Capacity Mapping Radiometer

The HCMR is a two-channel scanning radiometer, one covering the reflectance band from 0.5 to 1.1 μm, while the other channel monitors the thermal IR band between 10.5 and 12.5 μm. The two channels thus simultaneously provide measurements of reflected solar and emitted thermal IR bands. Measurement accuracy in the two channels is limited by the analog telemetry system to a noise equivalent radiance of 0.2 mW/cm^2 in the 0.5–1.1 μm channel and NETD of 0.4 K at 280 K in the 10.5–12.5 μm channel. Exclusive of the telemetry system, the thermal channel sensor itself has a measured NETD of 0.3 K at 280 K.

From the nominal orbit altitude of 620 km, the spatial resolution of the IR channel is approximately 600 m × 600 m at nadir, and the resolution in the reflectance channel is 500 m × 500 m. These values are masked by data processing, which generates registered data at a 481.5-m pixel size. Registration between channels is to 0.2 resolution elements. The swath of data coverage along the track is approximately 716 km wide.

The HCMR is composed of four major subassemblies mounted in a common housing. These subassemblies include a scan mirror and drive, optics, electronics, and radiant cooler. The scan mirror drive assembly provides a cross-course scanning of the instantaneous field of view with reference to the subsatellite ground track. The reflectance channel optics consist of a long-wavelength (greater than 0.55 μm) pass interference filter, focusing optics, and an uncooled silicon photodiode. The long-wavelength cutoff of the silicon detector limits the bandpass to wavelengths of less than 1.1 μm. The sensitive area of the detector is approximately 0.15 mm^2.

The thermal IR beam is focused onto the mercury-cadmium-telluride detector by using a germanium lens. Final focusing and spectral trimming is accomplished by a germanium aplanat located at the detector.

The reflected solar energy channel has no in-flight calibration capability and therefore relies on ground-calibration data. Since any degradation in the optical train or detector in this channel during the life of the mission is undetectable by the sensor, and since this degradation would appear as an error in the calibration of the data, the performance of the sensor is checked periodically by observing an instrumented ground-calibration site.

The thermal channel of the HCMR updates its calibration once each scan line (or 14 times a second) by using a combination of in-flight and preflight data. The optical configuration of HCMR is given in Figure 2-3-12.

Advanced Very High Resolution Radiometer

The advanced very high resolution radiometer (AVHRR) for *Tiros-N* and four follow-up satellites is a four-channel scanning radiometer that measures visible/near IR and thermal IR radiation.

The scanner module includes the 80-pole hysteresis synchronous motor, the motor housing, and the scan mirror. The scan motor continuously rotates the mirror at 360 rpm to produce a crosstrack scanning in orbit. The instantaneous field was chosen so that the satellite motion along its orbit would cause successive scan lines to be contiguous at the subpoint.

The optical system consists of a focal 20.3-cm aperture telescope combined with secondary optics that separate the radiant energy into discrete spectral bands, which are then focused onto their respective field stops.

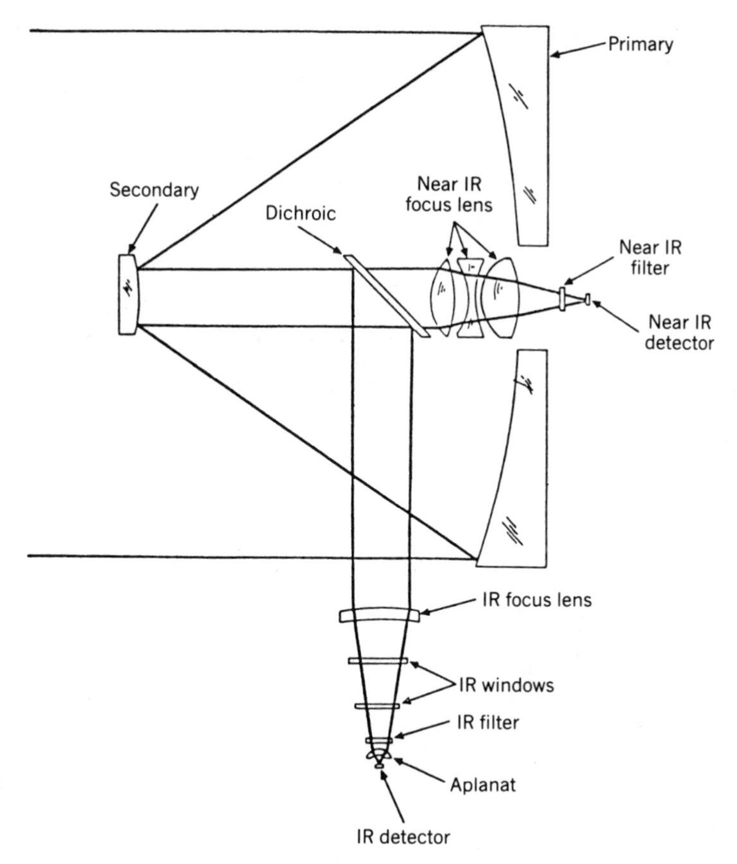

Figure 2-3-12. Optical diagram for HCMR. Courtesy of NASA, Nimbus Project.

The visible and near IR channels use silicon detectors to measure incident radiation; the IR channel uses detectors cooled to 105 K. The detector chosen for the 3.8-μm channel is indium antimonide, whereas the 11-μm channel uses mercury-cadmium-telluride. The indium antimonide has a 0.0173-cm^2 active area and is mounted with the aplanat lens forming a hermetic seal. An NETD better than 0.12 K (for a 300 K scene) is expected from this channel. The mercury-cadmium-telluride detector is optimized for best sensitivity between 10.5 and 11.4 μm. The detector is 0.0173 cm^2 and is bonded to the aplanat.

The approximate spectral bandwidths of the AVHRR channels for the *Tiros-N/ NOAA-A* through *G* series are given in Table 2-3-5. An example of the spectral response functions of two *NOAA 6* AVHRR channels is shown in Figure 2-3-13.

The AVHRR provides high-quality digital measurements that have a basic spatial resolution of 1.1 km at nadir in the visible (0.58–0.68 μm) and reflected IR (0.725–1.1 μm) bands and in two or three emitted IR window channels (3.55–3.93 μm, 10.3–11.3 μm, and 11.5–12.5 μm). *Tiros-N, NOAA 6,* and *NOAA 8* carried AVHRRs equipped with the first two window channels cited, whereas *NOAA 7* and *NOAA 9* have AVHRRs with all three. The full-resolution measurements are available locally by direct readout (high-resolution picture transmission) and by means of limited, temporary, onboard tape storage (local area coverage-LAC).

TABLE 2-3-5. Spectral Bandwidths (μm) of the AVHRR Channels for the *Tiros-N/NOAA* A through G Series

Channel No.	Tiros-N	NOAA-A, -B, -C, -E	NOAA-D, -F, -G	IFOV (mrad)
1	0.55 – 0.90	0.58 – 0.68	0.58 – 0.68	1.39
2	0.725– 1.10	0.725– 1.10	0.725– 1.10	1.41
3	3.55 – 3.93	3.55 – 3.93	3.55 – 3.93	1.51
4	10.5 –11.5	10.5 –11.5	10.3 –11.3	1.41
5	10.5 –11.5	10.5 –11.5	11.5 –12.5	not available

The instantaneous field of view (IFOV) of each sensor is approximately 1.4 mr leading to a resolution at the satellite subpoint of 1.1 km for a nominal altitude of 833 km. The analog data output from the sensors is digitized on board the satellite at a rate of 39,936 samples per second per channel. At this sampling rate, there are 1.362 samples per IFOV. A total of 2048 samples are obtained per channel per earth scan, which span an angle of ±56° from the nadir.
Courtesy of NOAA.

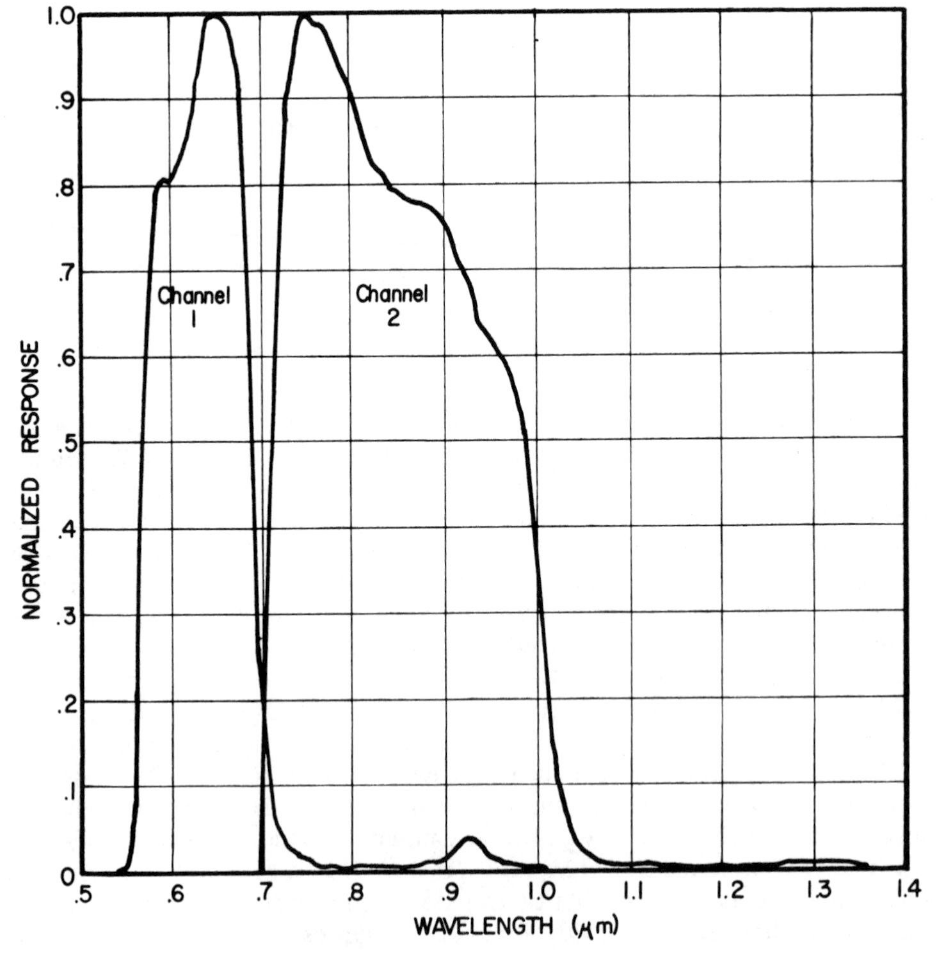

Figure 2-3-13. Spectral response curves for channels 1 and 2 for the AVHRR. Courtesy of NOAA.

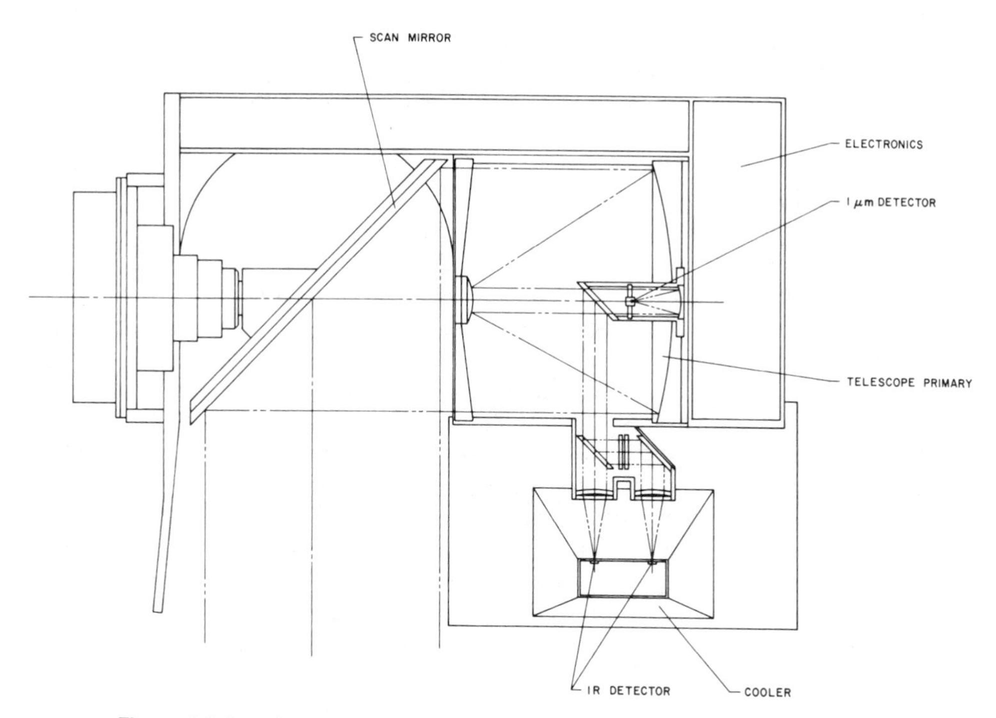

Figure 2-3-14. SCMR optical configuration. Courtesy of NASA, Nimbus Project.

Global area coverage (GAC) data is provided twice daily at a nominal resolution of 4 km by means of onboard data reduction and tape recording. There is provision for onboard calibration of the emitted IR channels.

The noise level has been exceptionally low (<0.1 K in the 11- and 12-μm channels). The 3.7-μm window data, however, tend to have an acceptable noise figure (<0.2 K) during the first 12 months or so after each satellite launch but then become increasingly contaminated by electrical interference thereafter (McClain et al., 1985).

Surface Composition Mapping Radiometer

The surface composition mapping radiometer (SCMR), experimentally flown on *Nimbus 5,* consisted of a three-channel instrument with thermal bands between 8.3–9.3 μm and 10.2–11.2 μm and a channel measuring the reflected solar energy between 0.8 and 1.1 μm.

Scanning of the field of view was in the direction of the spacecraft track with a mirror rotating at 10 rps with a scan width of about 800 km. The optical configuration of the instrument is given in Figure 2-3-14.

The 8.8- and 10.7-μm channels were selected to coincide with the Reststrahlen features of acidic (high silica content) and basic (low silica content) rocks, soils, and consolidated sediments. The object was to sense the radiation in both channels, convert them to equivalent blackbody temperatures, and determine if a temperature difference occurred. Since Reststrahlen manifests itself as a decrease in emissivity, variable in wavelength with silica content, the polarity of the temperature difference should indicate if the surface material is acidic or basic. The concept of differentiating surface material based on laboratory measurements of rocks and the resulting selection of appropriate wavelengths are shown in Figure 2-3-15.

Tiros Operational Vertical Sounder

The *Tiros* operational vertical sounder (TOVS) consists of three instruments: (1) the second version of the high-resolution infrared radiation sounder (HIRS-2), originally tested aboard the *Nimbus 6* satellite, (2) the microwave sounding unit (MSU), which is similar to the scanning microwave spectrometer (SCAMS) flown on *Nimbus 6* and 3) the stratospheric sound unit (SSU).

Table 2-3-6 provides the characteristics and purpose of the radiance observations provided by the various spectral channels of each instrument. Demonstration of the vertical resolution of the radiation observed within each channel is given in Figure 2-3-16. Each curve in the figure shows the sensitivity of the radiance observed in the spectral interval of the indicated channel to a local variation in atmospheric temperature. These are usually referred to as "weighting functions" because they appear as weights in the integral equations relating atmospheric temperature to the observed radiances. These weighting functions vary according to the water vapor and ozone content of the atmosphere and according to atmospheric temperature (Werbowetzki, 1981).

Calibrations of Infrared Sensors

All the radiometers flown on satellites undergo extensive prelaunch radiometric calibrations at the instrument manufacturer's facilities to establish their stability,

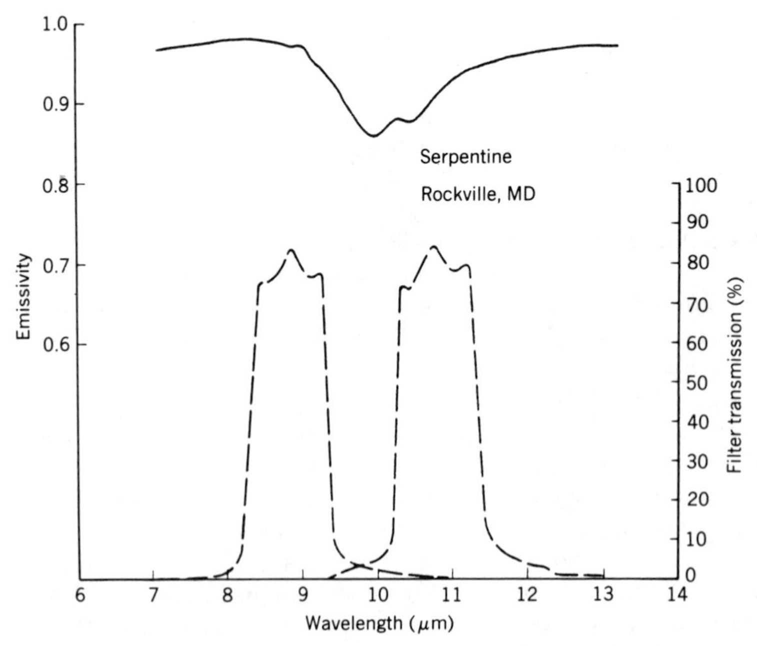

Figure 2-3-15. Laboratory measured emissivity of serpentine sample and the filter transmission of the SCMR. After Hovis (1972).

TABLE 2-3-6. Characteristics of TOVS Channels

HIRS Channel Number	Channel Central Wave Number	Central Wavelength (μm)	Principal Absorbing Constituents	Level of Peak Energy Contribution	Purpose of the Radiance Observation
1	668	15.00	CO_2	30 mb	*Temperature sounding.* The 15-μm-band channels provide better sensitivity to the temperature of relatively cold regions of the atmosphere than can be achieved with the 4.3-μm-band channels. Radiances in Channels 5, 6, and 7 are also used to calculate the heights and amounts of cloud within the HIRS field of view.
2	679	14.70	CO_2	60 mb	
3	691	14.50	CO_2	100 mb	
4	704	14.20	CO_2	400 mb	
5	716	14.00	CO_2	600 mb	
6	732	13.70	CO_2/H_2O	800 mb	
7	748	13.40	CO_2/H_2O	900 mb	
8	898	11.10	Window	Surface	*Surface temperature* and cloud detection.
9	1 028	9.70	O_3	25 mb	*Total ozone* concentration.
10	1 217	8.30	H_2O	900 mb	*Water apor sounding.* Provides water vapor corrections for CO_2 and window channels. The 6.7-μm channel is also used to detect thin cirrus cloud.
11	1 364	7.30	H_2O	700 mb	
12	1 484	6.70	H_2O	500 mb	
13	2 190	4.57	N_2O	1 000 mb	*Temperature sounding.* The 4-3-μm-band channels provide better sensitivity to the temperature of relatively warm regions of the atmosphere than can be achieved with the 15-μm-band channels. Also, the short-wavelength radiances are less sensitive to clouds than those for the 15-μm region.
14	2 213	4.52	N_2O	950 mb	
15	2 240	4.46	CO_2/N_2O	700 mb	
16	2 276	4.40	CO_2/N_2O	400 mb	
17	2 361	4.24	CO_2	5 mb	
18	2 512	4.00	Window	Surface	*Surface temperature.* Much less sensitive to clouds and H_2O than the 11-μm window. Used with 11-μm channel to detect cloud contamination and derive surface temperature under partly cloudy sky conditions. Simultaneous 3.7- and 4.0-μm data enable reflected solar contribution to be eliminated from observations.
19	2 671	3.70	Window	Surface	
20	14 367	0.70	Window	Cloud	*Cloud detection.* Used during the day with 4.0- and 11-μm window channels to define clear fields of view.

MSU	Frequency (GHz)	Principal Absorbing Constituents	Level of Peak Energy Contribution	Purpose of the Radiance Observation
1	50.31	Window	Surface	*Surface emissiity* and *cloud attenuation* determination.
2	53.73	O_2	700 mb	*Temperature sounding.* The microwave channels probe through clouds and can be used to alleviate the influence of clouds on the 4.3- and 15-μm sounding channels.
3	54.96	O_2	300 mb	
4	57.95	O_2	90 mb	

SSU	Wavelength (μm)	Principal Absorbing Constituents	Level of Peak Energy Contribution	Purpose of the Radiance Observation
1	15.0	CO_2	15.0 mb	*Temperature sounding.* Using CO_2 gas cells and pressure modulation, the SSU observes thermal emissions from the stratosphere.
2	15.0	CO_2	4.0 mb	
3	15.0	CO_2	1.5 mb	

After Werbowetzki (1981). Courtesy of NOAA.

linearity of response, and sensitivity in output digital counts per radiance unit. Instruments operating in the thermal IR and microwave regions of the spectrum are calibrated against precision blackbody sources. Instruments operating in the visible and near IR regions are calibrated against lamps whose output is viewed through the aperture of an integrating sphere.

Prelaunch IR calibrations are performed in a thermal/vacuum environment to simulate the environment of space. During the thermal/vacuum exposure, calibrations are performed at several instrument operating temperatures (nominal ± 10°C) to provide a measure of the deviation of the instrument's response as a function of temperature. Visible and near IR calibrations are performed at ambient temperature in air.

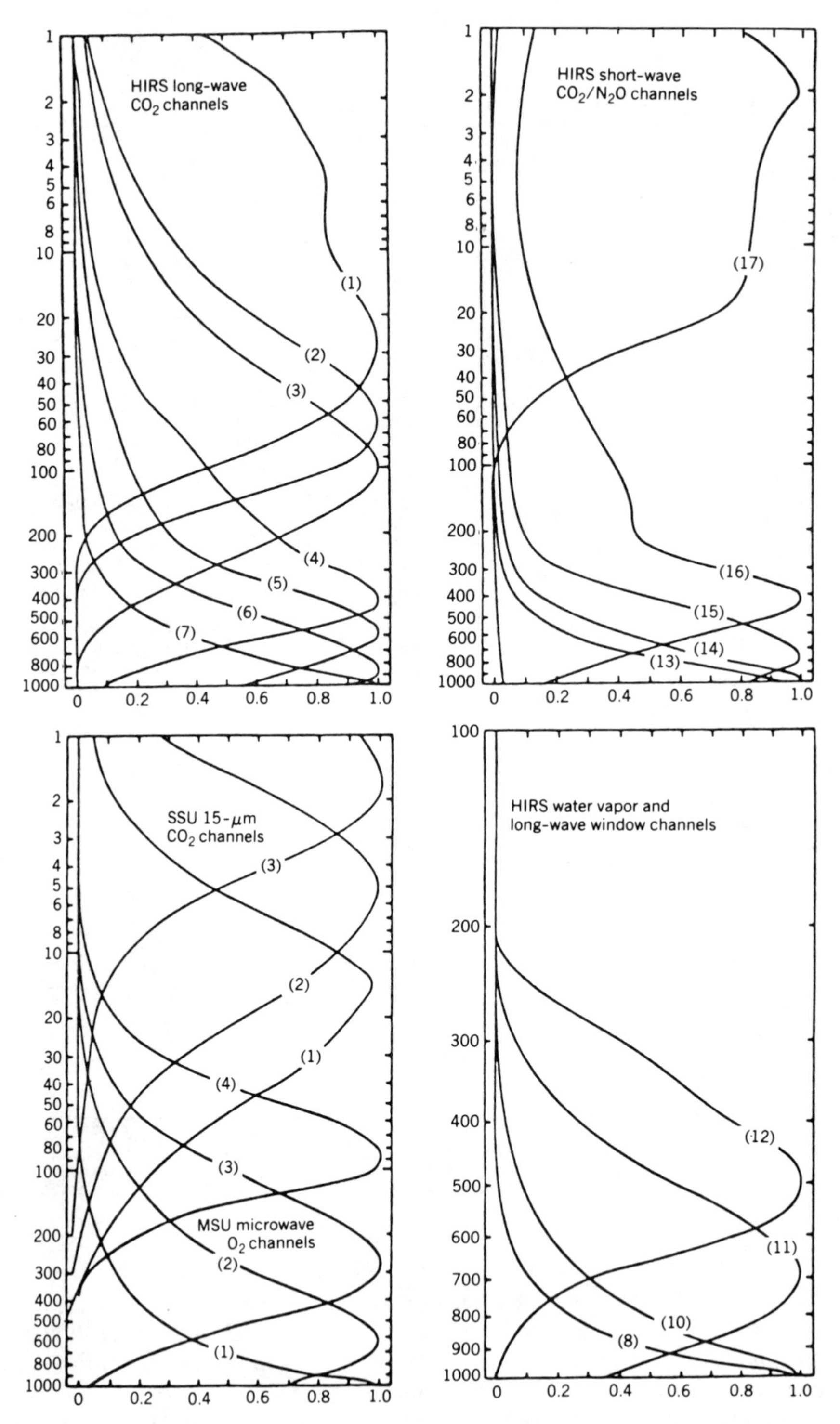

Figure 2-3-16. TOVS weighting functions (normalized). After Werbowetzki (1981). Courtesy of NOAA.

In-orbit calibration is accomplished by programming a given radiometer's scan mirror to view space (near-zero radiance) and part of its housing, which is a designed blackbody. This onboard blackbody is maintained at approximately the operating temperature of the radiometer (15°C or 288 K). Also, it is instrumented with temperature sensors whose outputs are multiplexed into the radiometer's telemetry. Thus, the zero radiance from the space look and the radiance from the 15°C onboard blackbody provide a two-point, in-flight calibration. Details of the calibration of the specific IR radiation sounders and MSUs are given by Werbowetzki (1981).

The main parameters for calibration of all electromagnetic radiation detection devices are essentially the same. Three fundamental quantities must be defined: the effective spectral response ϕ_λ, the effective radiance $\overline{N}$, and the equivalent blackbody temperature T_B. Here ϕ_λ is a composite function involving all of the factors that contribute to the spectral response of the instrument, such as filter transmission, mirror reflectances, and spectral responsivity of the detector.

In preflight laboratory calibrations, the field of view of the radiometer is filled by a blackbody target whose temperature can be varied and accurately measured from 150 to 340° K. From T_B, the spectral radiance of the target is determined by the Planck function B_λ. The integration of this function over ϕ_λ yields that portion of the radiance of the target to which the radiometer responds, $\overline{N}$,

$$\overline{N} = \int_0^\infty B_\lambda(T_\lambda)\phi_\lambda \, d\lambda \tag{2-1}$$

The effective radiance to which the orbiting radiometer responds may be expressed by

$$\overline{N} = \int_0^\infty N_\lambda \phi_\lambda \, d\lambda \tag{2-2}$$

where N_λ is the spectral radiance in the direction of the satellite from the earth and its atmosphere. It is convenient to express the measurement from orbit in terms of an equivalent temperature of a blackbody filling the field of view which would cause the same response from the radiometer. From Equations 2-1 and 2-2 we see that this "equivalent blackbody temperature" corresponds to the target temperature T_B of the blackbody used in the laboratory calibration. Therefore, the radiometer measurements can be expressed either as values of effective radiance $\overline{N}$ or as equivalent blackbody temperatures, T_B.

2-3-3 Microwave Sensors

Passive Microwave Sensors

Microwave radiometers perform passive microwave observations of the atmosphere; they measure the thermal emission from various broad layers between the earth's surface and the atmosphere. Physical processes that determine the observed emissions include

1. Line emission and absorption by molecular species, such as O_2, H_2O, and O_3, which typically are in thermal equilibrium at altitudes below ~100 km

2. Pressure broadening of spectral lines, which typically dominates line shape above ~50 km and which requires use of a matrix equation of radiative transfer
3. Nonresonant absorption and scattering by droplets and sometimes by water or ice particles, which may have nonuniform compositions and nonspherical shapes that scatter radiation in particular directions
4. Doppler thermal broadening and frequency shifts due to wind
5. Surface effects (Braun and Rausch, 1986)

Using scanning antennae, we can estimate the three-dimensional distributions of atmospheric temperature and composition by mathematically operating on the microwave spectrum observed.

Electrically Scanning Microwave Radiometer. A passive microwave sensor, the electrically scanning microwave radiometer (ESMR), was flown on board *Nimbus 5;* it has four principal major components (Wilheit, 1972):

1. A phased array microwave antenna consisting of 103 waveguide elements, each having its associated electrical phase shifter; the aperture area was 83.3 cm × 85.5 cm
2. A beam-steering computer, which determined the coil current for each of the phase shifters for each beam position
3. A microwave receiver with a center frequency of 19.35 GHz and a bandpass of 5 to 125 MHz, with a sensitivity to radiation from 19.225 to 19.475 GHz, except for a 10-MHz gap in the center of the band
4. Timing, control, and power circuits

The ESMR was arranged to scan perpendicular to the spacecraft velocity vector, beginning 50° to the left of nadir and scanning in 78 steps to 50° to the right of nadir every 4 sec. For a nominal orbit of 1100 km altitude, the resolution was 25 km × 25 km near nadir degrading to 160 km crosstrack by 45 km downtrack at the ends of the scan.

Brightness temperatures were measured at each scan position, which, when properly displayed, produced a microwave image of the portion of the earth near the satellite track.

Scanning Multichannel Microwave Radiometer. The SMMR, the only passive sensor of *Seasat*'s four microwave instruments, was flown earlier on *Nimbus 7*. The SMMR covered a 600-km swath to the right of the spacecraft. It received vertically and horizontally polarized radiation at five frequencies: 6.6, 10.7, 18.0, 21.0, and 37.0 GHz. From this matrix of 10 channels of information, the main goals for the instrument were the extraction of (1) sea surface temperature to 2 K absolute and 1 K relative over resolution cells approximately 80 km × 150 km; (2) ocean surface wind speed to 2 m/s; (3) integrated liquid water content and water vapor; (4) rain rate; and (5) ice age, concentration, and dynamics.

The SMMR measurements provided important information used to correct errors in the radar altimeter data due to atmospheric effects. A 10-channel instru-

TABLE 2-3-7. SMMR Performance Characteristics

Parameter	Channel 1	2	3	4	5
Wavelength	4.54	2.8	1.66	1.36	0.81
Frequency (GHz)	6.6	10.69	18.00	21.00	37.00
Rf bandwidth (MHz)	250	250	250	250	250
Integration time (ms) approximate	126	62	62	62	3
If frequency range (MHz)	10–110	10–110	10–110	10–110	10
Dynamic range (K)	10–330	10–330	10–330	10–330	10
Absolute accuracy (K rms)	<2.0	<2.0	<2.0	<2.0	
Temperature resolution, ΔT_{rms} (K) (per IFOV)[a]	0.9	0.9	1.2	1.5	
Antenna beamwidth ($\pm 0.2°$)	4.2	2.6	1.6	1.4	
Antenna beam efficiency (%)	87.0	87.0	87.0	87.0	
Scan cycle ± 0.4 rad ($\pm 25°$)/sec[b]	4.096	4.096	4.096	4.096	
Double-sideband noise (dB) (maximum)	5.0	5.0	5.0	5.0	

[a] IFOV are remapped to form equal-sized cells (150, 90, 50 km) across the swath prior to retrieval of geophysical parameters; the ΔT_{rms}'s are correspondingly lower.
[b] Add 2 msec (used for integer dump) for complete IFOV cycle time.

ment, the SMMR delivered orthogonally polarized antenna temperature data at the five microwave wavelengths indicated in Table 2-3-7. Six radiometers were utilized: those operating at the four longest wavelengths measured alternate polarizations during successive scans of the antenna; the others, at the shortest wavelengths, operated continuously for each polarization. A two-point reference signal system was used, consisting of an ambient rf termination and a horn antenna viewing deep space. A switching network of latching ferrite circulators selected the appropriate polarization or calibration input for each radiometer. The antenna, scan mechanism, rf module, and sky horn cluster were mounted on a bridgelike platform that was installed as an aligned and calibrated unit on the spacecraft. The electronics and power supply modules were mounted separately and were cabled to the instrument and spacecraft through connectors. Figure 2-3-17 shows the views of the instrument, and Figure 2-3-18 is the functional block diagram. The overall size can be visualized by noting that the elliptical antenna reflector was approximately 110 cm × 80 cm.

On *Seasat-A*, the SMMR scan pattern, covering a range of 50°, was aft viewing and biased toward the right of the flight path, so the center of the swath was 22° from the orbital track. This permitted overlap of the SMMR images with the data obtained by the *Seasat-A* imaging radar and microwave scatterometer. With a subsatellite velocity of 6.6 km/sec and a scan period of 4.096 sec, overlap coverage was provided at all wavelengths.

Active Microwave Sensors

A radar system is a ranging device that measures round-trip travel times and signal modification of a directed beam of pulses over specific distances. In this way, the directional location and separation distances from radar to reflecting target, as well as information about target shape and certain diagnostic physical properties, may be determined by the system. By supplying its own illumination, radar can

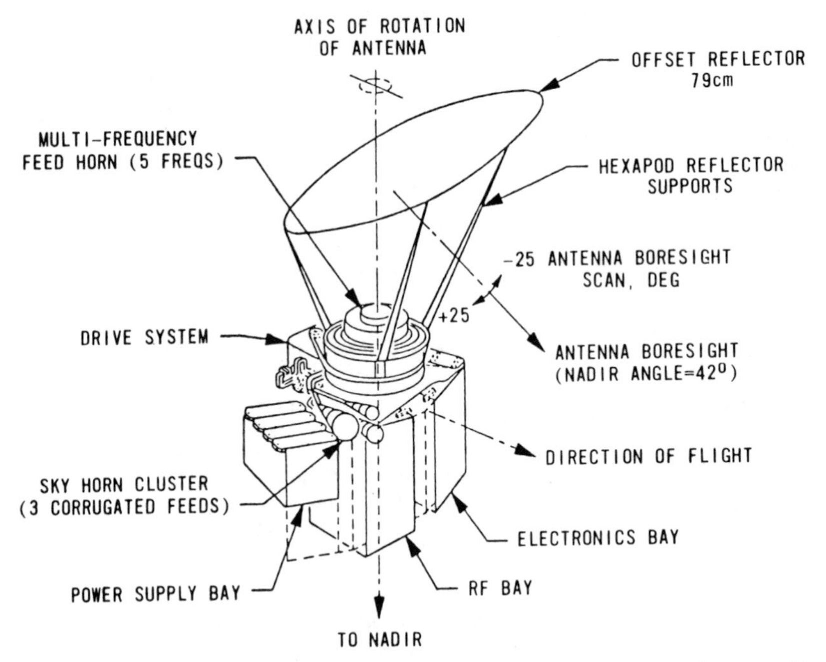

Figure 2-3-17. SMMR instrument configuration showing antenna, feed horn drive assembly, and electronic boxes. Courtesy of NASA, Nimbus Project.

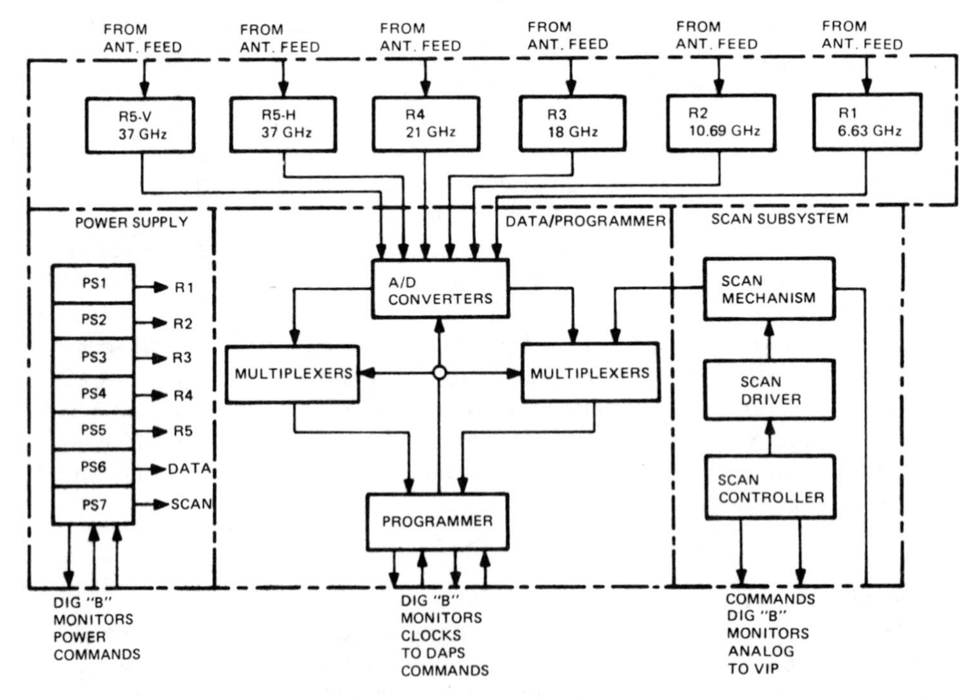

Figure 2-3-18. *Nimbus 7* SMMR functional block diagram. The functional units of the SMMR consist of the radiometer, the scan subsystem, the data/programmer, and the power supply. Courtesy of D. J. Cavalier, NASA.

function day and night and, for some wavelengths, without significant interference from adverse atmospheric conditions.

A radar system consists of (1) a pulse generator that presents time pulses of microwave/radio energy to (2) a transmitter, then through (3) a duplexer, to (4) an antenna that shapes and focuses the pulse stream as transmitted and then picks up returned pulses, which are sent to (5) a receiver capable of amplifying the weakened signals, which are then carried to (6) a real-time display device. Antennae for radars flown on aircraft for surveys are normally mounted on the underside and direct their beams to the side of the plane, that is, normal to the flight path, during operation.

Microwave Scatterometer. An active microwave sensor, a microwave scatterometer, can ideally be used to measure the wind vectors over the ocean surface.

The measurement principle of a microwave scatterometer derives from the fact that at microwave frequencies, the ocean surface roughness, which is a function of actual wind conditions, appears like a reflection grating. This results in a functional dependence between the normalized radar cross section of the ocean surface and the wind speed.

The radar cross section is anisotropic with respect to the angle between wind vector and incident radar beam. With the aid of several radar cross-sectional measurements of the same area from different measurement directions, the actual wind vector in terms of speed and direction can be determined. The conversion of the radar cross-sectional value into wind data is performed with a mathematical model, which defines the relationships among radar cross section, wind speed, wind direction, incidence angle of the scatterometer pulse, and the signal polarization. A typical microwave scatterometer configuration is described in Figure 2-3-19.

The scatterometer illuminates the sea surface sequentially for reflectivity measurements by rf pulses from different directions by three antennae. The nominal look angles of the antennae are 45° fore and aft as well as broadside to the satellite velocity vector. Reflectivity data are provided for a wide continuous swath along the satellite ground track to deduce wind speed and direction at resolution cells along and across the subsatellite track within the swath.

Each node is centered within a resolution cell with range direction determined by appropriate range or Doppler grating of the received echo signal and azimuth determined by averaging corresponding echo signals, being limited by the azimuth beamwidth of the antennae.

Wind Field Scatterometer. The wind field scatterometer, flown on *Seasat,* had an active radar that illuminated the sea surface with four fan-shaped beams with the intention of determining wind stress and wind direction. The angular effect of modulation at K-band frequencies corresponded with modifications in surface wind direction, and the radar backscatter coefficient was modulated by the change in small-scale roughness, which can be related to the wind stress at the air-sea interface.

Synthetic Aperture Radar. A SAR produces images of the earth's surface by virtue of the differing local normalized radar cross section (reflectivity) of the

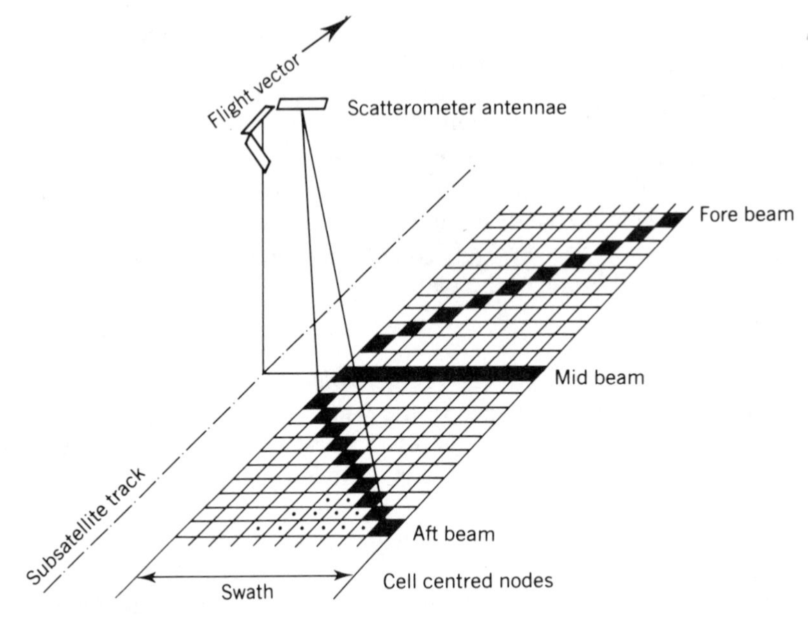

Figure 2-3-19. Microwave scatterometer geometry. After Braun and Rausch (1986). Reproduced with permission from Graham & Trotman, Ltd.: Szekielda, K.-H., 1986. *Satellite Remote Sensing for Resources Development*.

target. As the instrument provides its own source to illuminate the target, it can be operated independently of diurnal and weather-influenced solar illumination.

Pulses emitted from the SAR antenna are reflected from a wide swath. All objects within the swath bounce back signals with a strength according to their respective radar cross sections. The return from a pixel is characterized by amplitude, phase shift, time of arrival, and Doppler shift, the last being caused by satellite motion and earth rotation.

The pixel in Figure 2-3-20 is illuminated by many pulses during the SAR overflight and, consequently, reflects these echoes, which are summed up in the same range. They differ, however, by their respective Doppler shifts, which are known a priori by geometry and can be calculated.

The radar transmits pulses at the earth via an antenna with a 1° × 6° fan beam pointed 20° off nadir to the right side of the spacecraft. The 6° beamwidth illuminates 100 km swath centered about 300 km off the spacecraft track. Echoes from land and ocean are amplified, translated to an S-band telemetry frequency, and immediately transmitted to a ground tracking station. At the ground station, the radar echoes are digitized and recorded on magnetic tapes which are converted to film images or digital tapes by SAR processing adapted to orbital parameters.

A summary of the radar parameters of SAR is given in Table 2-3-8.

The main purpose of the SAR was to obtain high-resolution imagery of the sea surface and sea ice. The SAR operated at L band (1.4 GHz) and was designed to image a 100-km swath offset some 250 km to the right of the satellite's suborbital track. The geometric resolution was 25 m, and the respective SAR data rate was so high that no onboard recording was possible and the raw data were transmitted via a real analog channel to a ground station.

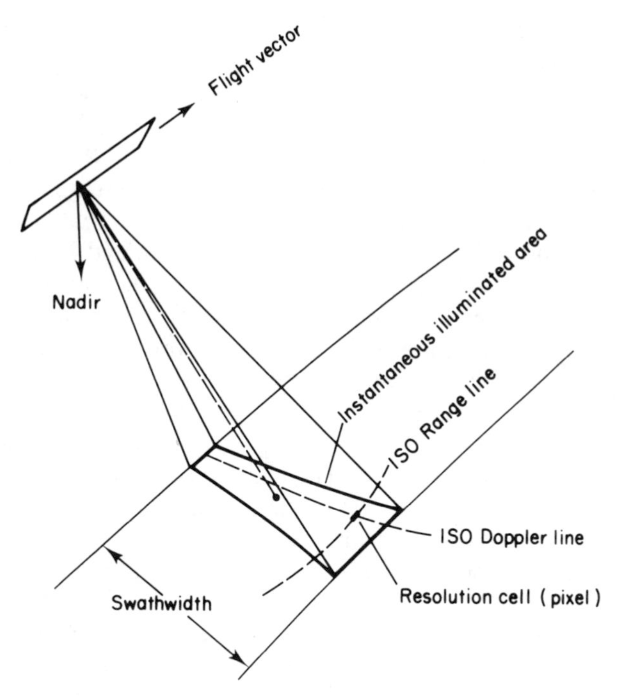

Figure 2-3-20. SAR measurement geometry. After Braun and Rausch (1986). Reproduced with permission from Graham & Trotman, Ltd.: Szekielda, K.-H., 1986. *Satellite Remote Sensing for Resources Development.*

TABLE 2-3-8. Radar Parameters of SAR

Frequency	1274.8 MHz
Wavelength	23.5 cm
Transmitted bandwidth	19 MHz
Pulse duration	33.8 μsec
Pulse time-bandwidth product	642
Transmitter rf power	800 W peak–46 W average
Transmitter type	Solid-state bipolar transistor
Antenna beamwidth, elevation	6°
Antenna beamwidth, azimuth	1°
Antenna beam center gain	34 dB, minimum
Antenna pointing angle	20° off nadir, right side
Surface resolution	25 m × 25 m
Image swath width	100 km
Image length	1000 to 4000 km
Downlink	20 MHz analog
Sensor weight	110 kg
Antenna weight	110 kg (est)
Data link weight	25 kg (est)
Sensor power	600 W, nominal operation

After Thompson and Laderman (1976).

The measurement objectives of the SAR were (1) to obtain radar imagery of ocean wave patterns in open ocean; (2) to obtain ocean wave patterns and water-land interaction data in coastal regions; and (3) to obtain radar imagery of sea and freshwater ice and snow cover.

The *Seasat-A* SAR antenna system consists of a 10.74 m × 2.16 m phased array system deployed after orbit insertion. This deployed antenna is configured to fly with the long dimension along the spacecraft velocity vector and boresighted at an angle of 20.5° from the nadir direction in elevation (cone) and 90° from the nominal spacecraft velocity vector. The antenna dimensions (10.74 m × 2.16 m) are dictated by a desire to limit range or Doppler ambiguities to acceptably low levels. At a nominal 20.5° look angle from nadir, a total beamwidth in elevation of 6.2° is required to illuminate a 100-km swath on the earth's surface from an 800-km-high orbit (Jordan, 1980).

Radar Altimeter. The radar altimeter is a nadir-looking active microwave instrument that can be operated over ocean, ice, and land. Over the oceans, it is used to determine the significant wave height, wind speed, and mesoscale topography. Over ice, it is used to determine the ice surface topography, ice type, and sea/ice boundaries.

The ocean radar altimeter is based on the electromagnetic backscattering of the ocean surface in response to a narrow pulse under near normal incidence. The average return echo (average over many return pulses) carries information on the sea surface, such as the mean sea level and the sea state (significant wave height and wind speed). The above power response can be modeled by the convolution of the following time functions: (1) the average flat-surface impulse response (decreasing exponential with a time constant related to the antenna beamwidth); (2) the height (converted into delay time) probability density function (pdf) of the sea surface scatterers (gaussian with standard deviation proportional to the rms wave height; and (3) the altimeter system point target response (gaussian with standard deviation proportional to the pulse width) (Braun and Rausch, 1986).

A short-pulse radar altimeter for the study of marine geodesy was flown on *Seasat* with the altimeter range of sea state up to about 20 m wave height with a precision of approximately 10%. This radar altimeter was based on similar altimeters designed for *Skylab* and GEOS-C experiments and allowed for studies of dynamic departures from the marine geoid corresponding to currents, storm surges, etc.

A listing of parameters for the *Skylab, GEOS-C,* and *Seasat-A* altimeters is given in Table 2-3-9. The variation in return signal strength with altitude and ground resolution required the use of pulse compression for *GEOS-C* and *Seasat-A,* whereas *Skylab,* which operated at a lower altitude, could use a 100-nsec uncompressed pulse. The 3.125-nsec pulse of *Seasat-A* defined a 1.7-km-resolution diameter and required a longer pulse and a higher antenna gain to give a per pulse signal-to-noise ratio equal to or better than *GEOS-C*.

Shuttle Imaging Radar (SIR-A and SIR-B). In November 1981, the shuttle *Columbia* carried the shuttle imaging radar (SIR-A), which successfully acquired radar images over numerous areas of the world. The objective of the experiment was to assess the capability of spaceborne radars for geologic mapping and earth resources observation.

TABLE 2-3-9. Comparison of Altimeter Parameters

	Skylab	*GEOS-C* (intensive)	*Seasat-A*
Mean altitude (km)	435	840	800
Antenna beamwidth (°)	1.5	2.6	1.6
Frequency (GHz)	13.9	13.9	13.5
Peak rf power (kW)	2	2	2
Average rf power (W)	0.05	0.24	6.5
Pulse width (uncompressed)	100 nsec	1 μsec	3.2 μsec
Pulse width (nsec) (compressed)	—	12.5	3.125
Repetition frequency (Hz)	250	100	1020
Footprint diameter (km)	8	3.6	1.7
Altitude precision (rms)	<1 m	<50 cm	<10 cm

Courtesy of NASA.

The SIR-A is a synthetic aperture imaging radar that uses the motion of the platform to synthesize a long aperture, which would allow high-resolution imaging. The SIR-A achieved a resolution of 38 m over an imaging swath of 50 km. The geometry was selected to achieve a 50° incidence angle, to acquire data where the scattering mechanisms are mainly due to surface roughness. This allows comparative analysis between SIR-A images and *Seasat* SAR images.

Space Shuttle Mission 41-G, launched October 5, 1984, carried as part of its scientific payload the SAR experiment SIR-B. Like *Seasat* and SIR-A, SIR-B operated in the L band (23 cm) and was horizontally polarized. However, SIR-B allowed digitally processed imagery to be acquired at selectable incidence angles between 15 and 60°. *Seasat* and SIR-A were both fixed-parameter sensors, acquiring images at constant look angles.

The SIR-B instrument is an upgraded version of its predecessor, SIR-A. Its new features include additional antenna electronics that allow selectable look angles, an increased bandwidth, a calibrator, and a digital data handling subsystem. The antenna and radar sensor have been carried over from SIR-A with only slight modification. The optical recorder remained unchanged. Table 2-3-10 is a list of SIR-B's operating characteristics.

The sensor works by generating microwave pulses at 1.28 GHz (L band) in the transmitter portion and radiating them outward through the antenna. The signals reflected from the earth are then collected by the antenna and sent to the receiver portion of the sensor where they are broken down to baseband to form an offset video output. This output is then sent to either the optical recorder or the digital data handling subsystem, or both.

Microwave Remote Sensing Experiment. The development of the spaceborne microwave remote sensing experiment (MRSE) was the first European step into radar remote sensing from space. The MRSE contained radar sensors of three kinds: a SAR, a dual-frequency scatterometer, and a microwave radiometer. This instrument failed on its first flight before SAR measurements could be made; hence, only few radiometer and scatterometer data are available. The measurement frequency of the MRSE lies within the X band, which makes the SAR mode

TABLE 2-3-10. SIR-B Operating Characteristics

Description	Value
Orbital altitudes	352, 274, 225 km
Orbital inclination	57°
Mission length	8.3 days
Wavelength	23.5 cm
Frequency	1.28 GHz
Polarization	HH
Pulse length	30.4 msec
Bandwidth	12 MHz
Optical recorder bandwidth	6 MHz
Minimum peak power	1.12 kW
Antenna dimensions	10.7 × 2.16 m
Antenna gain	33.2 dB
Look angles	15 to 60°
Swath width	20 to 50 km
Range resolution	14 to 46 m
Azimuth resolution	30 to 30 m (4-look)
Digital data rate	45.6 and 30.4 Mbit/sec
Total digital data	65 h
Total optical data	8 h

Courtesy of NASA.

of the MRSE interesting because it performed the first X-band SAR measurements from space for civil purposes.

2-3-4 Magnetic Sensors

Magsat

The *Magsat* spacecraft was launched in synchronous orbit with inclination 96.76°, perigee 352 km, and apogee 561 km. The cesium vapor scalar and flux gate vector magnetometers together measured the field magnitude to better than 2 nanotesla (nT) and each component to better than 6 nT. Two star cameras, a high-accuracy sun sensor, and a pitch-axis gyro provided the 10–20 arc-second attitude measurements necessary to achieve this accuracy. The magnetometers were located at the end of a boom to eliminate the effect of spacecraft fields. An optical system measured the attitude of the vector magnetometer and sun sensor at the end of the boom relative to the star cameras on the main spacecraft (Langel et al., 1982a).

The spacecraft (Fig. 2-3-21) consisted of an instrument module and a base module. The instrument module contained the spacecraft attitude determination system, the vector magnetometer axis determination system, and, on an instrument platform at the end of the magnetometer boom, the vector and scalar magnetometers. The base module contained the normal supporting systems, including the solar array/power system, the magnetic coil attitude control system, and redundant tape recorders to capture all of the *Magsat* data continuously. Two tape recorders were especially designed for *Magsat* to allow 10 hr of recorded data to be received by a NASA ground station during a typical 5-min pass.

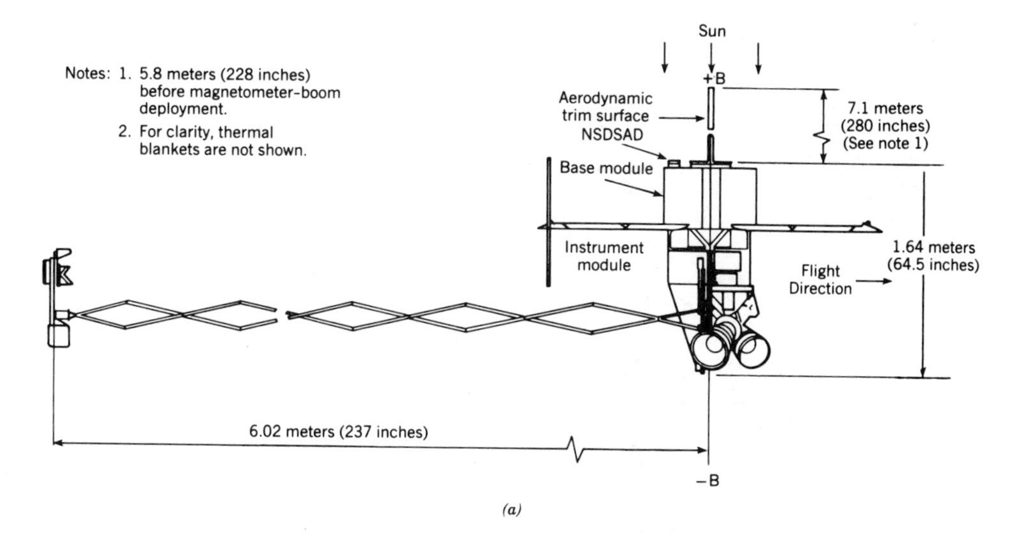

(a)

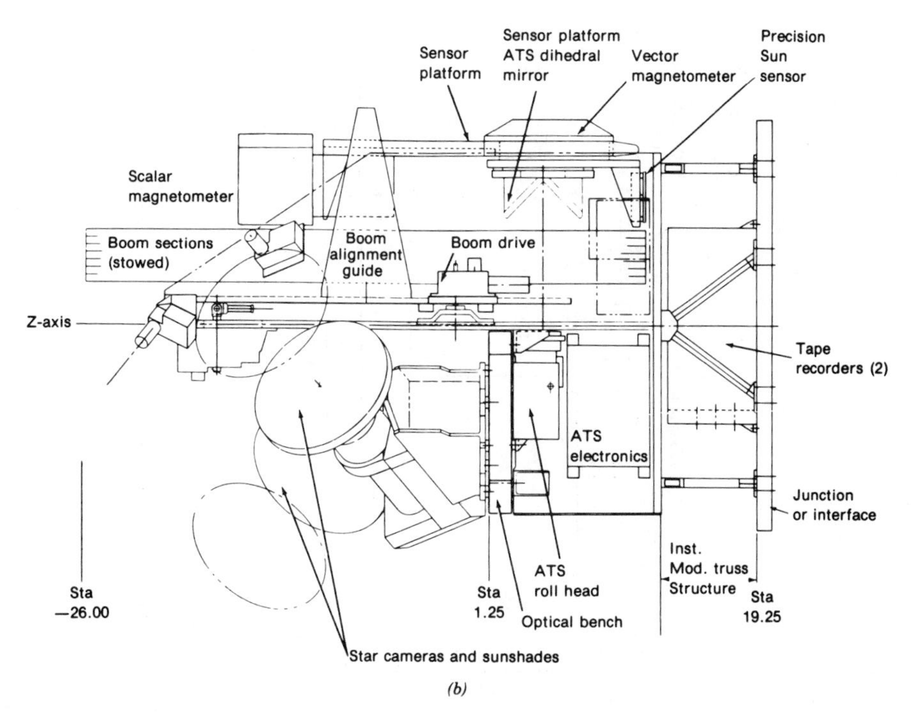

(b)

Figure 2-3-21. (*a*) *Magsat* orbital configuration; (*b*) *Magsat* instrument module. Courtesy of NASA.

The primary objectives of the *Magsat* mission were to obtain data for improved modeling of the time-varying magnetic field generated within the core of the earth and to map variations in the strength and vector characteristics of crustal magnetization.

The scalar magnetometer (cesium vapor type) (see Fig. 2-3-22) was accurate to about 1.5 nT when functioning properly. Unfortunately, it developed internal

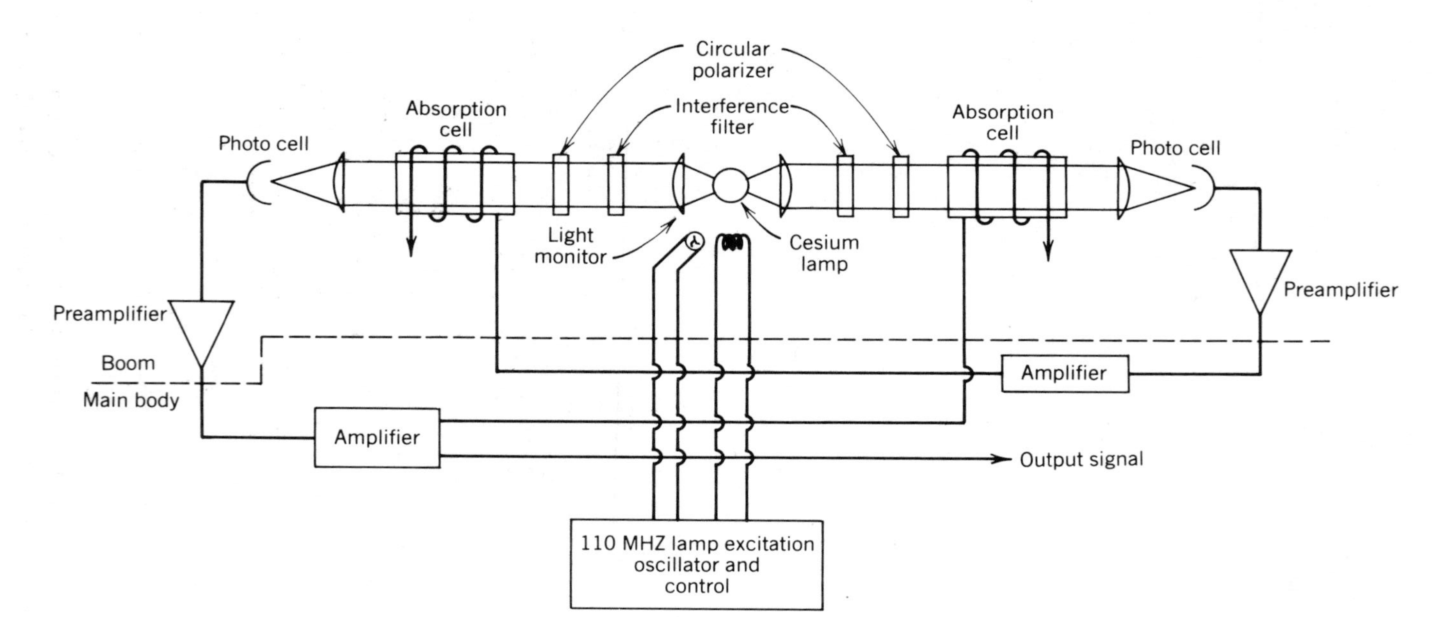

Figure 2-3-22. Dual-cell cesium vapor magnetometer sensor. Courtesy of NASA.

oscillations in its lamp circuitry shortly after launch, which prevented full data recovery and which, at times, slightly degraded its accuracy. Its data were, however, sufficient to provide in-flight calibration for the vector magnetometer (Lancaster et al., 1980).

The vector magnetometer (flux gate type) (Acuna et al., 1978) was built at the Goddard Space Flight Center. It performed flawlessly except for a gradual drift (about 20 nT over the mission lifetime), which was calibrated out by comparison with the cesium vapor magnetometer. The vector magnetometer consisted of three flux gate ring-core sensors mounted on a MACOR (Corning Glass Works tradename for a machineable ceramic) structure. The magnetometer construction was designed to minimize the effects of thermal expansion and contraction. The instrument accuracy, after in-flight calibration, was estimated to be within 3 nT in each axis (Langel et al., 1981).

The basic *Magsat* mission required knowledge of the magnetic field orientation to a total system accuracy of better than 20 arc seconds. The attitude determination system error budget for onboard spacecraft hardware was 14.5 arc seconds, requiring that the two star cameras and two attitude transfer system (ATS) optical heads (pitch/yaw and roll) be mounted on a temperature-controlled graphite-epoxy optical bench. Two mirrors (pitch/yaw and roll) were mounted on the back of the vector magnetometer that was on the end of a 6-m extendable boom, providing a reflected beam of light for accurate magnetometer axis determination (over a ±3-arc-minute field of view). In addition, redundant measurements were provided by a precision sun sensor mounted on the vector magnetometer for roll/yaw, and by pitch gyros mounted in the spacecraft. The star camera measurements were combined with the sun sensor/gyro measurements to provide a cross-check and to allow for attitude interpolation between star camera tracks. Finally, by utilizing the magnetic field measurements from the vector magnetometer, scientists developed an in-flight calibration system that ensured the successful accomplishment of the 20-arc-second mission requirement.

Instrument and attitude accuracies, together with other principal errors, are summarized in Table 2-3-11. Attitude determination was based on data from two star cameras, a sun sensor, a pitch gyro, and the attitude transfer system. In practice, attitude could be determined as long as data were available from any two of the three primary instruments (star cameras or sun sensor).

TABLE 2-3-11. Error Budget

Error Source	Scalar (nT)	Vector (nT)
Instrument	1.5	3.0
Position and time errors	1.0	1.0
Digitization noise	0.5	0.5
Attitude error (20 in at 50,000 nT)	—	4.8
Spacecraft fields	0.5	0.5
Random statistical sample (R.s.s.)	1.96	5.8

After Langel et al. (1982b).

2-3-5 Techniques for Geodynamics

Laser ranging and radio interferometry make it possible to detect crustal movements as small as 1–2 cm/yr over distances of many thousands of kilometers. These methods can be applied globally to measure simultaneously the movement of tectonic plates. They can also be applied regionally to study crustal deformation related to great earthquakes.

When radio antennas are located on different tectonic plates, the motion of these plates appears as a slow change in the base-line length between the antennas. Small and highly mobile, very long base-line interferometry systems are being used to measure regional crustal movements.

Very long base-line interferometry was initially developed by radio astronomers to study the structure of distant radio sources (quasars). Signals from these sources are recorded simultaneously by two antennas which may be thousands of kilometers apart. Atomic clocks, stable to better than 10^{-10} sec/day, provide accurate time for the signal recordings. The difference in arrival of the radio

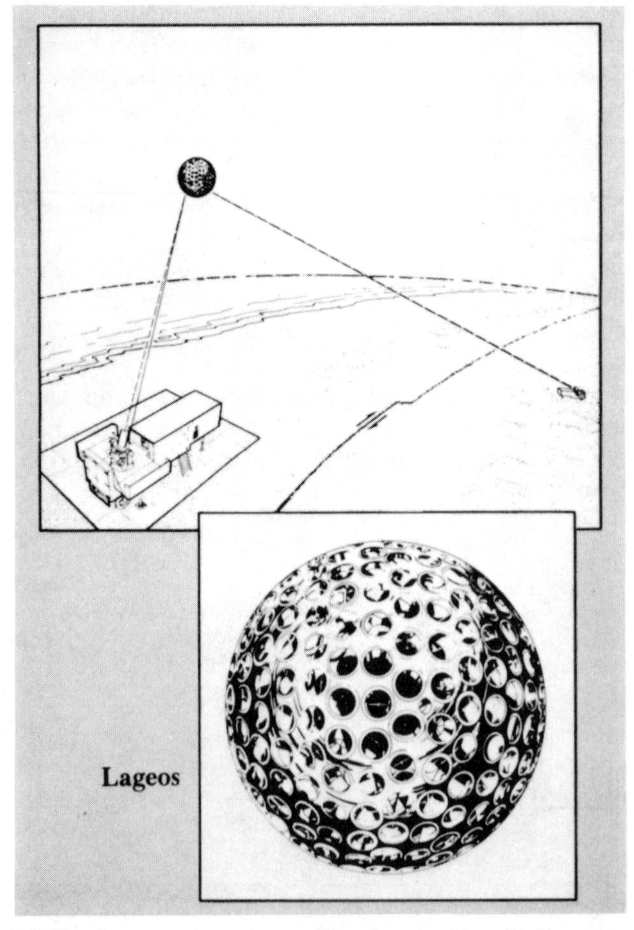

Figure 2-3-23. Laser geodynamics satellite, *Lageos,* 60 cm in diameter and weight of 411 kg.

signals at the two antennas is used to determine the distance between the radio antennas as well as the earth's rotational dynamics.

Ground-based lasers can transmit intense, short pulses of light to retroreflector-equipped satellites and record the round-trip travel time for the pulse to return. If the orbit of the satellite is well known, such ranging permits determination of the location of the laser station on the earth's surface. When two stations range to the same satellite, the distance between the stations can be accurately determined.

So far, 14 satellites equipped with retroreflectors have been launched by the United States and other countries. One of them is the laser geodynamics satellite (*Lageos,* see Fig. 2-3-23), which is only 60 cm in diameter. It was launched in 1976 into a high, extremely stable orbit. The satellite is completely passive; its only function is to reflect laser pulses.

3

ATMOSPHERIC CONSIDERATIONS

3-1 INTERACTION BETWEEN THE ATMOSPHERE AND RADIATION

Electromagnetic radiation emitted from the earth undergoes interaction within the atmosphere before it is detected at an orbiting platform. As a general rule, the attenuation increases as frequency increases or wavelength decreases. The interaction in the form of absorption and scattering may be caused by particulate matter, such as smoke, dust, haze, and condensed water, or gaseous components, such as N_2, O_2, O_3, CO_2, CO, H_2O, and CH_4. The first interaction is known as Mie scattering, and the second as Rayleigh scattering, which occurs when the radiation wavelength is much larger than the size of the scattering particles. In addition, nonselective scattering, which occurs when the scattering particle size is much larger than the radiation wavelength, must also be considered. In Rayleigh scattering, the volume-scattering coefficient σ_λ is

$$\sigma_\lambda = \frac{4\pi^2 N V^2}{\lambda^4} \frac{(n^2 - n_0^2)^2}{(n^2 + n_0^2)^2}$$

where N = number of particles per cm^3
V = volume of scattering particles
λ = radiation wavelength
n = refractive index of particles
n_0 = refractive index of medium

Mie scattering occurs when the radiation wavelength is comparable to the size of the scattering particles. For a continuous particle size distribution, the scattering coefficient is

$$\sigma_\lambda = 10^5 \pi \int_{a_1}^{a_2} N(a)K(a, n)a^2 \, da$$

where σ_λ = scattering coefficient at wavelength λ, km^{-1}
$N(a)$ = number of particles in interval a to $a + da$
$K(a, n)$ = scattering coefficient (cross section)
a = radius of spherical particles
n = index of refraction of particles

Mie scattering may or may not be strongly wavelength dependent, depending on the wavelength characteristics of the scattering area coefficient.

Measured radiation from the earth's surface at satellite altitudes is a composite of the contribution of a radiating target and the radiation from the gaseous compounds. The radiative transfer from a given test site may therefore be described by the following equation:

$$N = \frac{1}{\pi} \int_{\lambda_1}^{\lambda_2} \phi(\lambda) T_s(\lambda) E_\zeta \lambda) B(\lambda, T_s) \, d\lambda$$

$$+ \frac{1}{\pi} \int_{\lambda_1}^{\lambda_2} \int_{h=0}^{h_\infty} \phi(\lambda) \frac{\partial T(\lambda, h)}{\partial h} B[\lambda, T(h)] \, dh \, d\lambda$$

where λ = wavelength
$\phi(\lambda)$ = filter function (assuming that $\phi(\lambda) = 0$ outside the spectral envelope λ_1 and λ_2)
$T(\lambda, h)$ = transmissivity from a height h to satellite altitude h_0
$T_s(\lambda)$ = transmissivity from the radiating surface to the satellite
$E_\zeta(\lambda)$ = emissivity of the radiating surface
$B(\lambda, T)$ = Planck function
T_s = temperature of the radiating surface

For an ideal atmospheric transmissivity of $T = T_s = 1$, the second term on the right side of the equation disappears, and the signal received at the satellite then depends on the emissivity of the target, the temperature, and the angle of view. These assumptions may be made when IR imageries over cloud-free regions are interpreted. For absolute measurements the emissivity as well as the contribution of the atmosphere must be considered.

Total cloud cover absorbs the radiation emitted from the earth but reemits the energy according to the cloud top temperature. The response of a radiometer with partial cloud contamination in the field of view may, however, be described as a response to two radiation sources:

$$N_\lambda = \phi_\lambda \left\{ \tau_\lambda^2 [(1 - \alpha) R_\lambda + \alpha E_\lambda(T_c)] + \int_{P_c}^{P_s} E_\lambda(T) \frac{\partial \tau \lambda}{\partial p} \, d\rho \right\}$$

where ϕ = spectral response of the radiometer
τ_λ^2 = spectral transmissivity of the layer P_c to P_s
P_c = pressure at which the top cloud is found
P_s = satellite altitude

α = percent cloud coverage in the field of view
T_c = temperature at the cloud top
T = temperature
τ_λ = spectral transmission function
λ = wavelength
$E_\lambda(T)$ = Planck function
p = pressure
R_λ = spectral radiance

That part of the EMS where the atmospheric constituents have the highest transmission, allowing sensors in orbit to measure the outgoing signal from a test site through the atmosphere, is defined as an atmospheric window. At shorter wavelengths the atmosphere is almost opaque while in the visible and in some parts of the thermal IR, significant spectral regions with high transmission appear. Atmospheric windows are the sum of the spectral properties of all atmospheric gaseous constituents, the main absorption features of which are given in Table 3-1-1.

The high absorption of energy in the short wavelength region is based on the increasing energy of photons beyond the binding energies of molecules or atoms. Toward microwaves and radio wavelengths, photon energy is low and wavelengths are large compared with the distances in atoms and molecules. Well defined windows occur when there are no specific absorption bands of the atmospheric constituents. Qualitatively, the solar spectrum shows all the characteristic absorption bands of the gaseous constituents of the atmosphere. Broad absorption bands appear between 2.6–3.1 μm and 5–7 μm, and significant absorption appears

TABLE 3-1-1. Spectral Regions and Absorption Features of Atmospheric Constituents

Wavelength Region	Absorption
0.003 to 0.03 μm	Complete absorption due to short wavelength in relation to size of atmospheric particles and very high photon; high probability of interaction with atmosphere
0.03 to 0.13 μm	N_2 and O_2 electronic bands; almost complete absorption
0.13 to 0.22 μm	O_2 electronic bands; almost complete absorption
0.22 to 0.30 μm	O_2 electronic bands; strong absorption
0.30 to 0.40 μm	Rayleigh scattering
0.40 to 1.0 μm	Good transmission; few absorption bands except for water band at 0.9 μm
1.0 to 20 μm	Many rotational-vibrational bands throughout the region of H_2O and CO_2; electronic O_2 at 1.06 and 1.27 μm; rotational H_2O bands in 15–24 μm region window between 8 to 14 μm
24 to 1000 μm	Rotational lines, principally H_2O; strong absorption
1 mm to 10 cm	Widely spaced pure rotational lines, many clear windows
10 cm	Almost complete transmission

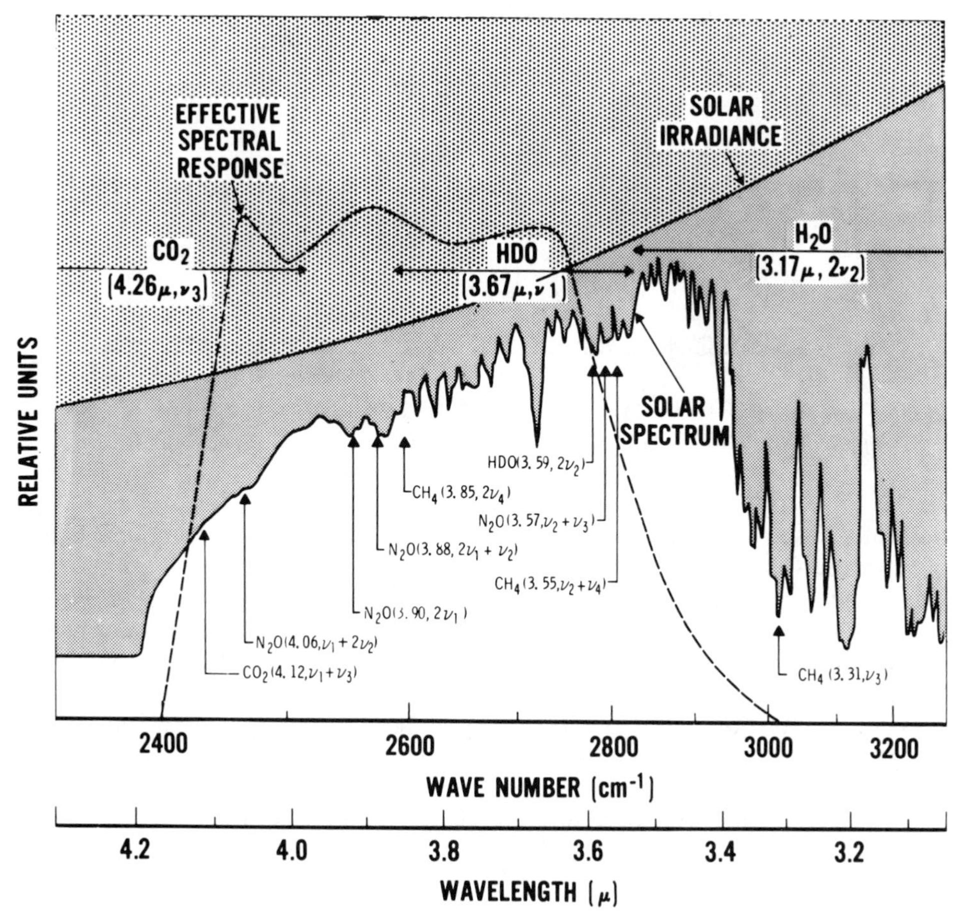

Figure 3-1-1. Atmospheric window between 3.2 and 4.2 μm. Courtesy of NASA.

at 9.4 and 9.6 μm, which is caused by ozone. Although of minor importance, hydrogen deuterium oxide, commonly written as HDO, shows a very well pronounced absorption maximum at 6.8 μm. Considering all molecular constituents, it is clear that only part of the spectrum may be used to measure from satellite altitudes the outgoing radiation from the earth's surface. The so-called atmospheric windows for thermal measurements, which show the minimum absorption of a cloud-free atmosphere, are located around 3–4 μm and 10–11 μm. With higher resolution, the spectrum of the atmosphere is more complicated and shows details that can also be annotated to the individual constituents of the atmosphere. Figure 3-1-1 shows a detailed description of the atmospheric window between 3.2 and 4.2 μm, including the spectral response of an IR radiometer that was flown on board *Nimbus 2*.

Similar high spectral resolution can be obtained for other portions of the EMS; for example, Figure 3-1-2 gives the ground level spectrum for the spectral region between 0.3 and 2.5 μm. It shows the high energy level transmitted through the atmosphere (or, in other words, the high transmittance) in the visible part, with

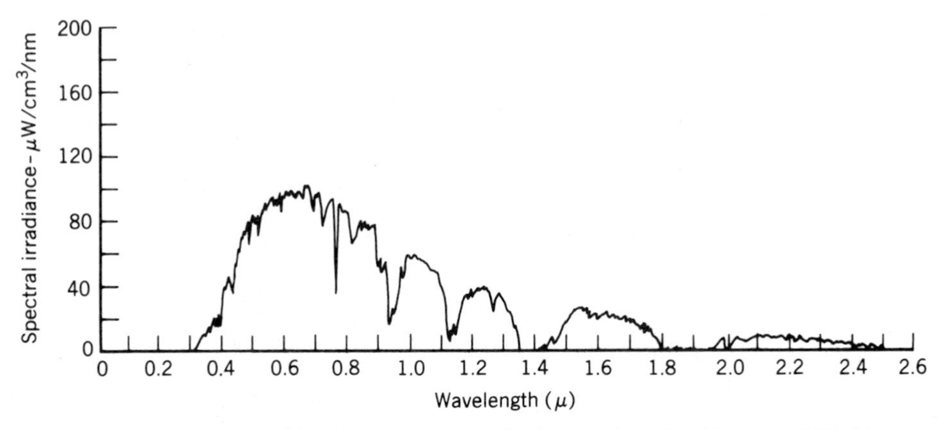

Figure 3-1-2. Ground level spectrum as obtained from clear sky. Courtesy of NASA.

the water absorption band at 0.9 μm and the effect at rotational-vibrational bands of H_2O and CO_2 in the wavelength above 1.2 μm, which, at this spectral part, makes the atmosphere almost opaque.

Water vapor and liquid water interact with radiation also emitted at microwave frequencies and modify the brightness temperature observed by a remote sensing device. Water vapor and liquid water in the atmosphere both absorb and emit microwave radiation and thus raise the brightness (resulting in temperature T_B) observed over the earth's surface. Wilheit (1972) showed that the effect depends on the temperature and pressure at which the water vapor or liquid water is located, and can be expressed for a wavelength of 0.81 cm by

$$T_B = A + BV + CL$$

where A = 125 K
B = 6.8 K cm²/g
C = 300 K cm²/g
V = net water vapor in a column
L = net nonraining cloud liquid water content

Hovis (1972) studied the optical properties of clouds, of which a sample in Figure 3-1-3 shows the reflectance of five clouds measured between 1.6 and 2.3 μm. The stratus and stratocumulus are water droplet clouds, whereas cirrus and cirrostratus are ice crystal clouds. In the region from 2 to 2.1 μm, ice has an absorption band, so one would expect the reflectance to be low for the ice crystal cloud. Water clouds vary in reflectance, the variation depending on the water droplet size: the smaller the droplet size, the more efficient the scattering and the higher the reflectance. Centered about 1.9 μm there is an absorption band due to water vapor. At the lower wavelengths of this band, all water droplet clouds reflect almost uniformly at about 1.72 μm. The ice crystal clouds vary in reflectance with particle size, the smaller particles reflecting more intensely than the larger particles. These spectra summarize results from a number of measure-

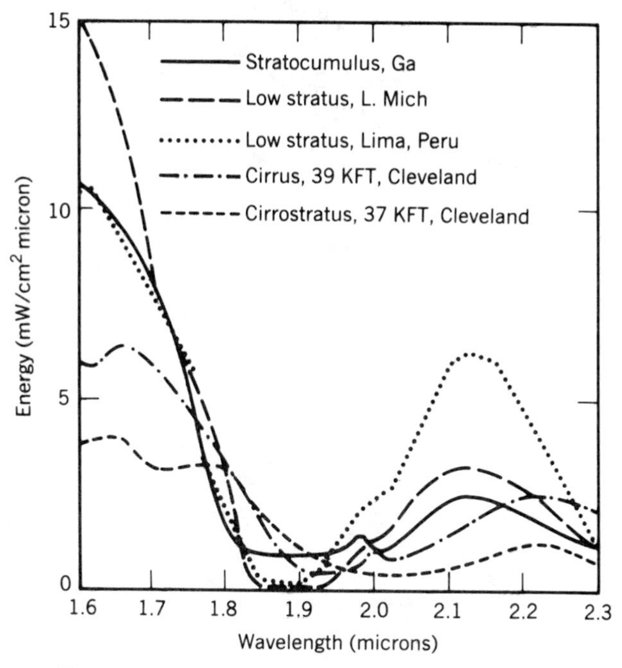

Figure 3-1-3. Reflectance of five clouds investigated.

ments. In general, they are true of all naturally formed clouds. They suggest that a two-channel radiometer with bands centered at 1.7 μm and 2.1 μm mounted on a satellite could be used to discriminate cloud type and determine to some degree the size of cloud particles.

Rain droplets, because of their larger size, comparable with microwave wavelengths, have an enhanced absorption effect. Ice in clouds has no measurable effect in the microwave region, which means that cirrus clouds or snow in the atmosphere do not affect the observed brightness temperature.

The observed brightness temperature, as influenced by water vapor, liquid water, and oxygen in the atmosphere can be expressed by

$$T_B = \int_0^H T(h) \frac{dt(h)}{dh} dh + t(H)[\varepsilon T_0 + (1 - \varepsilon)T_s]$$

where $t(H) = e^{-\int_0^h \gamma(x) \sec \theta \, dx}$
$T(h)$ = atmospheric temperature profile
H = altitude of the observation
ε = surface emissivity
T_0 = thermodynamic surface temperature

The magnitudes of atmospheric contributions due to liquid water and atmospheric water vapor are given in Figures 3-1-4 and 3-1-5.

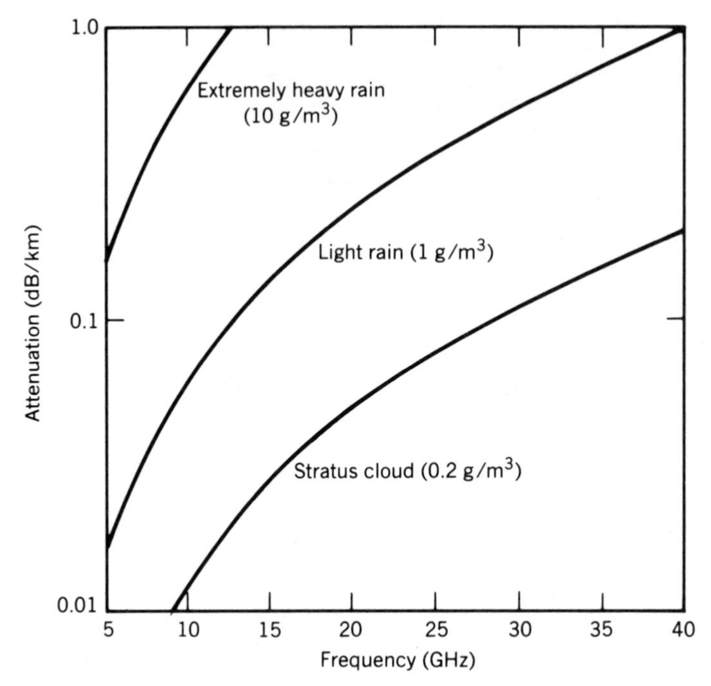

Figure 3-1-4. Microwave attenuation due to liquid water (approximate). Courtesy of NASA, Nimbus Project.

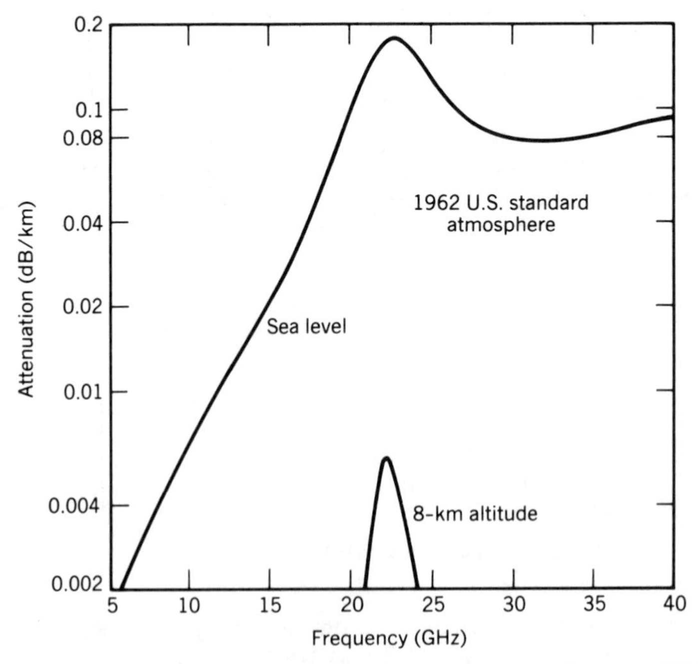

Figure 3-1-5. Microwave attenuation due to atmospheric water vapor. Courtesy of NASA, Nimbus Project.

3-2 GLOBAL DISTRIBUTION OF ATMOSPHERIC WATER

If quantitative measurements or radiance are required, the amount of aerosols and water in the atmosphere must be known. Atmospheric water, however, is not evenly distributed on the global scale. Humidity and cloud distribution, for instance, over the oceans are functions of sea surface temperature. Humidity may change from about 0.5 $g \cdot cm^{-2}$ for the high latitudes to about 6 $g \cdot cm^{-2}$ for the equatorial regions, and the water content in form of clouds may vary by a factor of 10.

The processes that control the distribution of clouds include (1) the introduction of water vapor into the atmosphere by evaporation of the oceans or by evapotranspiration from the continents; (2) transportation of water vapor by the atmospheric and local circulation currents; and (3) formation of clouds and precipitation by complex processes of cooling and condensation.

Continuous viewing of a satellite sensor over selected areas is restricted by the cloud coverage and its dynamics; that is, if the field of view is obscured by clouds, little or no information is obtained by the sensor. If changes in the distribution of cloud pattern occur, repetitive observations increase the probability of monitoring at least a selected percentage of an area.

Cloud statistics depend upon area size. For example, at a global area size, the amount of cloud cover is almost constant at near four tenths. At the other extreme, at a small point, the cloud cover may have only two values (i.e., 0 or 10). For areas between these extremes, it follows that (1) the smaller the area viewed, the more U-shaped the cloud frequency distribution; and (2) the larger the area viewed, the more bell-shaped the cloud frequency distribution.

Martin and Liley (1971) combined cloud data from the U-2, Apollo, and the Environmental Science Service Administration (ESSA) as illustrated in Fig. 3-2-1.

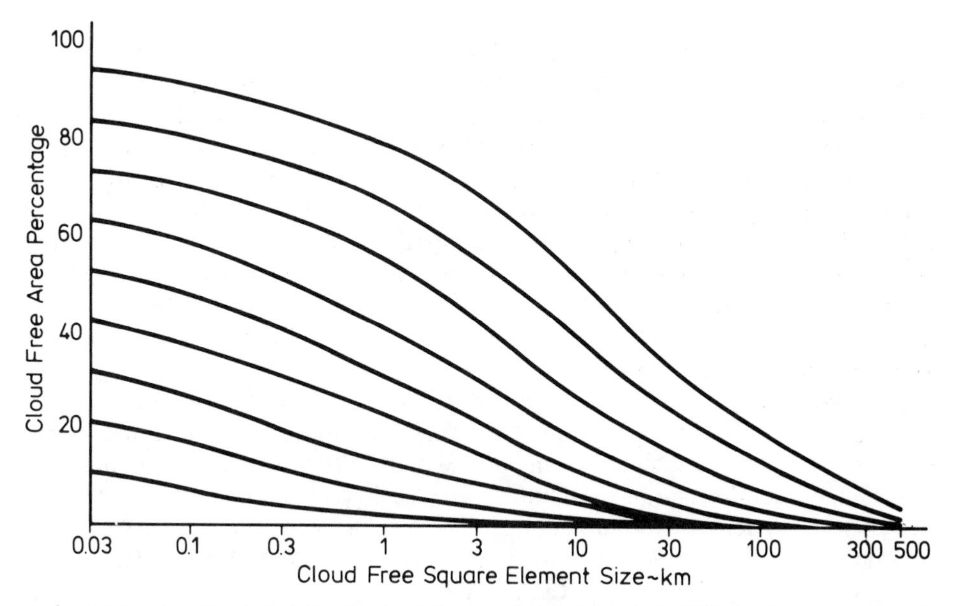

Figure 3-2-1. Combined statistics for cloud-free resolution elements in 10% cloud amount intervals. After Martin and Liley (1971). Reproduced with permission from Dr. E. Degens, University of Hamburg.

Figure 3-2-2. Infrared observations with *SMS-1* on April 1, 1975, 1730 Z. Black displays high IR thermal radiation, and white reflects low radiation of clouds. Courtesy of NOAA.

The curves represent the combined values representative of each cloud category and each 10% interval, respectively. These resultant statistics define the effect on the cloud-free area percentage of resolution element statistics for the range 0.3 to 200 km, and show the low probability for receiving cloud-free images over tropical regions by using a resolution of 1 km and covering large areas. On the other hand, monitoring over small sites has a much higher probability with high resolution.

Based on atmospheric circulation, the cloud distribution shows a pattern that reflects highly the outcome of the major circulation cells. For each hemisphere there are three circulation cells: (1) the tropical Hadley cell with convective character, (2) the terrel cell, which is a wind-latitude cell, and (3) the polar cell, which is also a convective cell.

The size of the cells decreases toward the pole and pressure belts result from the circulation and distribution of water and land into high- and low-pressure systems. The area of contact of air masses with different densities is normally indicated by condensation and clouds because lighter air masses rise over air masses with higher density and cooling. The displacement of an air mass by one with differing temperature leads to fronts that are marked by its specific cloud formation and sequence. Cold fronts may be recognized by a line of cumulus or cumulonimbus clouds with additional higher altostratus (ahead and behind). A warm front sequence may be characterized by cirrus, cirrostratus, altostratus, nimbostratus, and rain. Waves in a front develop to a cyclone or extracyclone, in

SMS-1 VISUAL
1 APRIL 1975
1700Z

Figure 3-2-3. Visible observations with *SMS-1* on April 1, 1975, 1700 Z. White shows high reflectance values of clouds. Courtesy of NOAA.

contrast to the tropical cyclone, which does not contain fronts. Other large circulations are the monsoons (winds that reverse their directions seasonally from the ocean to the continent) and the intertropical front, in the tropics at 5–10°N, where easterly trade winds converge. Superimposed on the larger wind systems are local winds due to gravitational air drainage, winds due to diurnal reversal of convective circulations, and sea and land breezes due to differential heating.

The distribution of clouds reflects the global wind pattern and is influenced by oceanic features such as currents and upwelling, the surface characteristics of the continents, and the interaction between wind systems in coastal regions. Figure 3-2-2 is an IR composite that reflects the cloud formation due to the major wind systems and convective circulations. The cloud distribution, as monitored by reflectance measurements in the visual channel by *SMS-1* at the same time (Fig. 3-2-3) gives a different cloud pattern than does the IR data. This is based on the fact that emitted energy from clouds is a function of height and the corresponding temperature. Consequently, it follows that the higher the cloud, the lower the radiation temperature, while the albedo for the same cloud may not undergo any significant modifications. This is summarized in Figure 3-2-4, which shows the cloud top temperature of part of Hurricane Gladys with a single scan line.

In addition to estimating the height through IR measurements, we can also quantify the percentage of cloud cover, an approach that has been demonstrated

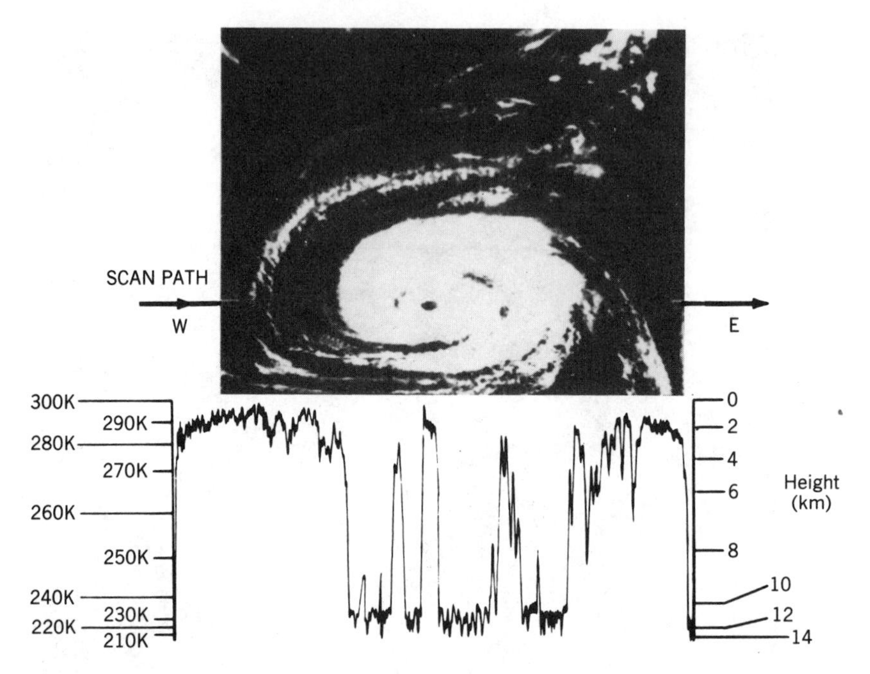

Figure 3-2-4. Analog trace of single scan through Hurricane Gladys, 18 September 1964. Courtesy of NASA.

Figure 3-2-5. Mean cloud cover and altitude for July 1979. Courtesy of NASA, Jet Propulsion Laboratory. See dust jacket for color representation.

to be feasible with the results shown in Figure 3-2-5. It is especially the high cloud deck, in connection with the Intertropical Convergence Zone during July, that is an outstanding feature. Figure 3-2-5 shows that the intertropical convergence (ITC) cloud band has different characteristics in different parts of the world. In the Atlantic, it is centered 3°N during the winter and moves to about 8°N by late summer (Anderson et al., 1969). There is some evidence of a second cloud band associated with the ITC in the eastern Pacific. Besides the main band north of the equator, a weak band appears at 5°S from January through March. The strength of this second band varies from year to year, and in some instances it fails to develop at all. The ITC cloud band in the Indian Ocean during the southern hemisphere summer is much broader than in either the Atlantic or eastern Pacific Oceans.

Mountains also affect the distribution of clouds. The Andes Mountains, for instance, are partly cloud covered throughout the entire year due to local convergence and convection over the mountain ridges.

The effects of local sea breeze circulations in clouds along the coastline are also apparent from satellite observations. This is particularly evident along the coastline of Brazil and the northeast coastline of Australia (Anderson et al., 1969). Besides the concentration of clouds onshore, the sea breeze circulation often produces a minimum of clouds offshore.

Microwave radiation emitted by the earth and the atmosphere in a wavelength band centered at 1.55 cm shows that emissivity of the earth and atmosphere varies considerably more than at IR wavelengths (Wilheit et al., 1976), and a minimum of interference from clouds exists, since most nonraining clouds are virtually transparent. Atmospheric moisture modifies the radiation emitted by the surface, and when cloud droplets reach precipitable size they enhance the radiation considerably over surfaces of low emissivity, thus making it possible to map areas of heavy cloudiness. Therefore, it is possible to determine the extent, structure, and intensity of rainfall and to identify over oceans the location of frontal rain, rain-snow boundaries, and the structure of tropical storms.

Over most of the microwave spectrum, raindrops absorb and scatter radiation, thus producing large changes in brightness temperatures relative to clear or cloudy conditions. Since the structure of rain varies substantially for different rain rates and climatological backgrounds, the raindrop size distribution, the rain-layer thickness, and the ice clouds above the rain layer are all important inputs to model computations. Calculations made by Burke et al. (1982) show that over ocean background, which is "cold" in the microwave region, lower frequencies are sufficiently sensitive to rain conditions. Over land background, higher frequencies have to be utilized in order to obtain the sensitivity to rain rates. It has also been demonstrated that by using discrimination tests of the radiometric data, the rain/no rain decision can be made and the rainfall rate can be retrieved from a statistical inversion technique.

As weather satellites provide continuously visible and IR data, cloud-covered areas as well as cloud-free surfaces can be monitored on an operational basis. This is important for agricultural applications in connection with rainfall estimates and daily total solar radiation on a horizontal surface.

Scofield and Oliver (1977) developed a technique to retrieve rainfall estimates from convective systems by using GOES thermal IR and visible data. The technique has been designed for deep convective systems that occur in tropical air masses with high tropopauses. Estimates of convective rainfall can be made by quantitative observation of changes between two consecutive observations, by using recordings in the thermal IR and in the visible bands. The technique is based on different steps, which include the identification of the active portion of the convective system, an initial estimate of expected average rainfall by using cloud top temperatures derived from enhanced thermal IR imagery, and successful pairs of visible and thermal IR images to identify additional clues that indicate heavier-than-average rainfall. These clues include overshooting tops, merging thunderstorms, merging convective cloud lines, and rapidly expanding anvils. The total rainfall prediction is computed by summing these two estimates (see also Yates et al., 1984).

Solar radiation is basic energy for growing plants, and the daily total solar radiation on a horizontal surface (or insolation) involves entering numerical models for estimating crop yield, potential evapotranspiration, and soil moisture. Therefore, knowledge of cloud coverage can be referred to radiation received at ground level.

The radiant energy incident at the top of the atmosphere, Q_0, is divided after interaction with the earth-atmosphere system according to

$$Q_0 = Q_R + Q_A + Q_G$$

where Q_R = incident energy reflected to space from the earth-atmosphere system
Q_A = energy absorbed in the atmosphere
Q_G = energy absorbed at the ground

are the components of the incident energy.

The incident radiation Q_0 can be predicted with acceptable accuracy for any location and time, and the reflected and absorbed components are determined by the state of the atmosphere. The reflected radiation term Q_R explains up to 80% of the variability in the portion of incident radiation that reaches the ground, and cloud cover determines the size of Q_R. Radiance in the visible spectrum measured by satellites is directly related to Q_R. All satellite insolation techniques amount to obtaining solutions to the above equations either by physical or statistical models.

The approaches are operational and provide daily estimates of insolation in Langleys (1 Ly = 1 cal$\cdot$cm^{-2}) on a 1° latitude-longitude grid (Yates and Tarpley, 1982). An example for the United States is shown in Figure 3-2-6.

Comparisons and ground truth pyranometer data for the eastern United States estimates have shown errors in the daily total insolation of 10 to 15% of the mean daily total when a sufficient number of pictures (at least four) are used in the daily estimate.

As early as 1960, radiometric observations from *Tiros-2* were made in the spectral region near the 6.3-μm water vapor absorption region (Allison, 1964). These observations were used to map relative humidity in the upper troposphere (Allison and Warnecke, 1965; Raschke and Bandeen, 1967) and to infer the location and extent of dynamic features in the upper troposphere (Nordberg et al., 1966; Beran et al., 1968; Martin and Salomonson, 1970; Steranka et al., 1973).

The distribution of moisture does not necessarily follow the location of clouds. The range in the level, from which the principal contribution in the 6.7-μm region arises, is approximately 3 km at a given station in mid-latitudes (Steranka et al.,

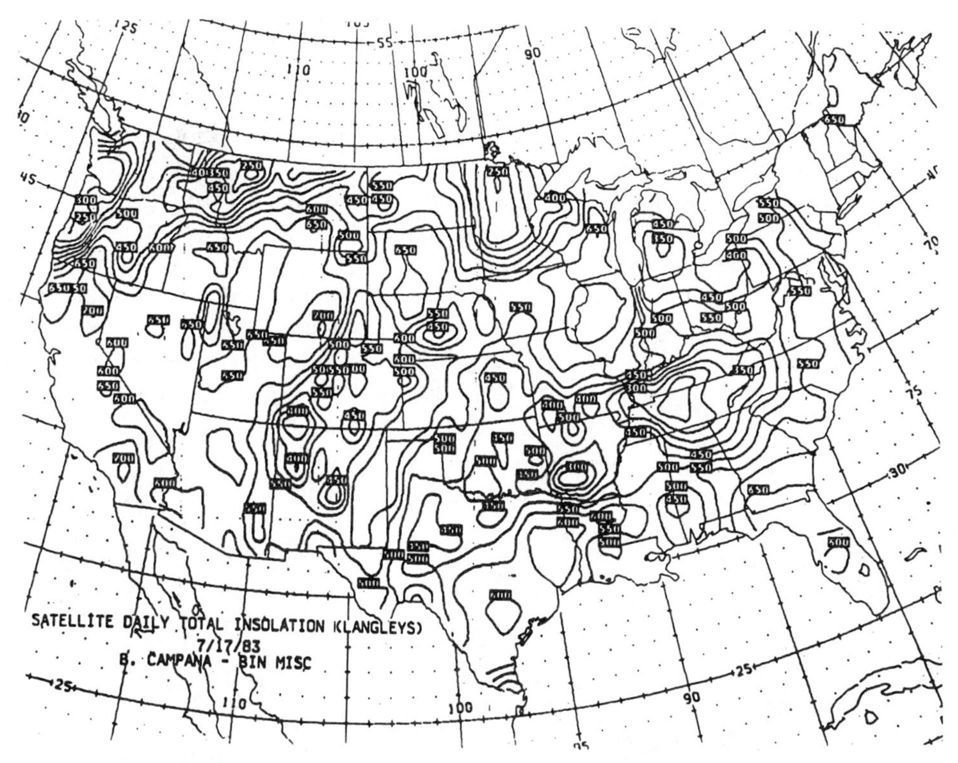

Figure 3-2-6. Daily total insolation (in Langleys) estimated from GOES data for 17 August 1983. Courtesy of Dr. J. D. Tarpley, NOAA. See also Justus et al. (1986).

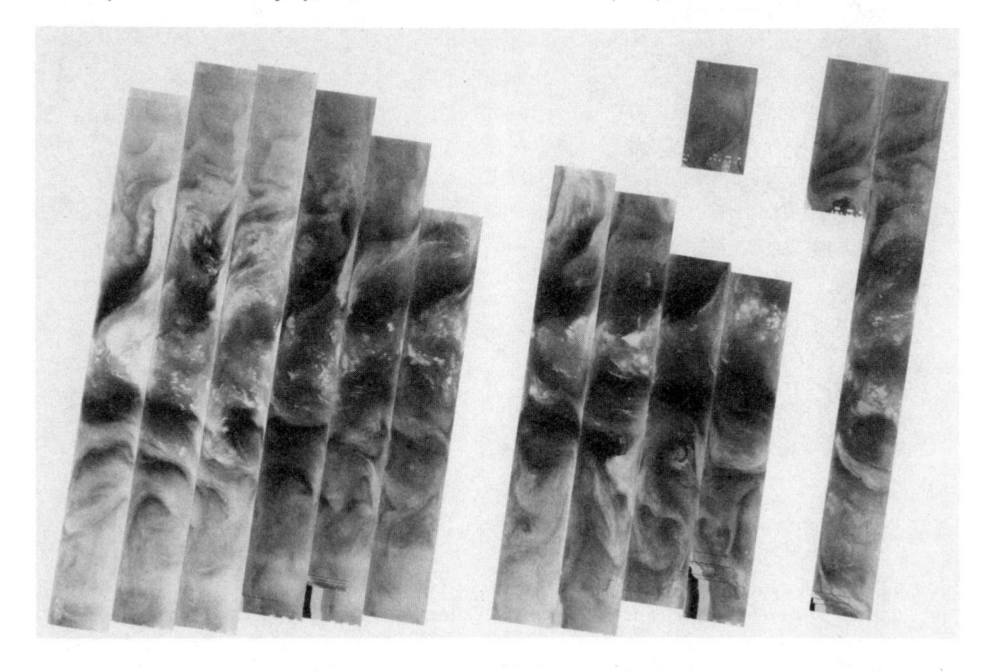

Figure 3-2-7. Composite of photofacsimile film strips (6–7 μm channel) *Nimbus 4* THIR nighttime measurements. Courtesy of NASA, Nimbus Project.

1973). This same range is somewhat larger in the tropics (4–5 km) and approximately one half as much (1–2 km) in the polar latitudes. The height of the mean contribution level goes from about 8 km in the tropics to near 4 km in the polar latitudes. An example of the global pattern of moisture distribution as measured by *Nimbus 4* is given in Figure 3-2-7.

A qualitative comparison of the moisture patterns on the 6.7-μm imagery, with the 400-mb level conventionally measured wind field, also indicates that moisture patterns are generally aligned with the wind field, which means that water vapor imagery can be used to infer the orientation of streamlines. Figure 3-2-8 shows a streamline analysis derived from water vapor channel imagery taken over the western United States. The direction applied to the streamline analysis requires knowledge of circulation patterns within meteorological systems such as low- and

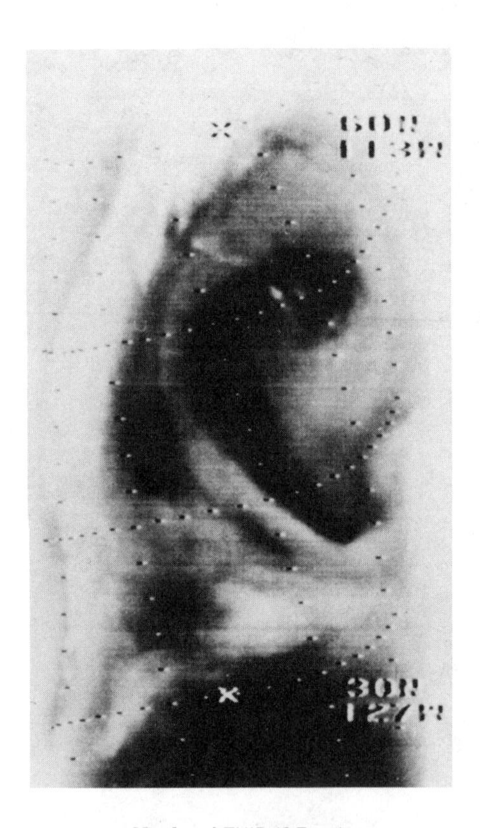

Nimbus 4 THIR (6.7 μm)
orbit 2565 (N)
16 October 1970

Derived streamline
analysis

Figure 3-2-8. A comparison of *Nimbus 4* THIR 6.7-μm imagery on orbit 2565 on October 16, 1970 and a detailed streamline analysis derived from the patterns shown in part (*a*) and conventional wind observations over the western United States. The arrowheads shown in part (*b*) were inferred from the light and dark patterns in part (*a*) that result from upper tropospheric dynamics affecting the distribution of water vapor. Courtesy of NASA.

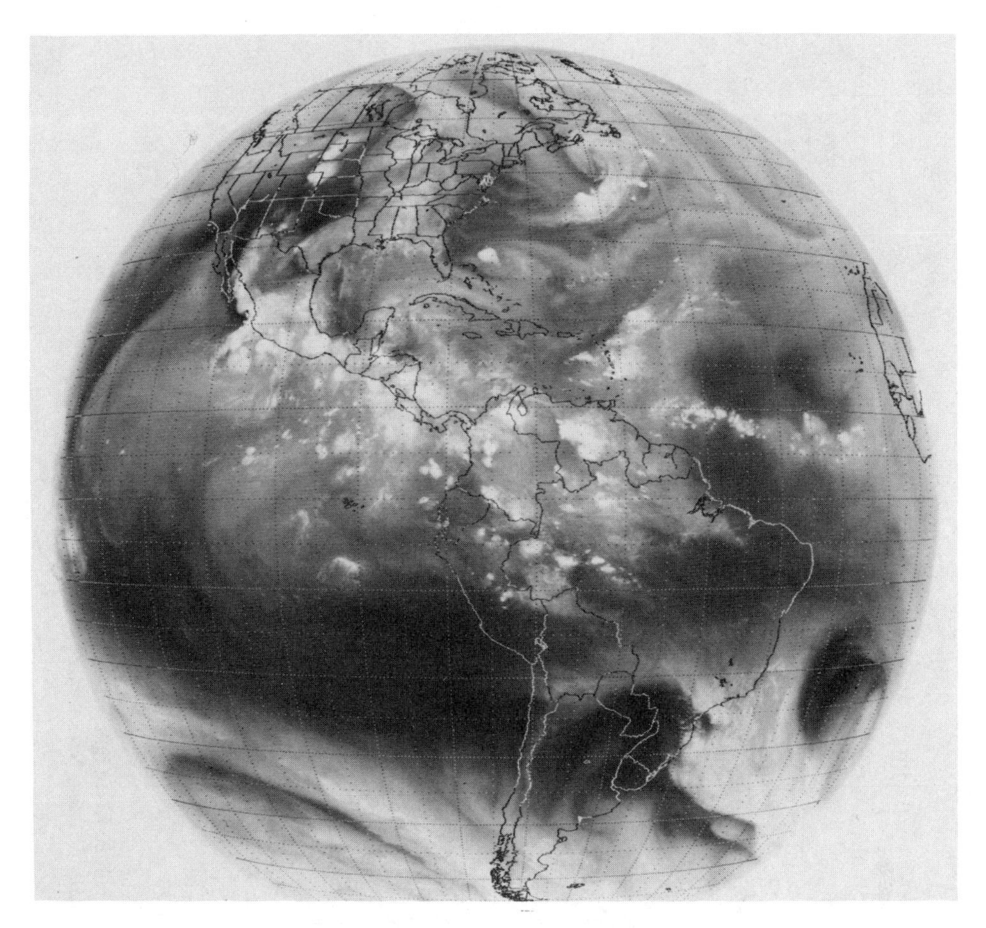

Figure 3-2-9. Water vapor distribution based on GOES measurements. Courtesy of NOAA.

high-pressure areas. Particular caution must be exercised in frontal regions with high, dense cloudiness, since frontal clouds are often not aligned with the upper tropospheric wind flow.

The newer generation of geostationary satellites also has the capability to monitor almost continuously the distribution of atmospheric moisture (see Fig. 3-2-9).

3-3 OZONE DISTRIBUTION

As shown earlier, ozone has a strong absorption in the thermal region, which prohibits measurements of the earth's surface features. The scattering of ultraviolet by ozone in the atmosphere, however, can be used to monitor ozone concentrations. This is of particular interest with respect to recent discussions on ozone depletion in the atmosphere.

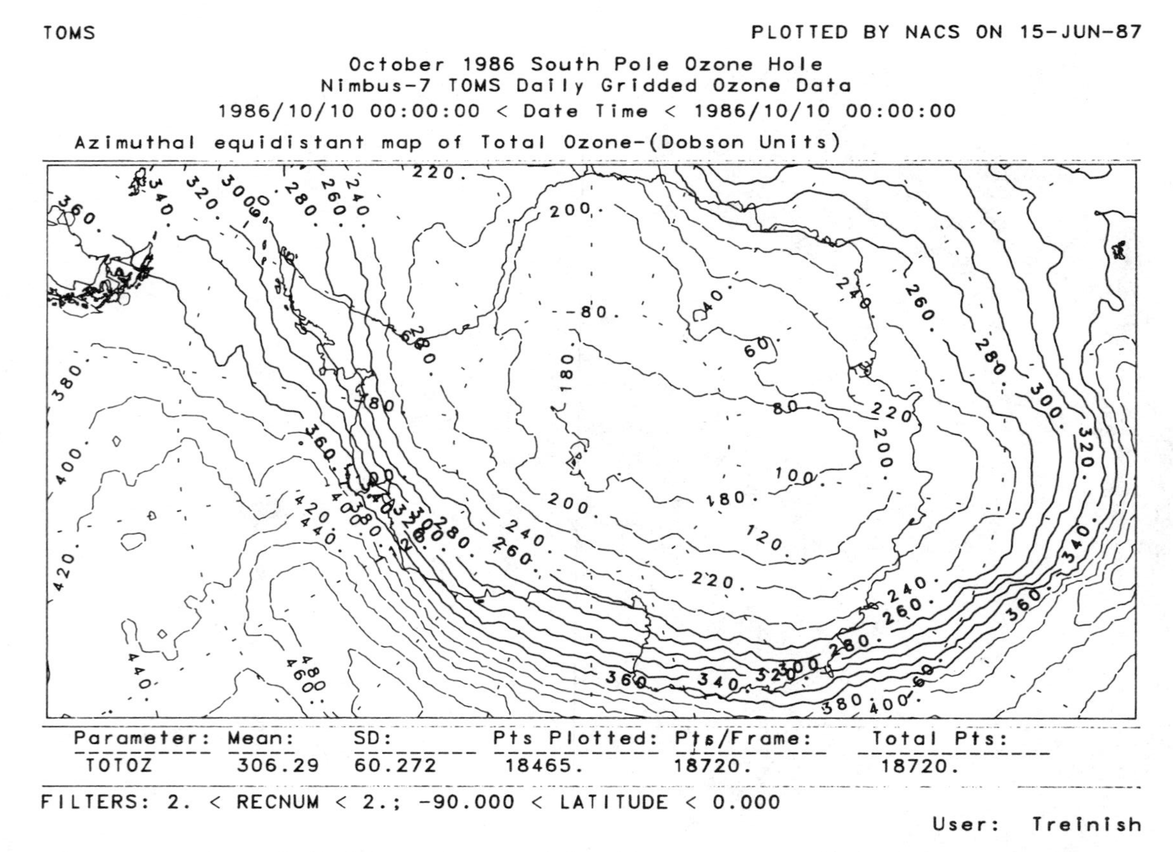

Figure 3-3-1. Ozone distribution over the Antarctic based on measurements with the total ozone mapping spectrometer during October 1986. Courtesy of Dr. P. A. Ross, NASA GSFC/NSSDC.

It is especially from studies of the Antarctic ozone hole that concentrations of ozone in the stratosphere have been observed to drop to half of its normal levels. Most of the depletion occurs in the first weeks after the return of sunlight to the southern latitudes, that is, during August-September (Pyle and Farman, 1987). At this time, levels over Antarctica may reach 180 Dobsons (matm·cm), of which one sample is shown in Figure 3-3-1.

4

SPECTRAL CHARACTERISTICS OF NATURAL SYSTEMS

4-1 GENERAL CONSIDERATIONS

Interpretation of remote sensing data obtained from space must take into account atmospheric conditions as well as the spectral characteristics of the surfaces to be investigated. Therefore, well-defined isolated spectral bands have to be used in order to properly monitor surface features or processes of the earth.

Spectral characteristics in the visible part of the EMS differ significantly between targets—for instance, between soil and vegetation. Although albedo in general can be used for identification of several natural surfaces, it is not sufficient for detailed information. However, the albedo over natural surfaces, as monitored by Hodarev et al. (1971) and outlined in Table 4-1-1 gives rise to sufficient gradients even working without wavelength-specific bands. For the identification and choice of proper wavelengths in monitoring natural surfaces with remote sensing techniques, it is rather important to know the reflection or emission spectra of the target to be investigated. Figure 4-1-1 gives the spectral reflectivity of several naturally occurring surfaces and indicates that with the selection of specific wavelengths a discrimination between different formations can be achieved. For instance, forests and limestones have a significant difference in their reflectivity over certain spectral regions, which can be used to distinguish them from each other by remote sensing if proper selection of the spectral signatures in the detection device or data processing has been provided.

Spectral characteristics depend not only on the composition of the surface but also on the conditions under which they are measured. As shown in Fig. 4-1-2, dry soil and damp soil give very different signals. The same material after weathering may also give a different spectral response (Fig. 4-1-3). This is true for all surfaces and demonstrates the importance of calibrating or verifying all remote sensing information with ground measurements.

Interaction between radiation and the atmosphere is severe at the short end of

TABLE 4-1-1. Albedo of Natural Surfaces Measured from Aircraft

Surface	Month	Sun's Elevation (°)	Albedo (%)
Tundra	VIII	24–36	11–23
Bogs	VIII	28–38	10–18
Wood (without foliage)	III	34	11
Coniferous wood (against a fresh snow-cover background)	III	19–23	35–40
Wood (leaf-bearing,	VI	39–55	14–17
coniferous	VIII	30–41	12–16
	IX	26–38	19–20
Leaf-bearing wood covered by fresh snow	XI XII	12–35	48–55
Steppe	VII VIII	40–62	17–89
Yellowy-gray sands	VIII	56	25
Light yellow sands	VIII	50	37
Gray sands	IX	36–45	20–28
Soils			
Black earth (humid)	V	41–45	5–6
Black earth (dry)	VI	41–48	10–15
Light gray podsol	VII	40	25–28
Dry fresh snow	VII	15	81–88
Light gray porous wet snow	III	18	43

After Hodarev et al. (1971). Reproduced with permission from the Environmental Research Institute of Michigan (ERIM).

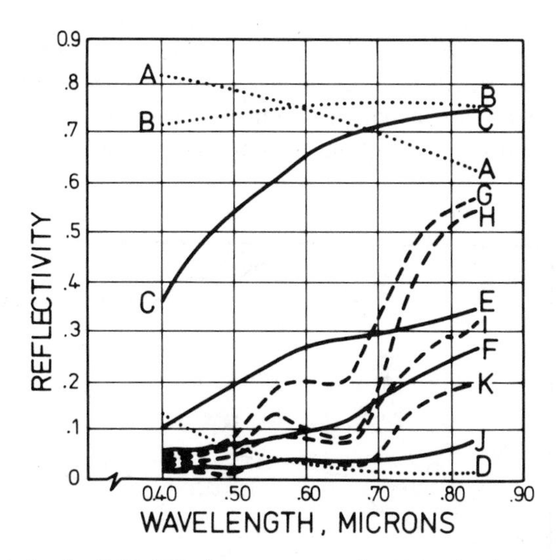

Figure 4-1-1. Spectral reflectivity from forest cover, soils, and other formations: A. Fresh snow; B. snow covered with ice; C. limestone, clay; D. water surface, viewed obliquely; E. desert; F. podsols, clay loam, paved roads; G. deciduous forests, autumn; H. deciduous forests, summer; I. conifer forests, summer; J. black earth, sandy loam, dirt roads; K. conifer forests, autumn. After Krinov (1970). Reproduced with permission from *Photogrammetric Engineering & Remote Sensing*, copyright by the American Society for Photogrammetry and Remote Sensing.

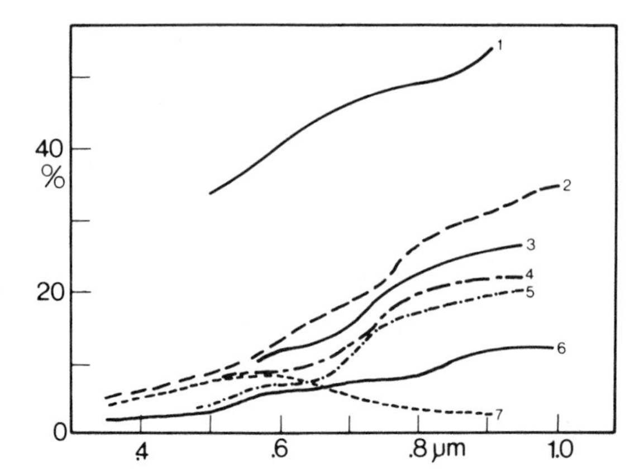

Figure 4-1-2. Spectral albedo for different surfaces. The sun's elevation is given in parentheses. Soils and road covers were characterized by the steady increase of albedo with increasing wavelength: 1. White river sand (47°); 2. stubble of cereals (35°); 3. road composed of dry soil (50°); 4. road composed of damp soil (50°); 5. soil after snow melting (24°); 6. black earth (40°); 7. lake (56°). After Hodarev et al. (1971). Reproduced with permission from the Environmental Research Institute of Michigan (ERIM).

the EMS. Since clouds are almost transparent for radiation at longer wavelengths, and the atmosphere has almost no effect on the long-wave radiation as emitted from an earth surface, the microwave region of the EMS is of particular interest for an all-weather monitoring system.

The measurements of the radiative characteristics in the very high frequency (VHF) band have been carried out at several wavelengths. The information on the emittance of various surfaces at $\lambda = 3.2$ cm is given in Table 4-1-2, which shows that wood and seafoam radiate practically as a blackbody, whereas other types of surfaces are characterized by noticeable departures from blackbody surfaces. Recently, attention has been paid to high-resolution spectroradiometry, a relatively new tool in remote sensing that is likely to be rapidly developed in the next

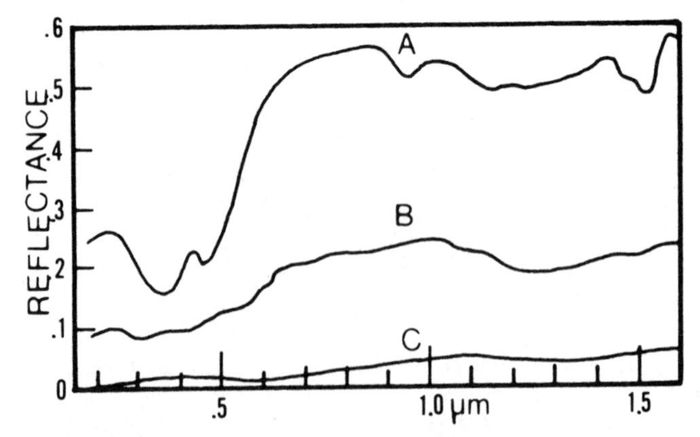

Figure 4-1-3. Spectroreflectance curves for common road surfacing materials: A. Weathered concrete; B. weathered asphalt; C. fresh asphalt. After Miller and Pearson (1971). Reproduced with permission from the Environmental Research Institute of Michigan (ERIM).

TABLE 4-1-2. Radiative Capacity of Various Surfaces in the Normal Direction for λ = 3.2 cm

Underlying Surface	Radiative Capacity (%)
Seawater	40
Wood	98–99
Bog	85
Field with thawed patches	75
Seafoam breakers	99–100
Soil	50

After Hodarev et al. (1971). Reproduced with permission from the Environmental Research Institute of Michigan (ERIM).

decade (Ferns et al., 1984). It is not only useful in defining the best wavebands for remotely sensed imagery, but direct spectral results from ground and airborne spectroradiometers have also proved their commercial value in mineral exploration. The Collins airborne spectroradiometer, for instance, has detected several mineral exploration targets chiefly by analyzing the red edge shift over biogeochemical anomalies (Chiu and Collins, 1978).

4-2 PROPERTIES OF WATER

4-2-1 Molecular Anomalies of Water

The water molecule has one oxygen atom that occupies the center and two hydrogen (H) atoms that build an angle of 104°31′ with respect to the oxygen (O) atom. The anomalies resulting from this angle are rather important for global aspects of, for instance, the heat budget and life on earth.

The regions of negative electrification based on the OH bond result in a strong dipole moment and may attract the positive partial charge of neighboring water molecules in order to build aggregates or clusters. The forces that bind the molecules together determine the physical properties of water and its aggregates. This is clearly shown if one compares the melting points of the hydrides of the elements in group VIA of the periodic table. This group is characterized by an increase in melting point as a function of increasing molecular weight. Water, on the other hand, has a boiling point 160°C higher than would be expected, from interpolation of the boiling points of the neighboring elements in the periodic table.

The lower density of ice, compared with liquid water, is based on the larger distance of H atoms from the O atom. Instead of being 0.96 Å, as is the case in the O—H···O bond in water vapor, the distance may be 0.99 Å or 1.77 Å.

4-2-2 Suspended Material in Water

Light penetration in water is affected by plankton and its decomposition products, as well as other dissolved and suspended matter. The composition of backscattered light from below the air-sea interface is therefore determined by spectral properties of water and the nature of the constituents in the water column. The main substances responsible for the absorption of light are photosynthetic pigments such as chlorophyll, carotenoids, and other pigments. Chlorophyll *a* is found in all terrestrial green plants and is also the predominant pigment in phytoplankton organisms. Chlorophyll *b* appears to be present only in the *Chlorophyceae,* whereas chlorophyll *c* is found in several algae, including all of the diatoms. Other major pigments are fucoxanthin in diatoms, phycoerythrin in red algae, and phycocyanin in red and blue-green algae. Spectral information relevant to remote sensing of ocean color is presented in Table 4-2-1. Major characteristics common to all species are found in the blue and red regions of the spectrum (Fig. 4-2-1). In contrast to the absorption spectrum of chemically pure chlorophyll in solution, algae suspensions absorb and scatter light more uniformly throughout the visible part of the EMS.

Most phytoplankton organisms show clearly the second absorption band of chlorophyll between 0.65 and 0.70 μm, but the first absorption band of pure chlorophyll at around 0.44 μm is not well defined, as can be seen from Figures 4-2-1 and 4-2-2. From the given spectra, it is also evident that the absorption signatures of different species are significantly different, which shows that the analysis of spectral response from natural populations, for the estimation of chlorophyll, for instance, must take into account the composition of the species.

When monitoring plankton or biomass from high altitudes, one must consider the fact that algae behave more like a suspension than like a pure solution of chlorophyll. As a result, solar light is scattered at the outer shell of plankton organisms. In addition, the absorption of incident irradiance by the cells depends on their outer structure and the optical density inside the cell. Variation in the optical density or the configuration of the cells may change the intensity of backscattered light even if the incident solar irradiance, sun angle, and chlorophyll concentration per unit of volume remain constant. Yentsch (1960) found that the red absorption band of chlorophyll has little influence on water color. This means that the signal obtained with the red band sensor would record primarily the effect of backscattered light from the organisms.

The size and concentration of particles or marine organisms seem to be the important contributors to the changes in backscattered light intensity as measured in the *Skylab* sensors (Szekielda et al., 1977). The scattering intensity of sus-

TABLE 4-2-1. Spectral Characteristics of Pigments (*in vivo*)

Wavelength (Peak Absorption, Å)	Pigment
6750–6950	Chlorophyll *a*
ca. 4400	
ca. 6500	Chlorophyll *b*
ca. 4650	
ca. 6400	Chlorophyll *c*
ca. 5850	
ca. 4700	Fucoxanthin
ca. 5650	Phycoerythrin
ca. 6200	Phycocyanin

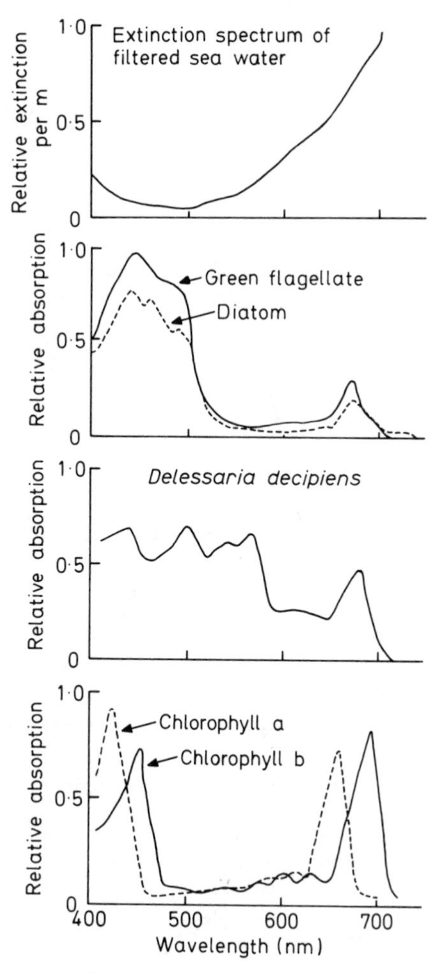

Figure 4-2-1. Spectral properties relevant to remote sensing techniques. After Szekielda and Duvall (1976).

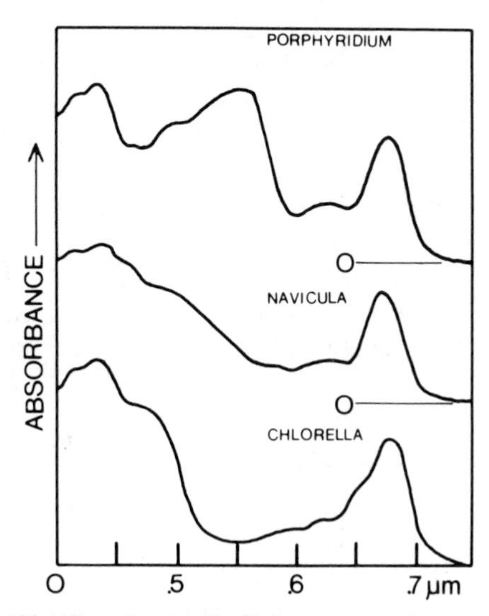

Figure 4-2-2. Spectra of *Chlorella vulgaris, Navicula minima,* and *Phorphyridium cruentum.* After Seliger and McElroy (1965).

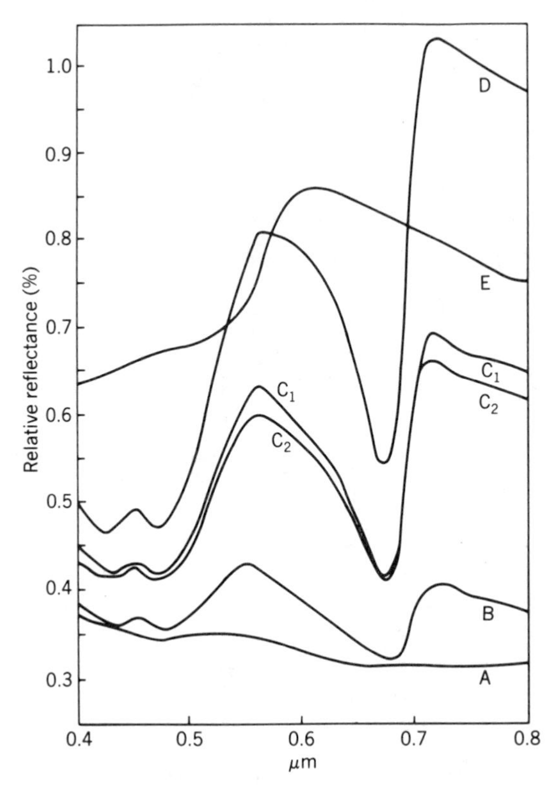

Figure 4-2-3. Examples of reflection spectra: A. distilled water; B. dense algae sample, *Carteria*, 4.86×10^4 mg/m^3, chl *a*; C_1, C_2, and D. mixed algae-clay suspensions; C_1 and C_2, sample B plus 1.0 mL of clay suspension 12.4×10^6 counts/mL; D. sample C plus 1.0 mL clay suspension; E. clay suspension 31×10^6 counts/mL. After Szekielda et al. (1977).

pended particles is proportional to λ^{-n}, where λ is the wavelength and n is the Rayleigh value, which may vary from 4 for pure water to 0 at high turbidity. In other words, the intensity of backscattered light increases with particle concentration. This shows the important influence of particles without chlorophyll on the backscattered light from below the sea surface. Laboratory studies found that backscattered light increases as a function of particle concentration, which is principally attributed to the scattering properties of both inorganic particles and algae in suspension (see Fig. 4-2-3.).

In near coastal regions human-made substances introduced into the marine environment can change the reflectance values of water, as can be shown with a comparison of spectra from major pollutants given in Figure 4-2-4.

4-2-3 Natural Water Bodies

Processes contributing to the radiance spectrum $N(\lambda)$ obtained at satellite altitudes can be written as

$$N(\lambda) = N_a(\lambda) + \gamma(\lambda)N_s(\lambda) + \alpha(\lambda)N_d(\lambda)$$

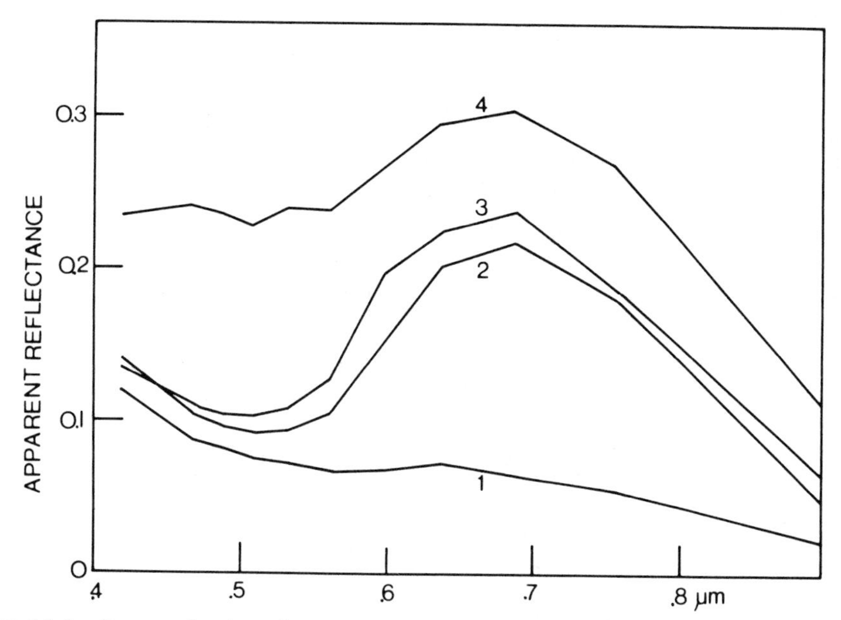

Figure 4-2-4. Spectra of major pollutants in the Rouge and Detroit Rivers: 1. Great Lakes steel 80-in strip mill; 2. Ford boat slip; 3. Great Lakes steel ecorse mill; 4. rouge oil scimmer. After Polcyn (1971).

where $N_s(\lambda)$ is the contribution at the surface due to reflection from the surface and whitecaps, $N_d(\lambda)$ is the diffused radiance just above the surface due to photons that have penetrated the ocean, $\alpha(\lambda)$ and $\gamma(\lambda)$ are atmospheric transmittance factors for $N_d(\lambda)$ and $N_s(\lambda)$, and, in general, are not equal, and $N_a(\lambda)$ represents the radiance scattered back without reaching the air-sea interface. From this formulation of the optical processes (see Fig. 4-2-5), it is evident that $N_d(\lambda)$ contains the information on concentration and composition of particulates and on dissolved material in the ocean, but represents only a fraction of 1% compared with the incident irradiance.

Maul and Gordon (1975) quantified the radiant energy interaction between the energy absorbed by water (w), suspended particles (p), and Gelbstoff (y) and the radiance penetrating the air-sea interface. Accordingly, the total attenuation coefficient c for these interactions can be written as

$$c = c_w + c_p + c_y$$

where $c_w = a_w + b_w$

$c_p = a_p + b_p$

$c_y = a_y$

are the beam attenuation coefficients of the water, particles, and yellow substance.

Incoming radiation from the sky and portions of the sun is partially reflected at the air-sea interface while other portions are absorbed, scattered, or reflected before reaching the ocean's surface. Part of the short-wavelength radiation is

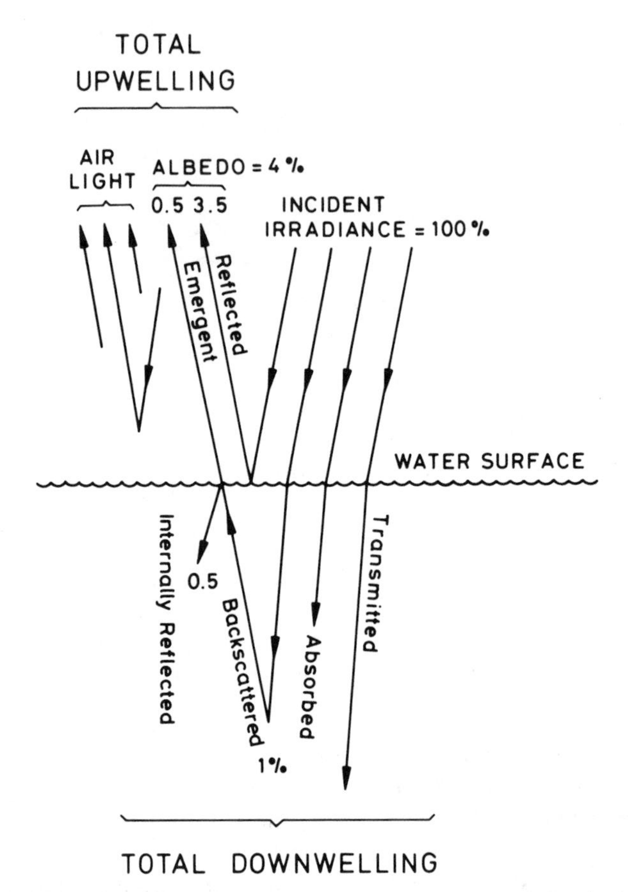

Figure 4-2-5. Schematic representation of the flux of the incident radiation in the air above the surface of the sea, at the surface itself, and in the water beneath the surface.

returned to the atmosphere, but its energy depends on the incident angle of sun or skylight, the refractory index, and the sea state, which is related to the foam formation at the surface. For practical purposes in remote sensing, the most important value is the albedo A, which can be expressed as the ratio of the incident radiation I_i and the outgoing energy I_o:

$$A\% = \frac{I_i}{I_o} \times 100$$

With respect to remote sensing of water properties, the absorption as well as the reflectance characteristics of water have to be considered. Figure 4-2-6 shows the absorption spectrum of seawater (expressed as the attenuation coefficient) and demonstrates that the lowest attenuation is measured in the visible part of the EMS while other parts absorb incoming energy to a high degree. This is especially true for the IR region at around 10 μm. Although the spectrum in Fig. 4-2-6 shows a well-defined transmission in the visible, it should be kept in mind that the dissolved and particulate components in seawater alter the spectrum significantly

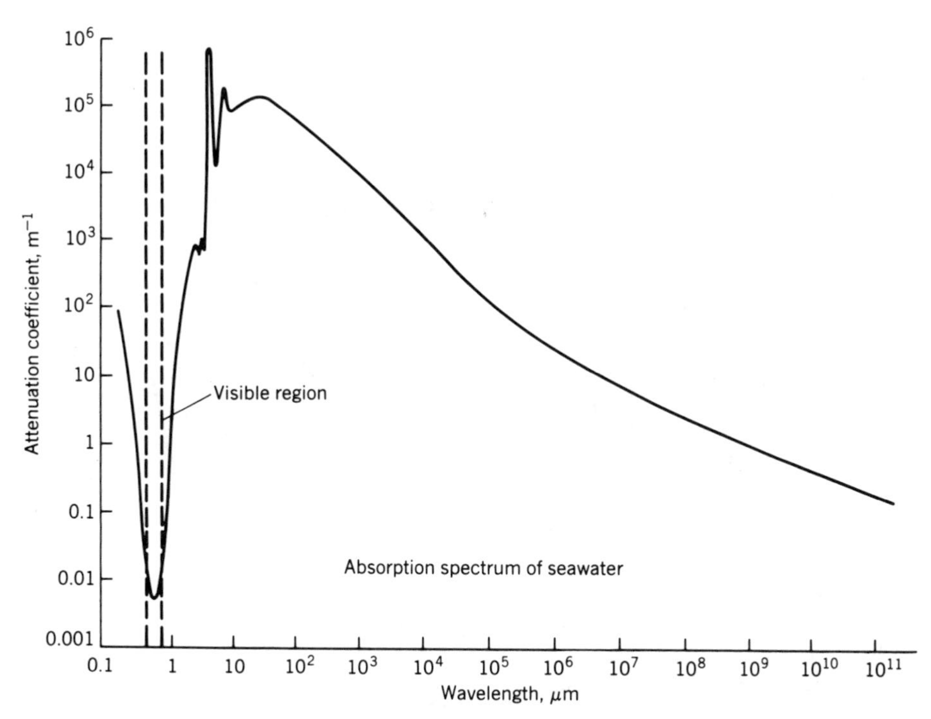

Figure 4-2-6. Absorption spectrum of seawater. After J. Arvesen (personal communication).

at specific wavelengths. This is shown in Fig. 4-2-7, where the absorption spectra of distilled water are compared with those of open ocean and near coastal water. Aside from the fact that the water with high particle load has a higher attenuation coefficient, it is also noticeable that the minimum attenuation coefficient shifts to shorter wavelengths as a function of increasing particle loads. Consequently, in the spectral absorption region of chlorophyll, one would expect above the air-sea interface to see decreasing reflectance from water bodies containing increasing levels of chlorophyll. This was also documented in the reflectance spectra in Fig. 4-2-8, expressed as percentage of incident light as a function of wavelength, and compared with the chlorophyll concentration in the surface water. From these recordings, it is evident that in the blue region the absorption of incident light may be interpreted as a function of the chlorophyll concentration, whereas in the green and red parts of the EMS increasing reflection is a function of increasing plankton and/or particle concentration.

Clarke et al. (1970) showed good correlation between shipboard measurements of chlorophyll and the spectra of backscattered light as measured from aircraft altitudes. Examining the spectral ratio 0.54 μm/0.46 μm, they found a decrease in reflectance at 0.46 μm and an increase in reflectance at 0.54 μm with increasing chlorophyll concentrations.

The intensity of backscattered light caused by plankton and dissolved matter from the ocean as a function of wavelength is given in Fig. 4-2-9, where Gulf Stream water was used as a reference water, with the assumption that the chlorophyll concentration of less than 0.02 $\mu g \cdot L^{-1}$ did not change the backscattered light

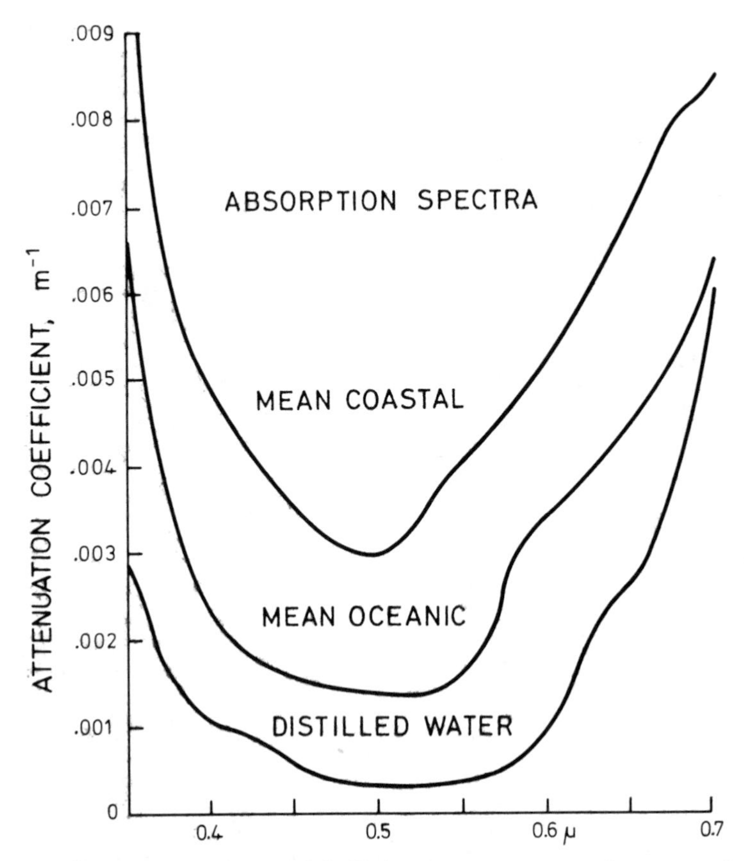

Figure 4-2-7. Absorption spectra of distilled water, open ocean, and near coastal water.

significantly compared with pure water. The spectrum is the difference in energy between the spectrum obtained in near coastal waters and the spectrum recorded over the Gulf Stream. It was assumed for the interpretation of the different spectra that sky conditions, sun angle, and sea state were the same over both sites. If both water masses had the same optical characteristics and oceanic conditions, the energy difference in both spectra should be equal to zero. Any differences between the two signals would thus be caused by dissolved or particulate matter in the sea.

Spectra reported by Noble (1972) also showed significant differences in reflectivity caused by different plankton species and different concentrations of biomass in ocean water. From reported data it is obvious that the spectral interval to monitor ocean color from space or from airborne radiometers is mainly between 0.5 and 0.8 μm. Since the reflected radiation is a function of the concentration of suspended matter, the relationship between the two parameters is highly pronounced in freshwater reservoirs (lakes, ponds, rivers) where high productivity leads to high concentration of primary standing stock. Data after Ritchie et al. (1974) in Fig. 4-2-10 demonstrate this clearly. In a quantitative approach they showed that solar radiation was best correlated ($r = 0.90$) with the concentration

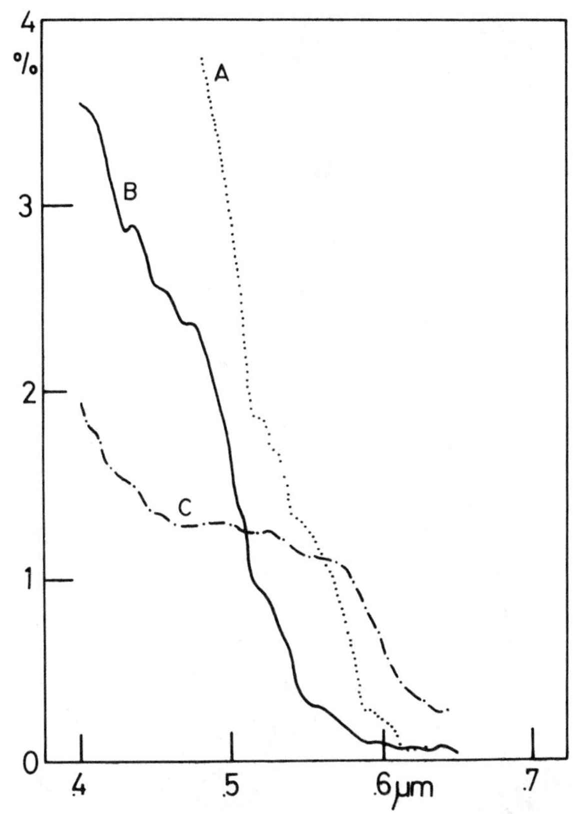

Figure 4-2-8. Spectra of backscattered light, in percentage of incident light, measured from an aircraft at 305 m on August 27, 1968. The spectrometer with polarizing filter was mounted at 53° tilt and directed away from the sun. Concentrations of chlorophyll *a* were measured from aboard a ship: A. Sargasso Sea, <0.1; B. slope water, 0.3; C. Georges Shoals, 3.0. After Ewing (1971).

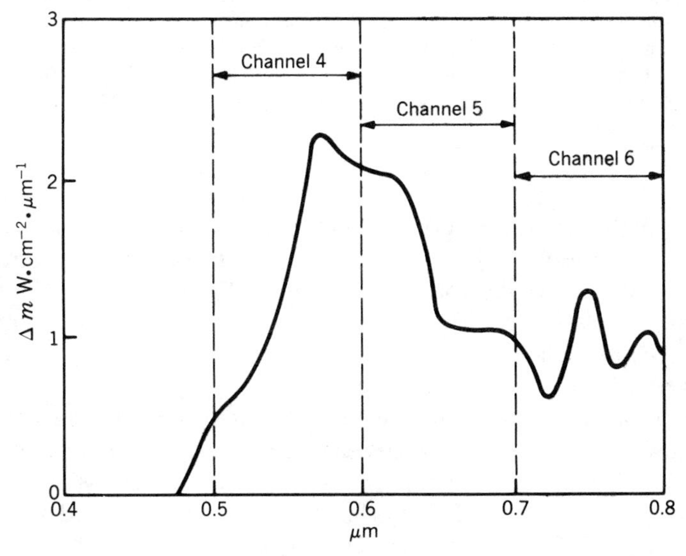

Figure 4-2-9. Spectral response of particulate matter in seawater based on overflights of the Gulf Stream and near coastal waters. Included are the spectral bands of *Landsat*.

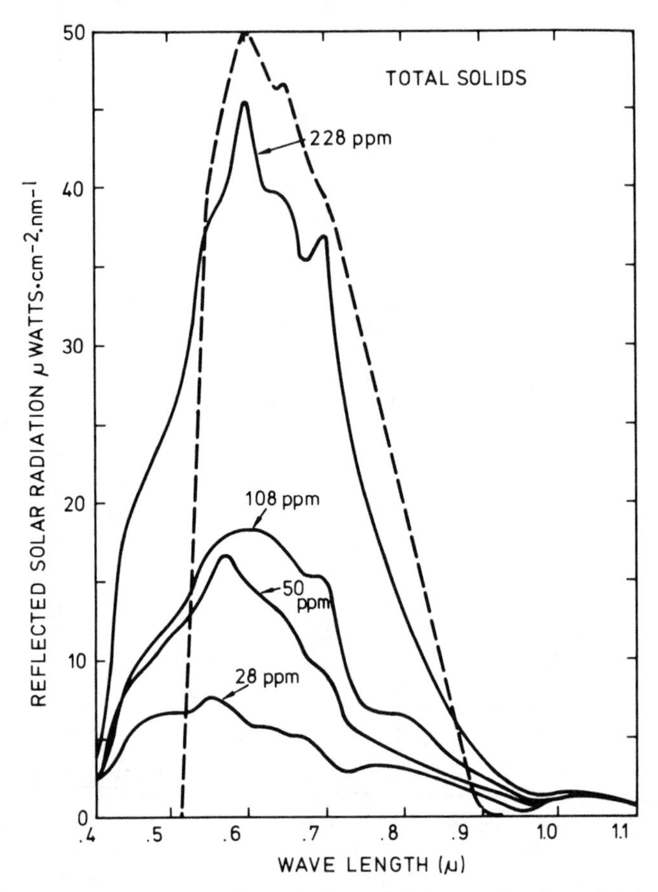

Figure 4-2-10. Relationship among reflected solar radiation, wavelength, and the concentration of total solids in surface waters. Reflected solar radiation was measured perpendicular to and approximately 20 cm from the surface water. Solar radiation measurements were made at 0.025-μm intervals from 0.400 to 0.750 μm and at 0.50-μm intervals from 0.750 to 1.550 μm for both reflected and incident solar radiation. Incident solar radiation was measured first. All solar radiation measurements were made on clear days within 2 hr of solar noon local time (after Ritchie et al., 1974). Dashed line includes the response of NOAA-AVHRR.

of total solids in the surface waters at wavelengths of 0.75 and 0.8 μm. The ratio of reflected solar radiation to incident solar radiation (reflectance) was also best correlated (r = 0.87) with the concentration of total solids in the surface waters at 0.8 μm. The correlation coefficient for the concentration of total solids in the surface waters with reflected solar radiation was greater (r = 0.87) for the 0.7–0.8 μm range than for the 0.6–0.7 μm range.

Regression equations were calculated that best fit the data for the concentration of total solids in surface water versus reflectance at all measured wavelengths. The highest coefficient of correlation (r = 0.90) for these regression equations for the concentration of total solids in surface water versus reflected solar radiation was at wavelength 0.75 and 0.80 μm. The highest coefficient of correlation (r = 0.87) for the concentration of total solids in surface water versus reflectance was at wavelength 0.80 μm (Ritchie et al., 1974).

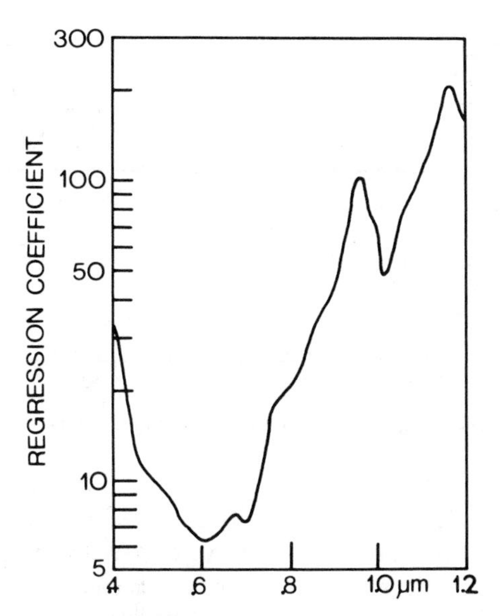

Figure 4-2-11. Relation between wavelength and the regression coefficient (slope) of the regression equations for reflected solar radiation and the concentration of total solids in the surface waters. After Ritchie et al. (1974). Reproduced with permission of Prof F. Sharokhi.

The regression coefficients (Fig. 4-2-11) for the calculated equations for the concentration of total solids versus reflected solar radiation showed that the highest sensitivity for the regression equations is at 0.6 μm. This is the wavelength where the maximum change in reflected solar radiation occurs for the minimum change in the concentration at total solids in the surface waters. The regression coefficients showed that the regression equation calculated for the 0.6–0.7 μm band was more sensitive than the regression equation for the spectral interval of 0.7–0.8 μm.

Relative signal changes of an algae bloom were reported by Grew (1973), which indicated that basically two spectral regions were useful in the monitoring of water color (Fig. 4-2-12).

Mueller (1972) used principal component analysis of ocean color spectra to develop an empirical equation for Secchi disk transparency.[1] Wezernak (1974) utilized reflectance data in the green and red regions of the spectrum in transparency analysis. The expression used to calculate transparency was

$$\log T_{\mathrm{RS}} = c + dR_2$$

where T_{RS} = Secchi disk transparency

c,d = constants

$$R_2 = \frac{0.50\text{–}0.54\ \mu\text{m band}}{0.62\text{–}0.70\ \mu\text{m band}}$$

[1] Secchi disk transparency is a function of absorption and scattering by materials in solution and suspension and is generally related to the volume attenuation coefficient in the green spectral region.

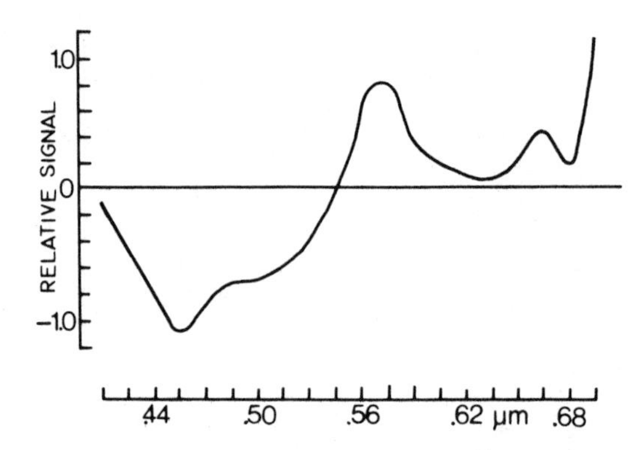

Figure 4-2-12. Spectral signature of an algae bloom in Clear Lake. After Grew (1973).

The green band used generally corresponds to the region of minimum attenuation in productive and moderately turbid waters. With increasing substances in suspension, the above ratio and transparency decrease.

In general, the derivative of a spectrum as a function of wavelength can be used to detect and resolve weak spectral features, overlapping lines, and subtle spectral differences. White (1972) pointed out that the second derivative of airborne data can provide spectra that are in close agreement with in situ measurements, and, when applied to chlorophyll analysis, the second derivative with respect to wavelength provides altitude-independent data of the water column. The second derivative of path radiance may be assumed to be a linear function of wavelength over the region of interest, thereby making the second derivative of path radiance zero. In practice, the technique is applied to a portion of the spectrum that exhibits an important inflection point.

Denoting wavelength and reflectance by λ and R, respectively, we write the first derivative of reflectance at a wavelength between λ_i and λ_j as

$$\frac{dR}{d\lambda} = \frac{R_j - R_i}{\lambda_j - \lambda_i}$$

Taking a portion of a reflectance curve described by reflectances R_1, R_2, R_3 at wavelengths λ_1, λ_2, λ_3, we write the first derivative at λ^2 (approximately) as

$$\frac{dR}{d\lambda}(\lambda_2) = \frac{1}{2}\left\{\frac{R_2 - R_1}{\lambda_2 - \lambda_1} + \frac{R_3 - R_2}{\lambda_3 - \lambda_2}\right\}$$

From satellite altitudes, rather broad bandwidths are received. However, Scherz (1977) showed that if high atmospheric effects are eliminated, "residual curves" from *Landsat* can be used to classify different types of lakes (Fig. 4-2-13). The satellite residual signal then can be expressed as

$$R_i'' = (\rho_{\mathrm{vi}} - \rho_{\mathrm{v_1}})H_0' \frac{\tau}{\pi}$$

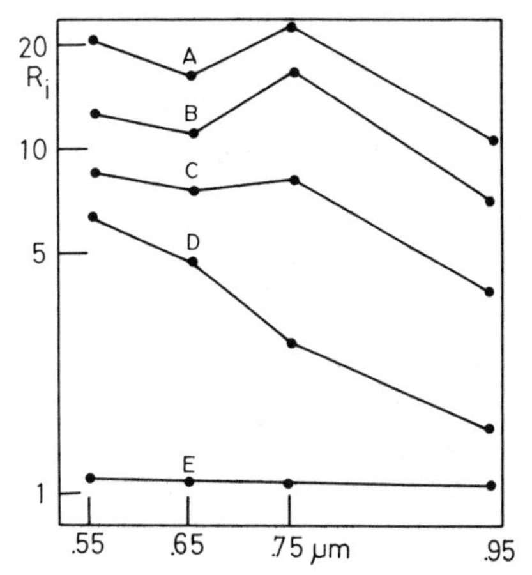

Figure 4-2-13. Satellite residual signal R_i for clear-type water lakes with various amounts of algae present: A. Prairie Lake; B. Yellow Lake; C. Wapagasset Lake; D. Shell Lake; E. Grindstone Lake. After Scherz (1977). Reproduced with permission from the Environmental Research Institute of Michigan (ERIM).

where H_0' *and* τ are total irradiance and atmospheric transmittance, respectively, ρ_{vi} represents the volume reflectance of the lake under investigation, and ρ_{v_1} the volume reflectance of a clear lake used as a "standard," with characteristics similar to distilled water. The residual R_i'' then is due to the material in the investigated lake.

In contrast to the spectral properties of geological features, the irradiance from water has to be interpreted in a three-dimensional model. For instance, the response of suspended matter varies with the depth, or the intensity of backscattered light might be the same for different concentrations if the particles are located at different depths, as shown, for instance, by Tomiyesu (1974).

The brightness temperature T_B can generally be expressed as

$$T_B = \varepsilon T_W$$

where e is the emissivity, and T_W is the absolute temperature of the sea surface. As shown by Nordberg et al. (1971), in the 1.55-cm microwave emission, the brightness temperatures in the nadir direction increased almost linearly with wind speed from 7 to 25 $m \cdot sec^{-1}$ at a rate of about 1.2°C$\cdot(m \cdot sec^{-1})^{-1}$. At 70° from nadir the rate was 1.8°C$\cdot(m \cdot sec^{-1})^{-1}$. This increase was directly proportional to the occurrence of white water on the sea surface. At wind speeds less than 7 $m \cdot sec^{-1}$, essentially no white water was observed and brightness temperatures in the nadir direction were approximately 120 K; at wind speeds of 25 $m \cdot sec^{-1}$, white water cover was on the order of 30% and average brightness temperatures at nadir were approximately 142 K. Maximum brightness temperatures for foam patches large enough to fill the entire radiometer beam were 220 K. This is based on the fact that emissivity of foam-covered water may change from about 0.4 to almost 1. Since foam cover and wind speed are correlated, it is possible to estimate the wind speed by microwave techniques.

Brightness temperatures (T_B) received by a radiometer are the result of three radiation components: radiation emitted to the sensor by the sea surface, radiation emitted to the sensor by clouds and atmospheric water vapor, and atmospheric radiation reflected toward the sensor by the sea surface:

$$T_B = \varepsilon T_W \tau_H + (1 - \varepsilon) T_S \tau_H + \int_0^H T_A(h) \frac{\partial \tau}{\partial h} \, dh$$

where T_W = water surface temperature
T_S = sky brightness temperature
T_A = atmospheric temperature
τ = atmospheric transmissivity
ε = surface emissivity
h = height above surface
H = aircraft altitude

Measurements of T_B over the Salton Sea by Conaway (1969) are shown in Figure 4-2-14. The observations show a considerably greater difference of brightness temperatures between smooth and rough seas than is indicated by theory. The measured differences are approximately 20°C at all nadir angles, in contrast to a theoretical difference of about 15°C at $\theta = 40°$, and a steady decrease of this difference with decreasing nadir angle. At nadir there should be practically no difference.

The value of T_B observed over the ocean by a satellite depends on the properties of the atmosphere:

$$T_B = \tau[\varepsilon T + R T_d] + T_u$$

where τ = atmospheric transmittance
T_d = downward-emitted atmospheric temperature
T_u = upward-emitted atmospheric temperature

The determination of this value is a complicated problem in radiative transfer. Atmospheric attenuation occurs in the presence of liquid water, and Beer (1980) showed that microwave temperature increases with greater precipitation rate over the sea (see Fig. 4-2-15). As the precipitation rate R increases, the signal depolarizes, and above a 5 mm/hr rainfall rate the underlying surface becomes immaterial.

4-2-4 Sea Ice and Snow

The physical properties of ice are more complex than those of water because, besides salinity and temperature, the age and fractionated crystallization of the dissolved constituents during the freezing and cooling process also modify the parameters of sea ice. For instance, the salt content in sea ice is a function of the age of sea ice; new sea ice has a salinity of 3–5‰, and old sea ice has a salinity of about 2‰.

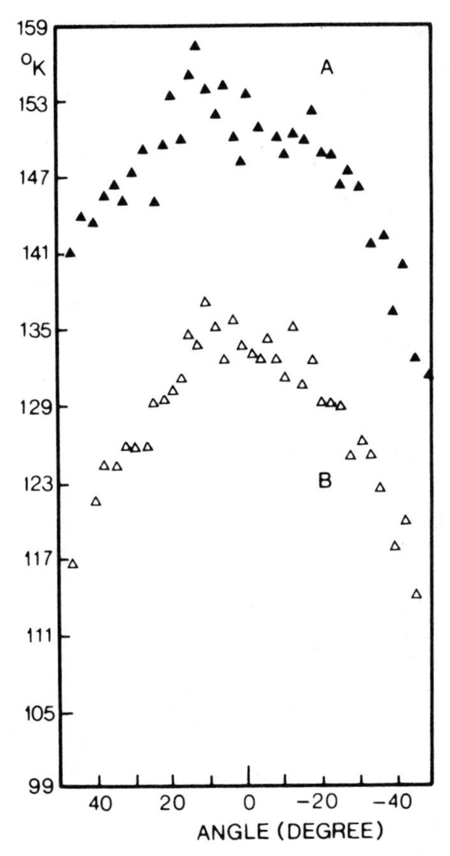

Figure 4-2-14. Observed brightness temperatures versus nadir angle at 1.55 cm over smooth (*A*) and rough (*B*) portions of the Salton Sea. Computations are for sea surface temperature of 290 K and a standard atmosphere. Observations were made with sea surface temperature of 294 K over the rough sea and 300 K over the smooth sea in a relatively moist atmosphere. Each point shown for the observed data represents an average of six consecutive scans at the respective nadir angle. After Conaway (1969).

The term salinity in connection with brine in sea ice should be considered as a comparative number, because the assumed constant ratio on which salinity measurements are based in the oceans cannot be equally applied to sea ice. This is based on the fact that pure water crystallizes from the solution at the onset of freezing, although some brine is found to be trapped between the ice crystals. When the temperature goes below the freezing point of seawater, the dissolved salts begin to crystallize as well. For instance, $Na_2SO_4 \cdot 10H_2O$ crystallizes from brines at −8.2°C, and $NaCl \cdot 2H_2O$ at −22.9°C.

The physical structure of sea ice is also affected by the duration within which freezing occurs: the slower the freezing process, the lower the salinity. This occurs because, as a result of the freezing, the brine diffuses from the ice to the water under it, which also explains why older sea ice has a lower salinity.

Air trapped in sea ice alters some of its physical properties, especially the density, which is also influenced by temperature and salinity. The relationship is

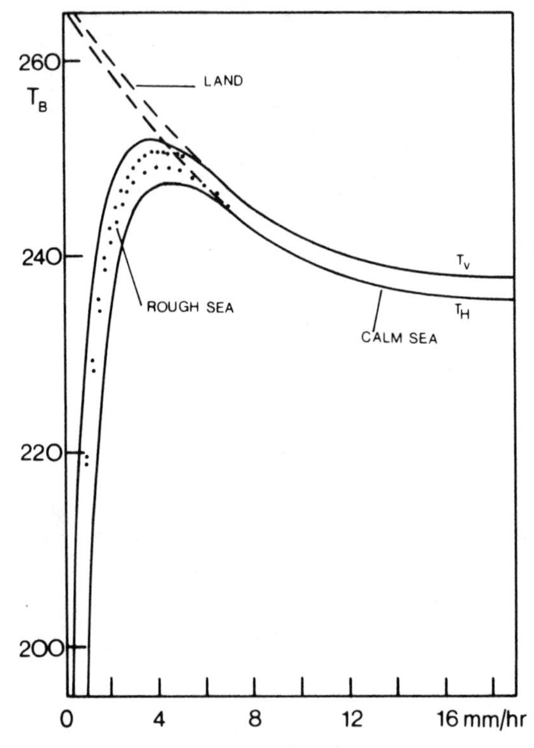

Figure 4-2-15. Theoretical T_B as a function of rainfall rate in mm/hr.

$$S_E = (1 - \alpha)\left(1 + \frac{4.56s}{T}\right) 0.917$$

where S_E = density, $g \cdot cm^{-3}$
α = air, % by volume
s = salinity, ‰
T = temperature, °C

During the freezing process small channels or pores provide a means for transporting the brine with a higher density to the layers below. As a consequence, additional air from the atmosphere replaces the brine, thereby modifying its density. Other parameters of sea ice are also changed by the concentration of salt and air. The specific heat, the mechanical behavior, and the thermal conductivity of sea ice, for instance, are influenced and modified by the temperature as well as by the trapped air.

The albedo A over ice-covered areas is a function of the ratio of the water and the ice-covered area. For short wavelengths, this relationship is

$$A = 0.59 - 0.32R$$

whereby only high sun angles are considered. This relationship may be important when the ground resolution of a sensor is less than the size of open water in the ice or the size of ice parcels floating in open water.

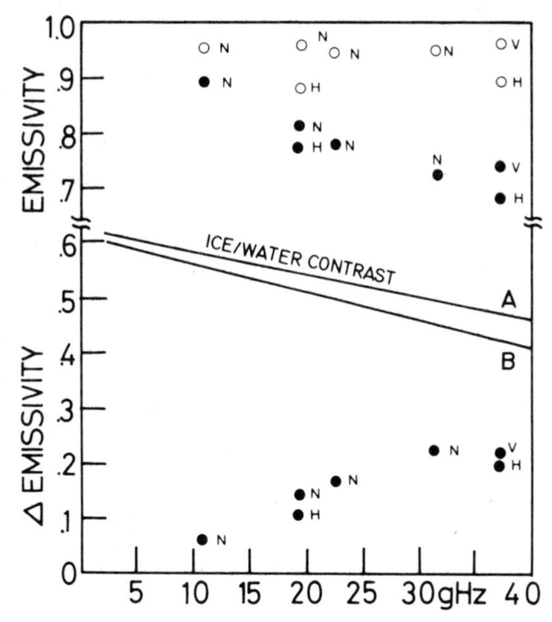

Figure 4-2-16. Microwave emissivity contrasts of old and new ice. Open circles represent data on new ice, and closed circles in the upper part of the graph show emissivity for old ice. Lower graph shows the difference in emissivities between new and old ice. Courtesy of NASA, Nimbus Project.

Ice in the higher latitudes may be derived from the freezing of seawater, or it may be transported by glaciers from continental ice. Therefore, based on its origin, ice also differs in its physical properties depending on the different electrolyte concentration and the incorported air.

Sea ice has a characteristic response in the microwave region of the EMS. It has been observed that the sea itself has emissivity variations, which are thought to be related to age, first-year ice being more emissive than older ice (Wilheit, 1972). The spectral nature of the differences between old and new ice is shown in Figure 4-2-16. The contrast between the two types of ice appears to increase linearly with frequency, amounting to about 12% at 19.35 GHz. This is about 20% of the contrast between new ice and open water.

Albedo of ice-covered sea surface depends on the percentage of ice at the surface as well as the region where it appears. This is shown in Table 4-2-2, which summarizes albedo measurements of different ice-covered regions.

TABLE 4-2-2. Albedo of the Ice-Covered Sea Surfaces with Various Ice Coverage (Summer Months)

	Ice Coverage (%)										
Region of Observation	0	10	20	30	40	50	60	70	80	90	100
Coastal waters	6	7	11	15	20	24	28	32	36	40	44
High seas	8	12	18	22	28	33	40	46	53	63	73

After Hodarev et al. (1971). Reproduced with permission from the Environmental Research Institute of Michigan (ERIM).

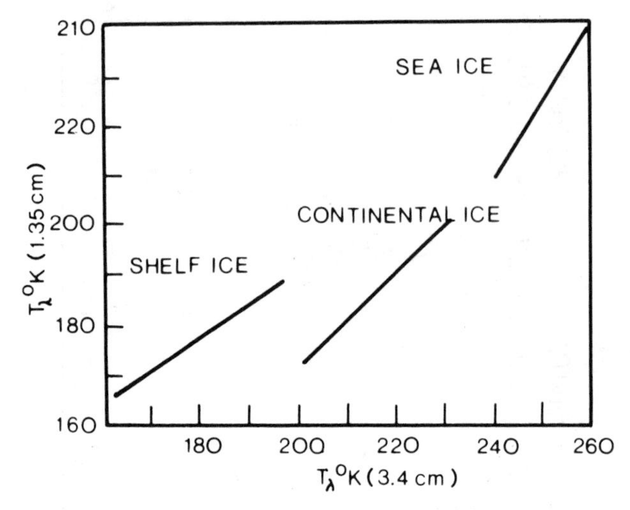

Figure 4-2-17. A correlogram of the Antarctic ice cover radio brightness temperature at wavelengths of 3.4 cm and 1.35 cm. After Basharinov et al. (1971). Reproduced with permission from the Environmental Research Institute of Michigan (ERIM).

Similar differences between coastal water ice and high sea ice in the visible can be shown in the microwave region. A correlogram of measurements at different wavelengths showed that continental ice barrier glaciers and floating ice are clearly recognized; see Figure 4-2-17 (Basharinov et al., 1971).

Because of the great changeability of snow properties, the spectral albedo of snow has a high variable value. Generally, in the range 0.5–0.7 μm, the reflectance has no structure. However, one can see some decrease of albedo at shorter wavelengths, with a sharper decrease at longer wavelengths, as shown in Figure 4-2-18. In addition, the albedo of snow is also a function of thickness.

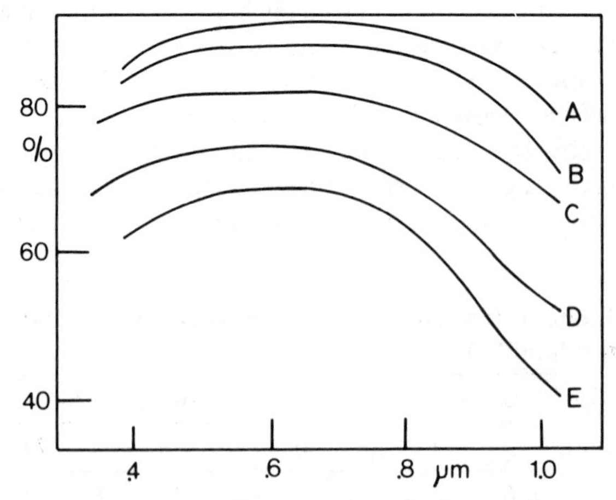

Figure 4-2-18. Snow albedo: A. Dry fresh snow; B. damp fresh snow; C. snow with ice crust; D. soil after snow melting; E. wet snow. After Hodarev et al. (1971). Reproduced with permission from the Environmental Research Institute of Michigan (ERIM).

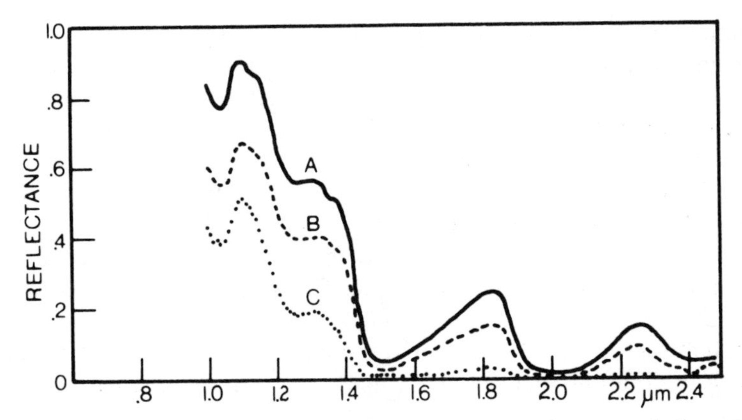

Figure 4-2-19. Changes in snow reflectance with natural aging. Reflectance relative to $BaSO_4$:

		A	*B*	*C*
Snow condition	Natural aging	14 hr	44 hr	70 hr
Snow density	g/cm^3	0.097	0.104	0.347
Snow hardness	g/cm^2	~4.5	~4	504

After O'Brien and Munis (1975).

Albedo is different also for different conditions of snow. As shown in Figure 4-2-19, dry fresh snow and wet snow have different albedos. Changes in snow reflectance are observed as a function of aging. O'Brien and Munis (1975) showed the reflectance properties of natural aged snow for the region between 1.0 and 2.5 μm. In this region snow has very characteristic bands, but snow also shows, especially between 1.0 and 1.3 μm, a decreasing reflectance as a function of age.

The spectral reflectance of water frost on ice as a function of temperature and grain size in the 0.65–2.5 μm wavelength region was investigated by Clark (1981). The 2.0-, 1.65-, and 1.5-μm solid-water absorption bands were precisely defined along with the 1.25-μm band and the 1.04-, 0.90-, and 0.81-μm absorption bands. It was found that the 1.65-μm component, which was thought to be a good temperature sensor, was highly grain-size dependent and poorly suited to temperature sensing.

4-2-5 Water-Covered Objects

The penetration depth of solar irradiation into water depends on the dissolved and particulate compounds and on the sun angle, which can be expressed by the relationship between the depth of penetration of solar rays into water and the height of the sun above the horizon:

$$I_1 = I_0 \left[1 + \frac{\cos^2(\phi + \psi)}{\cos^2(\phi - \psi)}\right]\left[\frac{\sin^2(\phi - \psi)}{2\, sin^2(\phi + \psi)}\right]$$

where I_1 = reflected light
I_0 = incident light
ϕ = angle of incidence of solar rays
ψ = angle of refraction of solar rays

The equation shows that a decrease in the height of the sun above the horizon results in an increase in light reflection at the air-sea interface. Reflection is also a function of sea state; in the presence of waves on the surface, reflection is greater even when the sun is at its zenith height, because, under those conditions, the surface is not completely normal to the solar rays. Therefore, water-covered objects are best observed at a high sun angle and with a calm sea surface, which would have high photon penetration.

Reaching the seafloor or other objects in the water, the reflected light is diffused in nature and depends on the spectral characteristics and structure of the objects.

Charnell and Maul (1973) calculated the percent of energy that penetrates the sea surface in different spectral intervals, by solving the intensity equation

$$I = \int_0^\infty \phi_\lambda I_{0\lambda} e^{-\alpha_\lambda z} \, d\lambda$$

where I is the intensity observed at a depth z through a bandpass filter ϕ_λ normalized over the region of wavelength λ for the spectral intervals, $I_{0\lambda}$ is the intensity at the sea surface, and α_λ is the spectral attenuation coefficient. The calculations show that a perfect reflector at a depth of 1 m would return 86% of the incident energy for the 0.5–0.6 μm spectral region, 55% for the 0.6–0.7 μm spectral region, 11% for the 0.7–0.8 μm spectral region, and 0.2% for the 0.8–1.1 μm spectral region.

The calculations on photon penetration also give a first estimate on the reflected energy of an object at a given depth. A perfect target at 5 m would reflect 50% in the green-yellow band, 7% in the orange-red band, and would have almost no reflection in the red or near IR bands.

Spectra of vegetation covered by water must also include the depth from which the signal is derived. Spooner (1969) reported directional underwater spectral reflectance curves of *Anacharis*. The spectra of the latter are shown in Figure 4-2-20. Although the levels of the curves vary by a considerable amount, the peak green and IR reflectance values and the minimum red reflectance value all occur at wavelengths that agree, respectively, within ± 0.01 μm. The wavelength is 0.55 ± 0.01 μm for the peak green reflectance for all curves, 0.8 ± 0.01 μm for the peak IR reflectance, and 0.67 ± 0.01 μm for the minimum red reflectance.

As is true for terrestrial plants, the overall levels of the curves for any water species also vary because of variations in plant leaf density, illumination, and water path lengths.

The plant leaf density factor may be defined both as the optical density of the leaf element and the placement density of small leaf elements. For instance, it was found that the old leaf of water sprite has a greater optical density than that of new leaves of the same plant. In addition, the variations caused by water path length are particularly significant at wavelengths greater than 0.8 μm. The peak values of the IR reflectance around 1.04–1.08 μm indicate the strong absorption of water in this spectral region.

Rock-type surfaces do not have pronounced reflectance features as vegetation. Nevertheless, the differences in reflection allows us to qualitatively differentiate surfaces through their spectral response. Selected samples of geological spectra

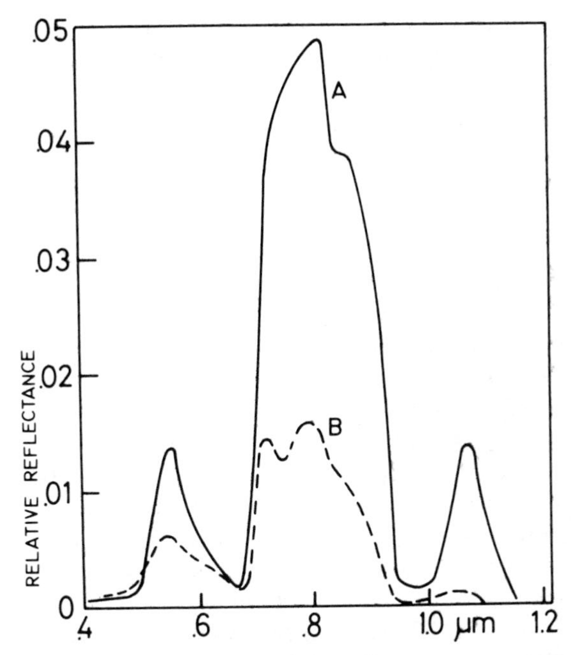

Figure 4-2-20. Relative spectral reflection of *Anacharis*. Curves are presented in order of increasing water path lengths, A having a shorter path than B. After Spooner (1969). Reproduced with permission from the Environmental Research Institute of Michigan (ERIM).

covered by water are given in Figure 4-2-21, including the spectra of mud and fine sand. The comparison between the last two spectra indicates that in flat ocean regions the composition of the bottom can significantly alter the signal obtained by a remote sensing system over shallow water.

Detection of fish schools by remote sensing is possible by using the proper

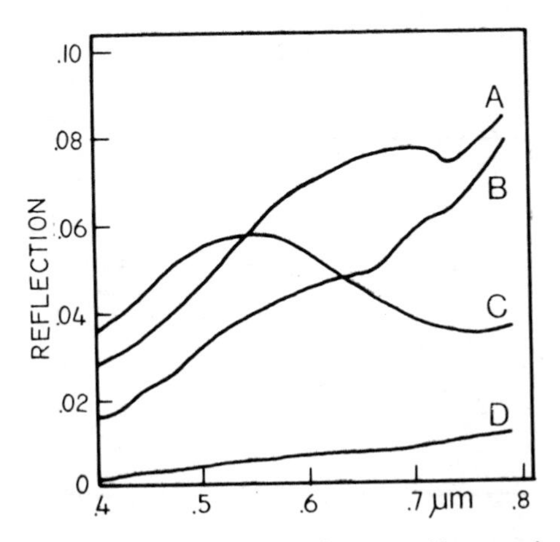

Figure 4-2-21. Data are reduced relative to a loose barium sulfate standard: A. High surface beach sand; B. fine sand with phosphorite olites (5.1% P_2O_5); C. jadelike green rock; D. green mud. After Spooner (1969). Reproduced with permission from the Environmental Research Institute of Michigan (ERIM).

spectral bands according to radiation properties of the fish. Calculations by Drennan (1971) showed that the minimum density in which fish could be detected under these conditions is about 10%. Benigno (1970) also gave spectral reflectivities for different fish species and showed that they differ significantly from each other. Chub mackerel has the highest reflectivity, at around 0.5–0.6 μm, and bluefish and menhaden show rather low reflectivity at this wavelength.

4-3 SPECTRAL CHARACTERISTICS OF GEOLOGICAL TARGETS

Type and composition of minerals characterize the albedos and the emissivity over the whole spectrum. Over small spectral intervals in the thermal part of the EMS, significant temperature differences exist due to the different emissivity. In the microwave region slight changes of the emissivity can be recognized easily, but the sensitivity due to changes of the physical temperature is much less.

Differences in spectral shape can be used to distinguish among geologic materials and, in some cases, to estimate their bulk composition. Several factors, including the crystal lattice, the surface of the rocks, and atmospheric conditions, have to be taken into account for such an approach.

In the visible part of the EMS, sedimentary rocks do not display distinct spectra, making it difficult therefore to determine rock types. Short and Lowman (1973), however, reported that several individual rocks displayed greater variation between spectra from fresh and weathered surfaces than between different rock types. A difference in reflectance values is normally seen in both the visible and the IR part of the EMS.

With a constant chemical composition, it has been shown that for soil, the reflectance increases as particle size decreases. Absorption increases with particle size, since reflection occurs at each particle-air interface. As a consequence, it can be stated that the smaller the particle, the more multiple scattering or multiple reflections contributing to the total reflectance can be predicted. The variation of total reflectance for various sieve sizings of clay was investigated by Piech and Walker (1974) and is shown in Figure 4-3-1.

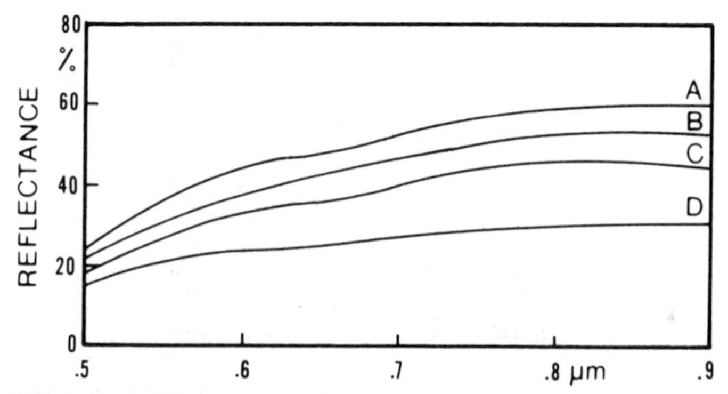

Figure 4-3-1. Variation of total reflectance for various sieve sizings of a Yaucoa clay: A. <0.062 mm; B. 0.082–0.125 mm; C. 0.125–0.25 mm; D. 2 mm. After Piech and Walker (1974). Reproduced with permission from *Photogrammetric Engineering & Remote Sensing,* copyright by the American Society for Photogrammetry and Remote Sensing.

Reflectance properties of soil also depend on the moisture, with lower reflectance indicated with higher soil moisture. In addition, the concentration of organic matter, iron oxide, and roughness may influence the spectral characteristics of soil.

Singer (1980) measured the near IR spectral reflectance of mineral mixtures containing pyroxenes, olivine, and iron oxides (see Fig. 4-3-2).

Reflectance values at the spectral interval between 0.2 and 1.6 μm for travertine and basalt show that significant differences in the absolute reflectance in the visible and near IR regions can be used to identify and detect geological targets having different chemical composition (Miller and Pearson, 1971) (see Fig. 4-3-3).

Spectra of hematite, limonite, which are hydrated ferric iron minerals, and jarosite have a strong blue-green absorption edge that can be attributed to the ferric ion. The shape and location of the absorption edge in the visible region also change according to the mineralogy. Ferric absorption in the IR appears in a broad band centered at about 0.9 μm. The minimum for jarosite is 0.5 μm farther into the IR than is the minimum for hematite. The relative variations in the visible and IR absorption bands cause the apparent peak in the red region to shift between 0.70 and 0.75 μm.

Generally speaking, rocks have a rather low reflectance at shorter wavelengths, with increasing reflectance toward longer wavelengths, and reaching maximum reflectance between 1 and 3 μm (Fig. 4-3-4.).

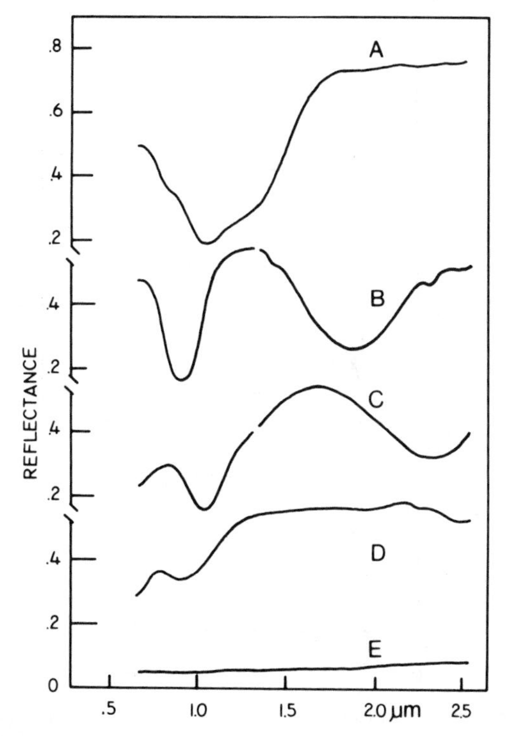

Figure 4-3-2. Bidirectional reflectance spectra of five mineral samples. All spectra presented were obtained with an illumination angle of 10° and an emission angle of 0° (normal): A. Olivine; B. orthopyroxene; C. clinopyroxene; D. limonite; E. magnetite. After Singer (1980).

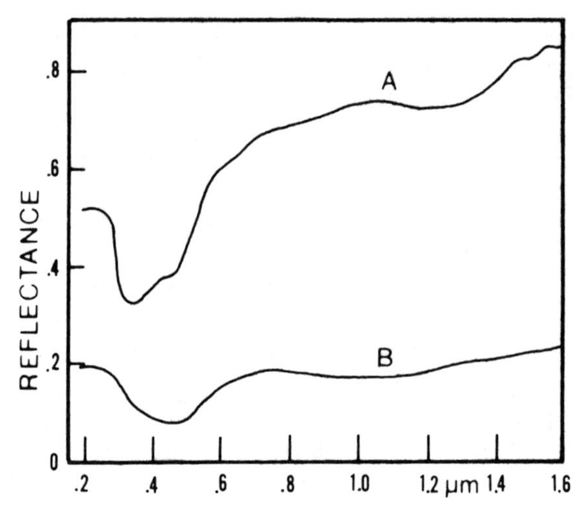

Figure 4-3-3. Spectroreflectance curves for geologic materials: A. Travertine; B. basalt. After Miller and Pearson (1971). Reproduced with permission from the Environmental Research Institute of Michigan (ERIM).

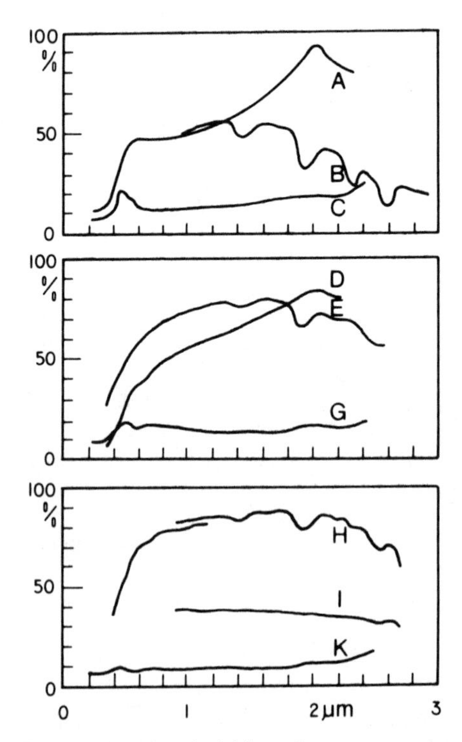

Figure 4-3-4. Spectral reflectances and emissivities of common rock types: A. Shale; B. coral; C. basalt; D. yellow sandstone; E. chert; G. gray feldspar; H. limestone, I. siltstone; K. lava. After Suits (1978).

The spectral reflectances of rocks measured on the ground and those measured in the laboratory suggest higher reflectance values for laboratory specimens, although the nature of the spectral response in different wavelength intervals is similar. Average spectral reflectance values of individual rocks at an interval of 0.1 μm for both field and lab measurements appear to be distinctive (Guha and Mallick, 1985).

More significant differences in the spectra of rocks appear in the thermal IR between 9–12 μm, due primarily to silica content which causes a shift in the minimum emission in this spectral part. The shift may be 0.49 μm for gypsum sand compared with that of silica sand (Bell and Eisner, 1956; Bell et al., 1957; see Fig. 4-3-5).

Absorption and reflection spectra were compared for various silicon dioxide modifications in the wavelength region between 7 and 24 μm (Sevchenko and Florinskaya, 1956). The shift in wavelength is mainly based on silicate. Lyon and Patterson (1966) used this shift to develop an analytical tool for determining the mineralogical and chemical composition of surface rocks. They determined the shift to be approximately 2.0 μm, from 8.8 to 10.8 μm between rock compositions of granite and dunite. Another marked spectral band for silicates appears between 20 and 25 μm, which, as the first band, shifts further in matic and ultramatic rocks due to an increase in the FeO/(FeO + MgO) ratio. With finely powdered material, however, the effect of surface roughness of the sample loses much of its information.

Samples of spectra of common terrestrial rocks, tektites and chondritic meteorites, between 8 and 13 μm, as measured in the laboratory, are shown in Figure 4-3-6. Although the spectra are based on laboratory measurements, they demonstrate that a specific emissivity spectrum can be assigned to each rock type.

Spectral peaks in the thermal IR depend on bulk composition of the rock and do not change with increasing grain size (Lyon and Burns, 1963). In addition, regardless of whether a given composition of rock is in the physical form of glass, felsite,

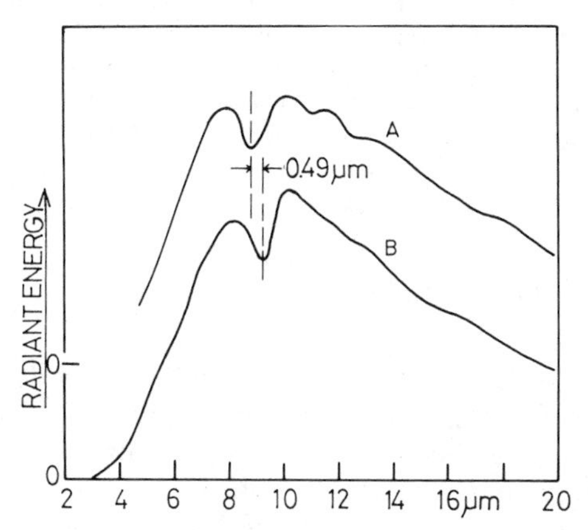

Figure 4-3-5. Spectral emission from gypsum sand (A) compared with that for quartz sand (B), showing a 0.49-μm peak shift between sulfate and silicate functional groups. After Bell and Eisner (1956) and Bell et al. (1957).

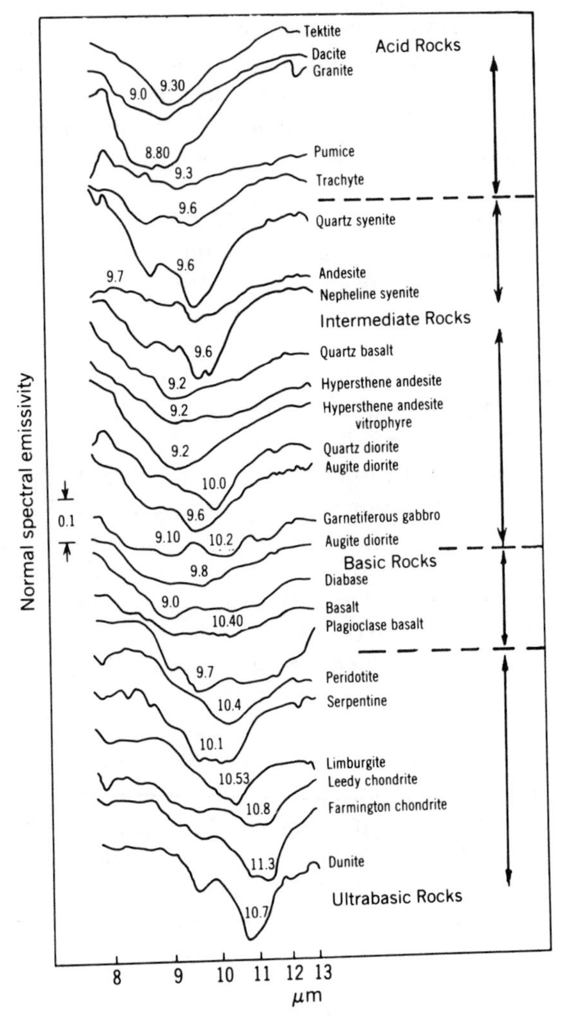

Figure 4-3-6. Spectral normal emissivities of common terrestrial rocks, tektites, and chondritic meteorites, between 8 and 13 μm. Spectra arranged in decreasing SiO_2 content from "acidic" to "basic" rock types. Numbers refer to the wavelength of prominent minima. Specimen surfaces were either sawed or broken naturally. A relative scale of emissivity is given along the vertical axis. Samples were heated to 640 K. After Lyon and Patterson (1966). Reproduced with permission from the Environmental Research Institute of Michigan (ERIM).

fine- or coarse-grained flow, or plutonic rock, the spectral maximum of reflection remains fixed in wavelength.

Hovis (1972) showed the equivalent blackbody temperature, measured in the nadir, when flying over a serpentine quarry. As shown in Figure 4-3-7, over trees and fields on either side of the quarry, the two temperatures agree, but over the quarry the shorter-wavelength channel produces a warmer equivalent blackbody temperature than does the long-wavelength channel. Since both measurements were made simultaneously, the difference is due only to the emissivity variation with wavelength.

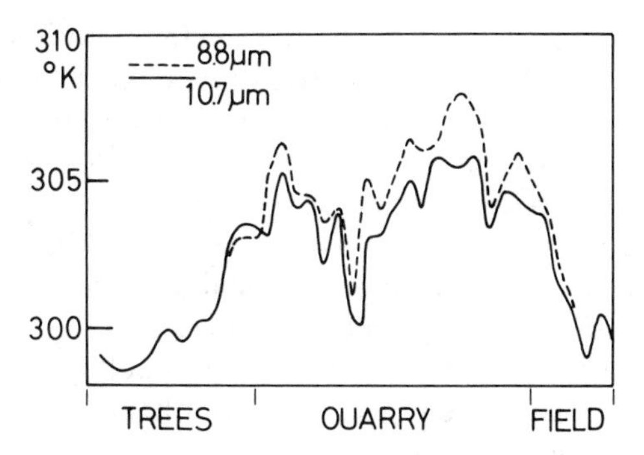

Figure 4-3-7. Flight recordings of the 8.8- and 10.7-μm equivalent blackbody temperatures over a serpentine quarry. After Hovis (1972).

In an effort to monitor the composition of different rock types from space, one has to identify the proper wavelength in connection with the atmospheric properties. For acidic rocks, this coincides with the Reststrahlen features, which correspond to high silica content such as granite and ultrabasic rocks with low silica concentrations, as for example, in dunite. However, the strong ozone absorption band between 9 and 10 μm in the atmosphere allows only the selection of few spectral regions where one can take advantage of specific emission spectra of different terrestrial rock types. For a better illustration of the spectral interval, Figure 4-3-8 outlines the spectral emission of granite and dunite as well as the radiated energy from ozone (Hovis, 1972). Since Reststrahlen indicates a decrease

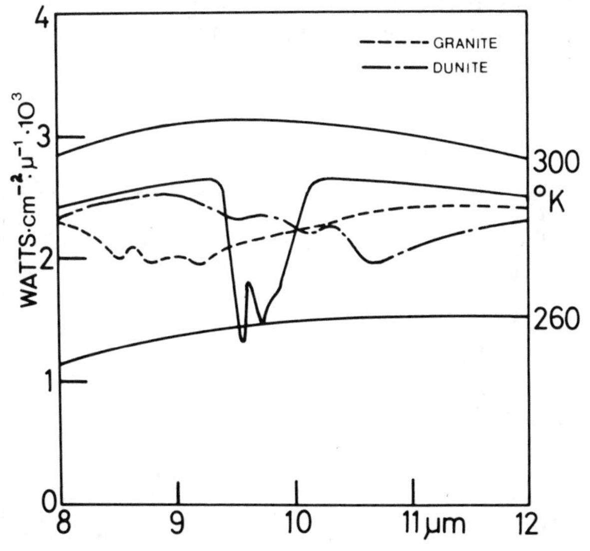

Figure 4-3-8. Effects of Reststrahlen on radiated energy from a granite, representing the acidic case, and from a dunite, representing the basic case, where both are at a true temperature of 290 K. It can be seen that the minimum for granite occurs near 8.8 μm and for dunite near 10.7 μm. The 9.6-μm ozone band is shown to illustrate why the Reststrahlen for intermediate materials cannot be observed. After Hovis (1972).

in emissivity as a function of silica concentration in the rock sample, the temperature difference can be used as a measure to estimate the qualitative composition of rocks. Note that from the spectral characteristics of granite, dunite, and ozone, intermediate rocks, as outlined in Figure 4-3-6, cannot be differentiated from space in this particular spectral envelope.

Thermal measurements have been used to estimate thermal inertia by remote sensing. Thermal inertia, also called the thermal contact coefficient, is a measure of the rate of heat transfer at the interface between two dissimilar media. As Figure 4-3-9 shows, rocks have different rates of cooling and heating, and these rates can be used to identify them, if their rate of heat transfer can be estimated.

Thermal inertia P is given by

$$P = (dc\lambda)^{1/2}$$

where λ = thermal conductivity, $cal \cdot cm^{-1} \cdot sec^{-1} \cdot {}^\circ C^{-1}$
d = density, $gm \cdot cm^{-3}$
c = heat capacity, $cal \cdot gm^{-1} \cdot {}^\circ C^{-1}$

Materials with a low P are relatively insensitive to temperature perturbations at their boundaries.

Watson (1975) showed an approximately linear relationship between density and thermal inertia for dry soils and rocks. However, this does not mean that a unique interpretation of soil and rock type can be achieved, because soils and rocks with different chemical composition may have the same bulk density. Thermal inertia values for moist soils and rocks change rapidly, depending upon their water saturation and porosity. The reason for the rapid change is the large difference in the thermal properties of water compared with the air that it is replacing. Therefore, it is possible, at least in arid and semiarid environments, to determine thermal inertia by remote sensing.

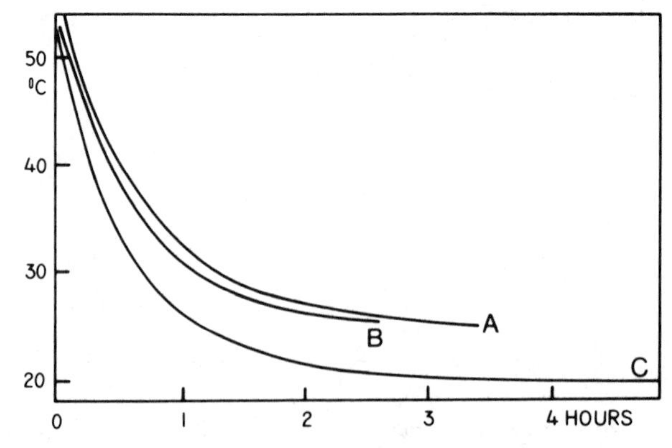

Figure 4-3-9. Rate of cooling of various rock samples. Equivalent blackbody temperatures (°C) measured in the spectral range 10.5–12.5 μm. Angle of view, 2°: A. Wet, coarse-grained granite; B. wet marble; C. wet clayish kaoline. After Pouquet (1969). Reproduced with permission from the Environmental Research Institute of Michigan (ERIM).

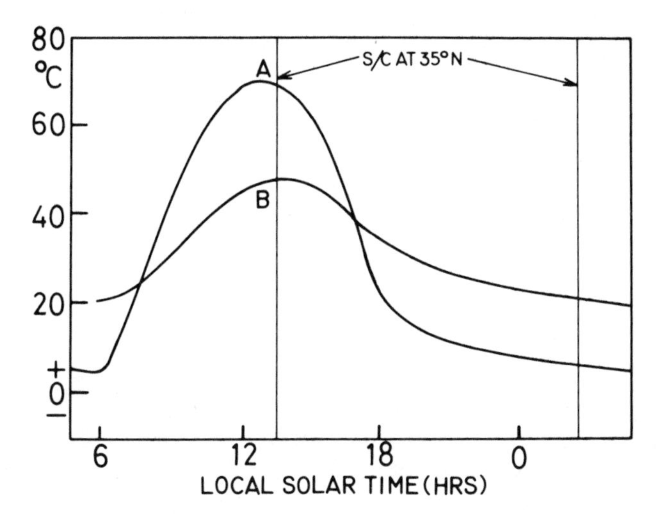

Figure 4-3-10. Diurnal surface temperature variation as a function of thermal inertia. A has a value of 0.01, and B a value of 0.05.

The maximum diurnal temperature variation ΔT, as shown in Figure 4-3-10, is primarily a function of the ground thermal properties and is essentially independent of the surface emissivity. The fact that it is independent of emissivity is extremely important, because this eliminates one of the major unknown variables involved in the interpretation of thermal remote sensing data for thermal inertia.

In the microwave region the emissivity of a land surface is greater than 0.9, which shows that a strong gradient between land and water features may be expected and the brightness temperature contrast between them may reach 100 K. Values of emissivity at 37 GHz for some selected surface materials are given in Table 4-3-1, and brightness temperatures for different substances as a function of angle of incidence are given in Figure 4-3-11. The brightness temperature as seen from a satellite (T_B) can be calculated as follows:

$$T_B = \tau[\varepsilon T_g + (1 - \varepsilon)T_d] + T_u$$

where τ = atmospheric transmittance
ε = emissivity
T_g = ground temperature, K
T_d = downward-emitted atmospheric temperature, K
T_u = upward-emitted atmospheric temperature, K

For different surfaces with emissivities ε_1 and ε_2, respectively, but with the same ground temperature, the difference in the brightness temperature is

$$\Delta T_B = \tau(T_g - T_d)(\varepsilon_2 - \varepsilon_1)$$

At 19.35 GHz, for instance, with a dry atmosphere, $\tau = 0.5$, and $T_g = 310$ K, the brightness temperature for sand would be 285 K and for limestone it would be 237 K, giving a ΔT_B of 48 K (Allison, 1976).

TABLE 4-3-1. Emissivities of Selected Surface Materials at 37 GHz

Surface Materials	Emissivity
Desert sand (quartz, dry, medium fine)	0.93
Sandstone	0.93
Granite	0.90
Limestone	0.75

After Allison (1976).

In the microwave region, it was shown that the brightness temperature at 1.55 cm is also correlated with soil moisture in the top 1 cm. This has been demonstrated, for example, by Tinney et al. (1975).

The unique dielectric properties of water afford two possibilities for the remote sensing of the moisture content in the surface layer of soil (Schmugge, 1980). The dielectric constant for water is an order of magnitude larger than that for dry soils at the longer microwave wavelengths, approximately 80 compared with 3 or 4 for dry soils. As a result, the surface emissivity and reflectivity for soils are strong functions of their moisture content. The changes in emissivity can be observed by passive microwave approaches, and the changes in reflectivity can be observed by active microwave approaches.

Allison et al. (1979) gave examples of aircraft results at 1.55-cm wavelengths, which are presented in Figure 4-3-12, showing the dependence of T_B as a function

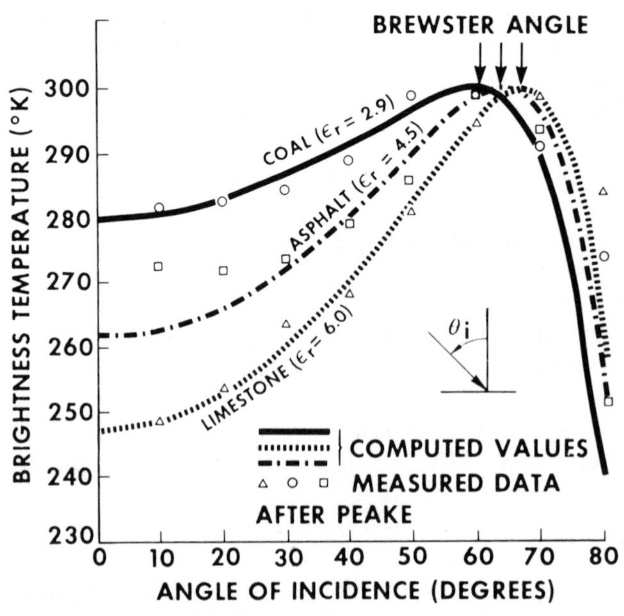

Figure 4-3-11. Computed and measured brightness temperatures of limestone, asphalt, and coal at 10 GHz, vertical polarization. After Edgerton and Trexler (1970).

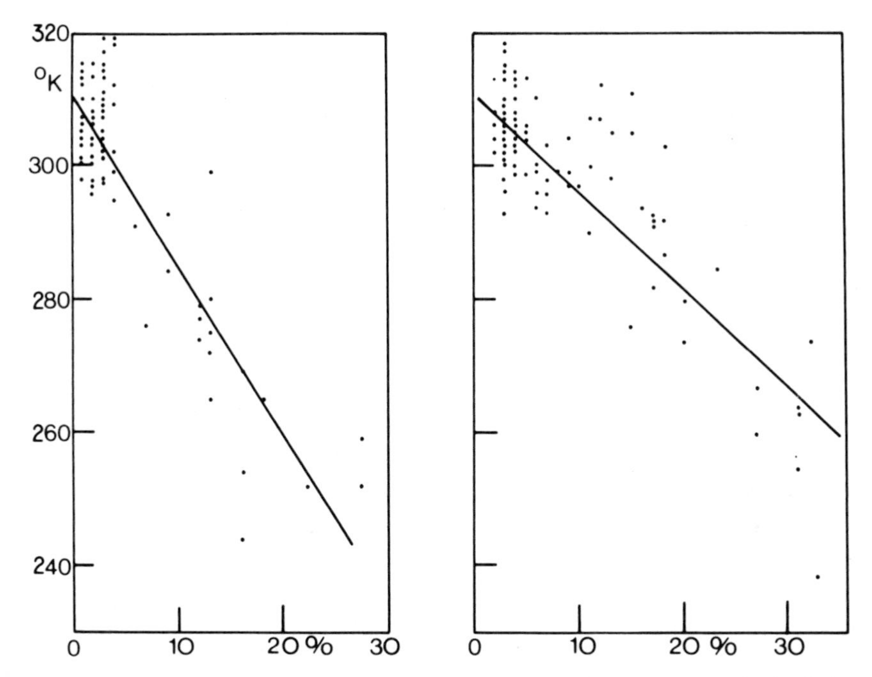

Figure 4-3-12. 1.55-cm brightness temperatures versus soil moisture by weight in the 0–1 cm soil layer. Left data show light soils and right data, heavy soils. After Allison et al. (1979). Courtesy of NASA, Nimbus Project.

of soil moisture. It was found that microwave emissivity of heavy soils (high clay content) showed a lesser sensitivity (slope) to soil moisture content than did the light soils (sandy loam). The dependence on soil type can be removed by expressing soil moisture content in units of percentage of field capacity (FC) by use of the following empirical equation:

$$FC = 2.51 - 0.21 \times \text{sand} + 0.22 \times \text{clay}$$

where sand and clay represent respective soil fractions in percentages (Schmugge et al., 1976).

Soil moisture content expressed in volume basis is a better parameter to use than the moisture expressed as a percentage of dry weight, because of its unique relationship to dielectric permittivity for a given soil. Normalized brightness temperatures T_{NB} are a function of volume basis (W_V) and dry weight (W_W) in data from both 1.4-GHz and 5-GHz frequencies at 20° incidence angle and horizontal polarization, as shown in Figure 4-3-12. The W_V and W_W are evaluated at two different layers, 0–2.5 cm and 0–0.5 cm, for 1.4-GHz and 5-GHz measurements, because of the difference in the sampling depth. The results shown in Figure 4-3-13 define T_{NB} as

$$T_{NB} = \frac{T_B}{T_S}$$

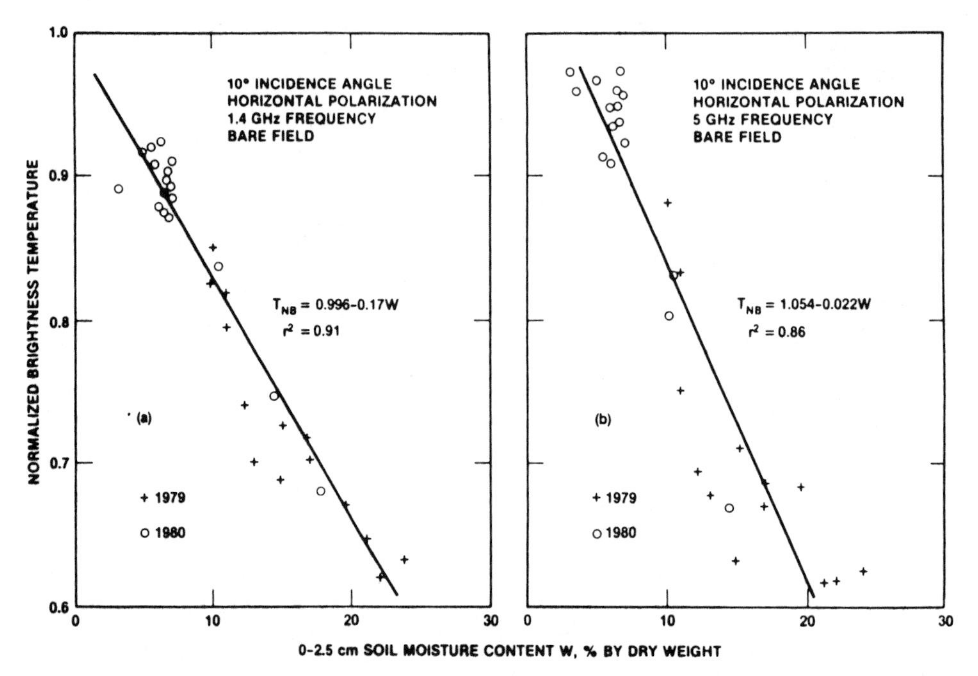

Figure 4-3-13. The variation of normalized brightness temperatures with soil moisture content in the top 2.5-cm layer; (*a*) 1.4-GHz frequency and (*b*) 5-GHz frequency. The measurements were made over bare fields in both 1979 and 1980 at 10° incidence angle and horizontal polarization. After Wang et al. (1982a). Reproduced with permission of Elsevier Publishing Co., Inc.

where T_B is the measured brightness temperature and T_S is the soil's thermal temperature measured at the two different layers, as indicated in the figures corresponding to the two frequencies. In an evaluation of different data sets collected over several years, Jackson et al. (1984) concluded that the optimal surface soil moisture sensor using passive microwave techniques was a 21-cm wavelength radiometer.

4-4 SPECTRAL CHARACTERISTICS OF VEGETATION

Spectral characteristics of vegetation are far more complex than those of geological features, due to the radiation interactions with chlorophyll and auxiliary pigments. In pure solutions the individual pigments of plants have distinct absorption bands over a bandwidth of 0.25 μm; however, in living plants the interactive effects of the different pigments in leaves broaden the absorption bands due to overlapping of the specific absorption bands. As a result, vegetation shows only a slight rise in the visible except in the green part of the reflection spectrum at 0.55 μm. In the near IR, leaf material is transparent to the incoming long-wave radiation and is scattered by multiple reflection at optical density boundaries within the mesophyll structure, which results in an increased reflection in the near IR region.

From satellite altitudes the leaves rather than the stems, roots, flowers, and fruits are responsible for the signals obtained by a sensor, which means that the foliage and its spectral characteristics are the prime concern in investigations of vegetation by remote sensing. Reflection or emission from the leaf's cuticle and epidermis is relatively minor; most of the light from the blue and red ends of the visible spectrum is absorbed by chlorophyll, whereas green light is largely reflected by it. Energy from the near IR part of the spectrum is slightly affected by the chlorophyll, but is greatly affected by the gross structure of the spongy mesophyll tissues deep within the leaf.

The peak reflectance in the green and the low reflectance in the blue and red ends of the visible spectrum are primarily attributable to the interactions of light with chlorophyll. The peak reflectance in the near IR is primarily due to interactions of IR radiant energy with the spongy mesophyll in them. Figure 4-4-1 shows that the spectrum of different vegetation is specific for the species and also shows that highest differences in spectra are in the near IR.

If the spectra of leaves are to be considered, one has to keep in mind that incident energy is partly reflected, absorbed, and transmitted, as shown in Figure 4-4-2. The transmittance spectrum has the same shape as the reflectance spectrum, whereas the absorptance spectrum is opposite in character.

The reflectance from a canopy is considerably less than that from a single leaf because of a general attenuation of radiation by variations in illumination angle, leaf orientation, shadows, and nonfoliage background surfaces such as soil (Knipling, 1970).

A nearly continuous broadleaf canopy typically might be about 3–5% for the visible and 35% for the near IR (Steiner and Gutermann, 1966), whereas the corresponding values for a single leaf are about 10% and 50%. The relatively smaller reduction in IR reflectance is due to the fact that much of the incident IR energy transmitted through the uppermost leaves is reflected from lower leaves and retransmitted up through the upper leaves, thus enhancing their reflectance.

The strong absorption by a leaf in the IR beyond 1.3 μm is due to water, with two absorption maxima around 1.4 and 1.9 μm. Using the absorption characteristics of water, Allen et al. (1969) and Gausman et al. (1970) derived the equivalent water thickness to indicate the thickness of a water layer that can account for the

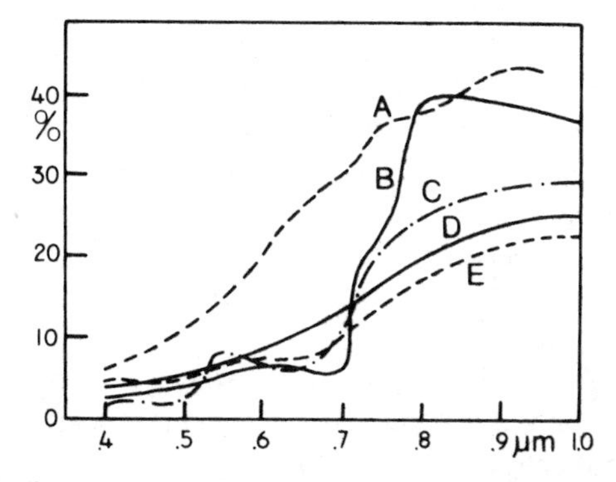

Figure 4-4-1. Spectral albedo for selected vegetation cover. Sun's elevation is in parentheses: A. Straw (56°); B. thick green grass (56°); C. sunflower (52°); D. dry grass (28°); E. shoots of winter wheat (43°). After Hodarev et al. (1971). Reproduced with permission from the Environmental Research Institute of Michigan (ERIM).

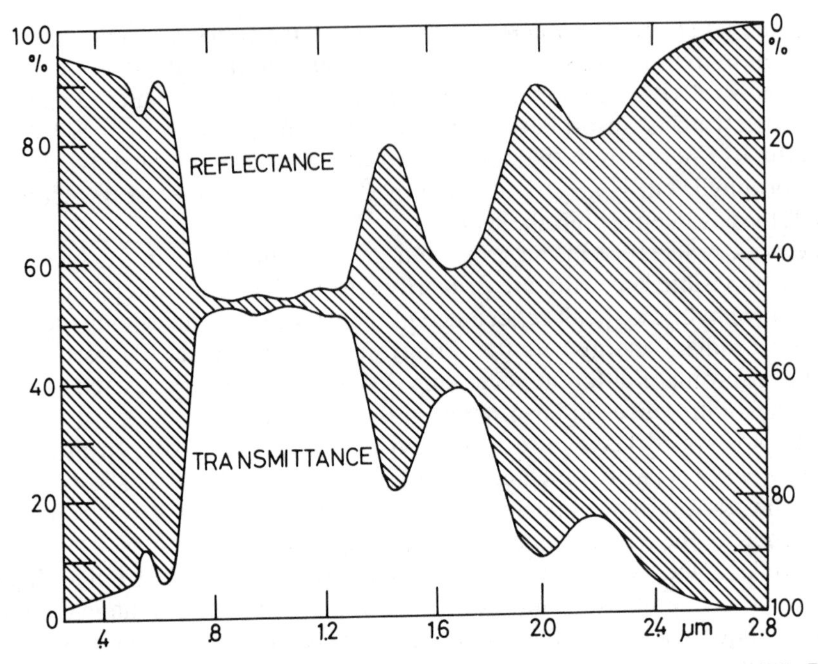

Figure 4-4-2. Reflectance and transmittance spectra of a plant leaf. After Knipling (1970). Reproduced with permission of Elsevier Publishing Co., Inc.

absorption spectrum of a leaf in the 1.4–2.5 μm spectral range. Turbid and mature corn and cotton leaves indicated a theoretical layer of about 150 μm and were in close agreement with the measured amounts of water in the leaves.

Small amounts of IR radiation are absorbed internally between 0.7 and 1.3 μm, but about 40–60% of it is scattered upward through the surface of incidence and is designated reflected radiation, whereas the remainder is scattered downward and is designated transmitted radiation. This internal scattering mechanism accounts for the similarity in the shape of the reflectance and transmittance spectra, as shown in Figure 4-4-2. The high reflectance values of visible and IR beyond 1.3 μm from white and dehydrated leaves suggest that the interaction of radiation of these wavelengths with the leaf structure is not really different from the interaction of near IR energy (Knipling, 1970). However, when chlorophyll and water are present, much of the radiation energy is absorbed before it escapes the leaf.

Basically, the following regions exist between 0.35 and 2.50 μm where different physiological variables control the resulting canopy spectral reflectance, and which are important for the interpretation of remote sensing data on vegetation (Tucker, 1978):

1. The *0.35–0.50 μm* region is characterized by strong absorption by carotenoids and chlorophylls. At this spectral region, a strong relationship exists between spectral reflectance and the plant pigments present.
2. Between *0.50–0.62 μm* a reduced level of pigment absorption is found. This results in a weaker relationship between spectral reflectance and the plant material.

3. The *0.62–0.70 μm* region is characterized by strong chlorophyll absorption, and a high correlation exists between spectral reflectance and the chlorophyll concentration in the vegetation.
4. The *0.70–0.74 μm* region is characterized by the transition from strong chlorophyll absorption and the high reflectance of green vegetation starting at around 0.75 μm. Consequently, there is a low relationship between the amount of green vegetation and reflectance (Tucker and Maxwell, 1976).
5. Between *0.74–1.10 μm* high levels of reflectance occur in the absence of major absorptance. A strong relationship exists between spectral reflectance in this region and the amount of green vegetation present (Knipling, 1970; Woolley, 1971).
6. The *1.30–2.50 μm* region is characterized by the strong absorption of water in the vegetation. At this spectral interval a strong relationship exists between the reflectance and the amount of water in the leaves (Knipling, 1970; Woolley, 1971).

A summary of the spectral regions that are useful for monitoring vegetation is given in Table 4-4-1.

Although vegetation in general has typical high reflectance levels in the near IR, these levels may differ significantly between species. A common vegetation in salt marshes, for instance, is *Spartina alterniflora,* of which the community reflectance is shown in Figure 4-4-3. Pfeiffer et al. (1973) found that the community reflectance of tall *S. alterniflora* was higher in the visual range and almost twice as high in the near IR as the intermediate height stand of the same species. This is partially due to the presence in creek bank stands of wider leaves, which have a large horizontal surface and form a highly reflective canopy. The reflectance values are lower in the short stand than in the tall *S. alterniflora* community, which is probably due to the lower biomass and vertical leaf orientation.

Although *Juncus roemerianus* forms very dense stands, relatively low reflectances were recorded. This may be primarily due to the foliage angle of the plants. The cylindrical leaves rise almost vertically from the soil and since the leaves

TABLE 4-4-1. Ordered List of Spectral Regions in Descending Usefulness for Monitoring Green Vegetation

Number	Wavelength (μm)	Utility for Vegetation
1	0.74–~1.1	Direct biomass sensitivity
2	0.63– 0.69	Direct in vivo chlorophyll sensitivity
3	~1.3 – 2.5	Direct in vivo foliar water sensitivity
4	0.37– 0.50	Direct in vivo carotinoid and chlorophyll sensitivity
5	0.50– 0.62	Direct/indirect and slight sensitivity to chlorophyll
6	0.70– 0.74	Indirect and minimal sensitivity to vegetation;
7	~1.1 – 1.3	perhaps valuable nonvegetational information

After Tucker (1978).

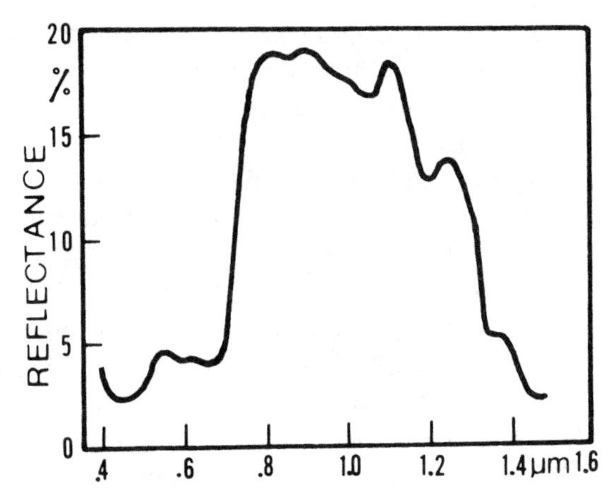

Figure 4-4-3. Community reflectance of 1.5-m-long *Spartina alterniflora* as measured perpendicular to the soil surface. After Pfeiffer et al. (1973). Reproduced with permission from *Photogrammetric Engineering & Remote Sensing,* copyright by the American Society for Photogrammetry and Remote Sensing.

start to die from the tip downward, the canopy surface has many brown components; incident radiation is therefore scattered and reflectance is low. When the living leaves of *J. roemerianus* were oriented horizontally, the resulting reflectance of incoming radiation was higher than any of the natural communities.

A near IR reflectance similar to the tall *S. alterniflora* signature has been recorded for the *Salicornia virginia* community. The higher reflectance in the yellow to red region (0.55–0.65 μm) of the *S. virginia* spectrum can be the result of B-cyanin which is the source of the red coloration of portions of the vegetation. Much of the reflectance is found to be at the outermost layer of the waxy cuticle covering the plant.

From the investigations of the spectral reflectance of salt marsh vegetation, it can be concluded that the spectral signature of vegetation depends on the interaction of numerous plant factors, such as leaf area index, leaf thickness, stem density, leaf orientation, leaf-to-stem ratio, leaf pigmentation, and water content. The composite of all these factors may be recorded instantaneously by remote sensing and utilized to study the distribution of various species and related ecological problems.

Hodarev et al. (1971) showed that the change of vegetative phases can be monitored through their spectra. As corn grows, the profile of the chlorophyll absorption band changes, and within the spectral portion of 0.45–0.7 μm albedo increases by 2–3%. From 0.75 to 1.0 μm, albedo decreases by 2–5%. At the ripening period, albedo increases even more in the 0.4–0.7 μm range, and the chlorophyll band disappears due to the changes in the vegetation pigments (see Fig. 4-4-4).

Miller and Pearson (1971), after investigating various grassland constituents, showed that the amount of vegetation can be measured by its characteristic spectroreflectance. As shown in Figure 4-4-5, the individual spectroreflectance curves collected indicate the differences in various species in the fall. The curves also indicate how various plants with different spectroreflectances couple with the

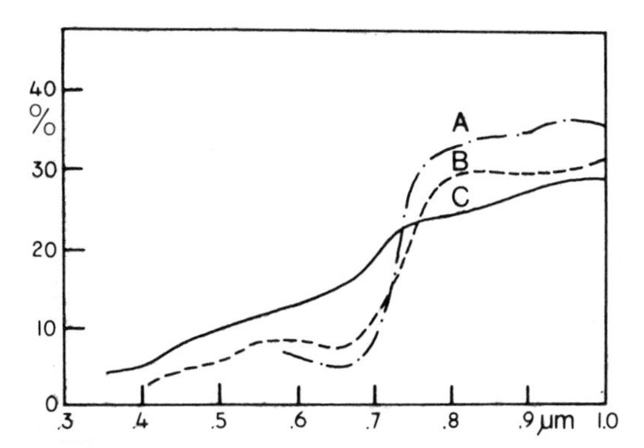

Figure 4-4-4. Dependence of a corn albedo on vegetative phases: A. Silo corn; B. high green corn; C. yellow corn. After Hodarev et al. (1971). Reproduced with permission from the Environmental Research Institute of Michigan (ERIM).

spectral distribution of incoming solar energy to differentially control the flow of solar energy into the plants and soils of the biosystem. In most cases grass will give a "mixed spectrum," which includes the spectral response of the vegetation as well as that of covered soil. This is demonstrated in Figure 4-4-6, where laboratory measurements on *Poa pratensis* are compared with field measurements.

Besides seasonal changes, single species spectra can also be influenced by the conditions under which they grow. Yost and Wenderoth (1971) showed that reflec-

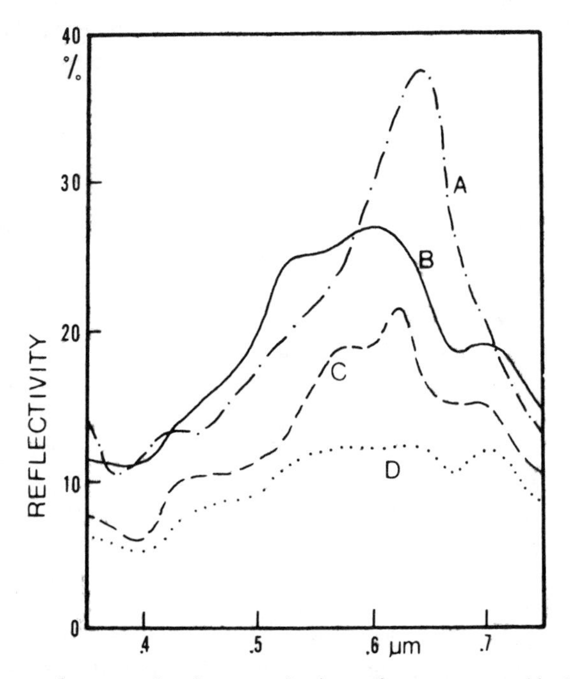

Figure 4-4-5. Spectroreflectance of various grassland constituents measured in the fall. All measurements were made on October 3, 1970, normal to the underlying soil surface. After Miller and Pearson (1971). Reproduced with permission from the Environmental Research Institute of Michigan (ERIM).

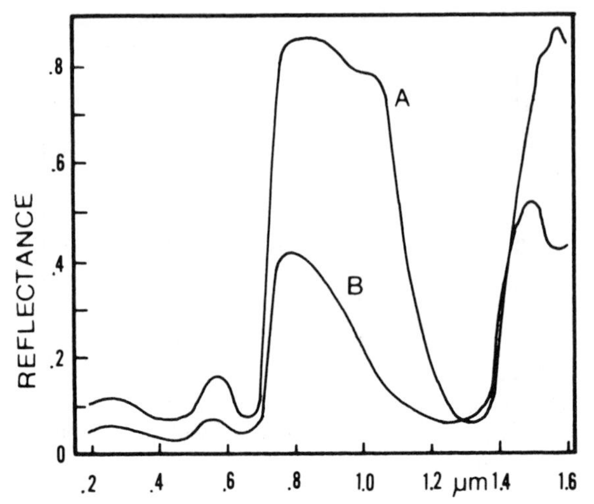

Figure 4-4-6. Smoothed spectroreflectance of *Poa pratensis:* A. Comparison of curves obtained for crisscrossed mats of grass taped to black paper and illuminated by a quartz iodine source; B. curves measured vertically into grass plots illuminated by the sun near noon. The "in situ" spectroreflectance measurements of undisturbed biological materials interacting with incoming solar irradiance yield measurements similar to those of a multispectral scanner viewing the same patch of grass at the same angle, and are different from laboratory curbs. After Miller and Pearson (1971). Reproduced with permission from the Environmental Research Institute of Michigan (ERIM).

tance anomalies of red spruce and balsam fir are correlated with the metal content of the tree needles and soil chemistry. Statistical computations indicated that at the 95% confidence level, there existed at certain wavelengths a significant difference in the spectral reflectance of the mineralized balsam fir that grew in the anomalous soil area compared with the other tree groups. As shown in Figure 4-4-7, at every wavelength the anomalous fir group had a higher reflectance than did the background group. The statistical analysis of these data showed that in the 0.525–0.750 μm region the difference was significant at the 95% confidence level. Note that the balsam fir group was rooted in soils that contain distinctly more copper and molybdenum than do the spruce group soils.

Working in the region of 0.35–1.84 μm, Rao et al. (1978) reported field investigations with a field spectroradiometer. Experimental field plots of wheat, oats, barley, fava beans, soybeans, and rapeseed were analyzed statistically to assess discriminability among the crops. It was found that at early stages of plant growth, interference from soil reflectance was dominant.

Certain specific narrow bands of the spectral range 0.35–1.85 μm can be used to distinguish between selected cereal crops and their cultivars, and the fluctuations around the mean values at specific wavelengths have the potential to be used as a quantitative measure for possible separation of the crops and cultivars.

Spectral reflectance studies were made by McClellan et al. (1963) on seedling, vegetative, and mature soybean and wheat plants grown under controlled conditions with high and low levels of fertilizer and moisture, thin and normal stands, and dark and light soil backgrounds. Reflectance measurements were also made on disease-free wheat plants and those infected with black stem rust as well as on healthy and mildewed barley plants.

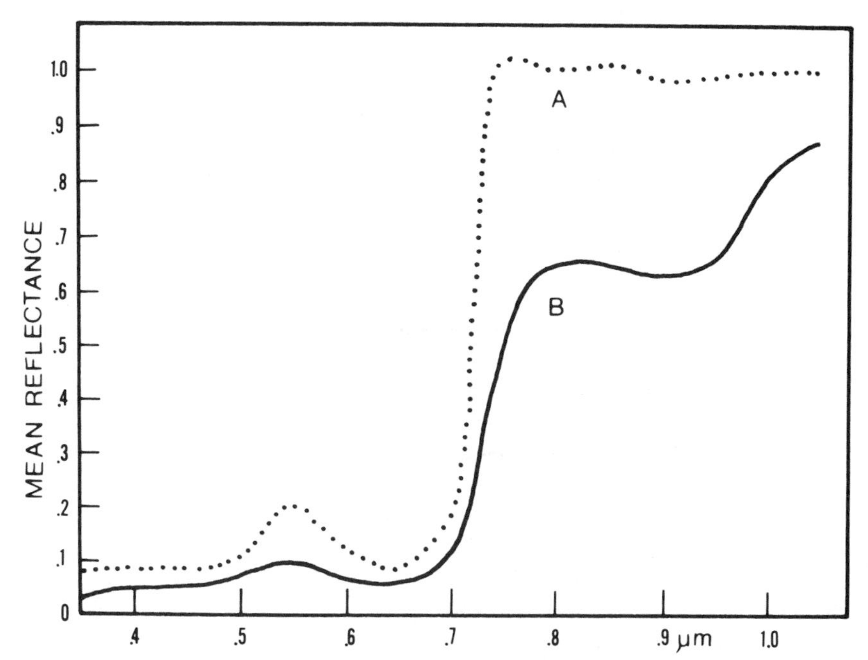

Figure 4-4-7. Reflectance spectra of balsam fir. Averages of percentage of directional reflectance for anomalous (A) and background balsam fir (B). Data obtained during summer of 1968. After Yost and Wenderoth (1971). Reproduced with permission from the Environmental Research Institute of Michigan (ERIM).

Examples of selected spectra are given in Figure 4-4-8, which shows the spectral differences between different species and the changing signatures as a function of growth stage and healthy conditions.

Light reflectance as a function of leaf maturity was reported by Gausman et al. (1969). Many factors, such as water and salinity, affect leaf growth. With limited water supply prior to and during the leaf expansion period, expansion is suppressed, intervascular intervals are reduced, and leaves are thinner, since leaf expansion precedes leaf thickening. In cotton, sulfate salinity limits cell enlargement much more than cell division, whereas chloride salinity inhibits cell division but stimulates cell extension. Chloride salinity retards the formation of leaf initials and their differentiation (Strogonov, 1962).

Chiu and Collins (1978) showed the spectral development of young wheat in several stages of growth (Fig. 4-4-9). The different radiance curves are due to both plant maturation and increasing canopy density, whose effects cannot be separated by the broadband spectral features alone. However, a far-red shift in the position of the chlorophyll absorption edge between 0.70 and 0.75 μm is characteristic of plant and pigment system maturation (Collins, 1978). This shift is visible only under a spectral resolution of 0.02–0.10 μm and can be distinguished even under moderate variations in canopy density. A very pronounced red shift occurs in the wheat spectrum at heading. This spectral shift and information in other spectral regions may enable wheat classification, canopy density evaluation, and grain yield estimation to be carried out.

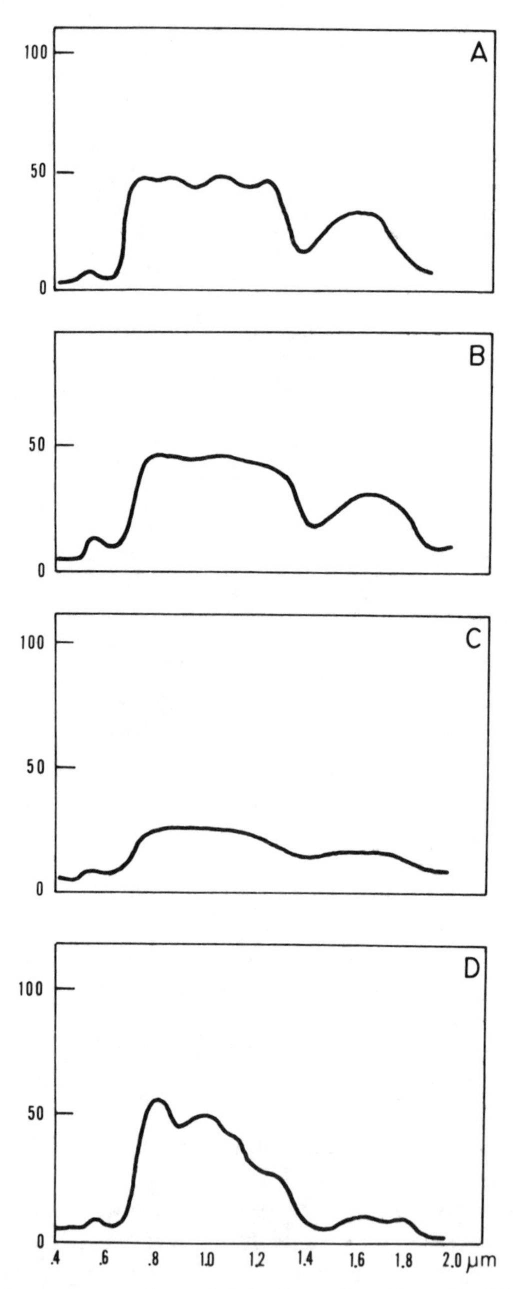

Figure 4-4-8. Reflectance from soybeans (*A*), barley at flowering stage (*B*), healthy barley (*C*), and tomato (*D*). After McClellan et al. (1963). Reproduced with permission from the Environmental Research Institute of Michigan (ERIM).

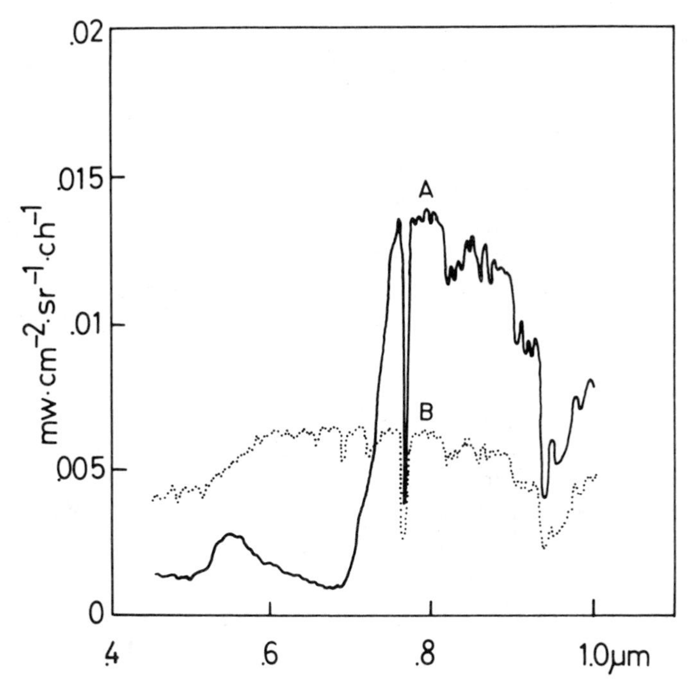

Figure 4-4-9. Spectral difference of wheat in different stages of growth: A. 20-in wheat; B. 2-in wheat. Reproduced with permission from *Photogrammetric Engineering & Remote Sensing*, copyright by the American Society for Photogrammetry and Remote Sensing.

Collins (1978) showed also that a red shift in the chlorophyll absorption edge of heading wheat and grain sorghum is clearly visible in high spectral resolution measurements. The position of the absorption edge shifts progressively toward the longer wavelengths during the crop growth cycle and reaches a maximum in the fully headed preripening stage. The red shift of 0.07–0.1 μm can be measured by using 0.01-μm-wide spectral bands centered at 0.745 μm and 0.785 μm. These two bands, plus a band in the pigment absorption region at 0.670 μm, contain enough information to identify wheat and grain sorghum in the heading stage, to indicate the degree of heading, and to indicate canopy density during the heading stage.

Physiological damage on vegetation is apparent in the reflected near IR. Chronic damage or a low-level effect over a long period eventually causes a deterioration of chloroplasts. This change in physiology is seen as a yellowing of the foliage. Therefore, the change in spectral reflectance is the shifting of the green peak towards the red wavelength.

The incoming spectral irradiance interacts with the vegetation canopy and, depending upon the vegetation density, with the soil background. These interactions with the soil background become less frequent as the vegetational density or biomass increases until the asymptotic spectral radiance or reflectance is reached (Tucker, 1977a). Increases in the vegetational density or biomass cause no change in the canopy spectra when the asymptotic spectral radiance or reflectance is reached. When the vegetation is dense enough to produce asymptotic spectral reflectance, there is no soil spectral reflectance contribution to the composite canopy spectral reflectance.

Soil background influences were found to seriously hamper the assessment and characterization of vegetated canopy covers. Both a soil brightness and a soil spectral effect were found to influence greenness measures, not only at low vegetation densities but also at canopy covers approaching 75%. Soil brightness influences were of greatest concern, because even for complete, bare-soil spectral normalization, soil influences became greater with increased vegetation canopy covers (Huerte et al., 1985).

Theoretical considerations indicate that the soil spectra can be extracted by regressing canopy spectral reflectance against some measured biophysical characteristic of the canopy such as total biomass, chlorophyll, and leaf water concentration.

The limitation of soil-water availability to plants is a common environmental occurrence. This limitation of soil-water availability to plants is referred to as water stress or drought stress. The terms "water stress" or "drought stress," as used here, refer to the combination of abiotic conditions that produce serious internal plant water deficits, which limit photosynthesis and restrict plant growth.

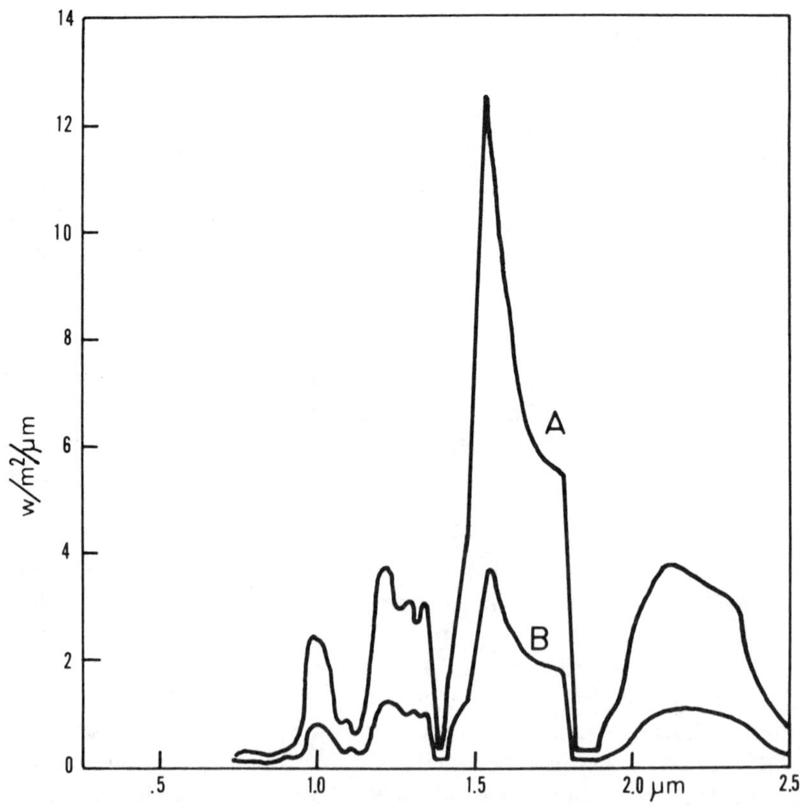

Figure 4-4-10. Spectral radiance differences between leaf water contents (A. 90–60%; B. 90–80%). The spectral radiance differences resulted from multiplying the simulated spectral reflectance differences by a solar spectral irradiance distribution. Note the superiority of the 1.55–1.75 μm region when the spectral reflectance–spectral irradiance product is considered. Reproduced with permission from *Photogrammetric Engineering & Remote Sensing*, copyright by the American Society for Photogrammetry and Remote Sensing.

Tucker (1980b) suggested that leaf water content changes were more easily remotely sensed at wavelengths where moderate water absorption occurred (i.e., 2.1–2.35 μm, 1.42–1.56 μm, 1.83–1.88 μm, and 1.57–1.82 μm) than at wavelengths where strong water absorption occurred (i.e., 1.90–2.05 μm). Considering the solar spectral irradiance and atmospheric transmission characteristics, he concluded that the 1.55–1.75 μm spectral interval was the most suitable band in the 0.7–2.5 μm region for monitoring plant canopy water status from space platforms (see Fig. 4-4-10).

4-5 APPARENT RADIANCE

Spectral data obtained from spacecraft altitudes represent the reflected or emitted energy from the target as well as the absorption characteristic of the atmosphere. In addition, according to the spectral response of the radiometer, only a selective part of the EMS is recorded. Although these "residual spectra" can be estimated from the knowledge of the spectral characteristics of the target, the atmospheric absorption and the spectral response of the radiometer, the residual reflectance signatures of the targets as may be seen from space have to be considered separately for data interpretation.

The combined effect of atmospheric absorption on radiation emitted from a target and its specific radiation can be demonstrated with a continuous spectrum recorded by *Nimbus 4* for Sahara Desert sand. The spectral band between 6.7 and about 20 μm (500–1500 cm^{-1}), as shown in Figure 4-5-1, also compares the SiO_2 Reststrahlen with the curves of constant brightness temperatures, where the apparent differences in radiance between the curves of constant brightness and observed radiance are due to the atmospheric absorption of radiance.

So far, most residual spectra or apparent radiance have been based on data monitored by *Landsat* and are useful for interpreting the different signals obtained from the MSS of *Landsat* either for data interpretation of the images or for supervised classification.

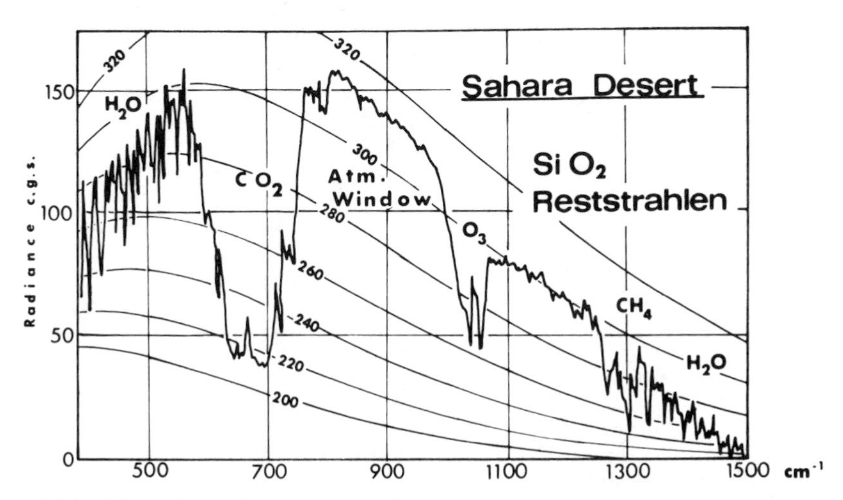

Figure 4-5-1. Thermal emission spectra compared to curves of constant brightness temperatures (K).

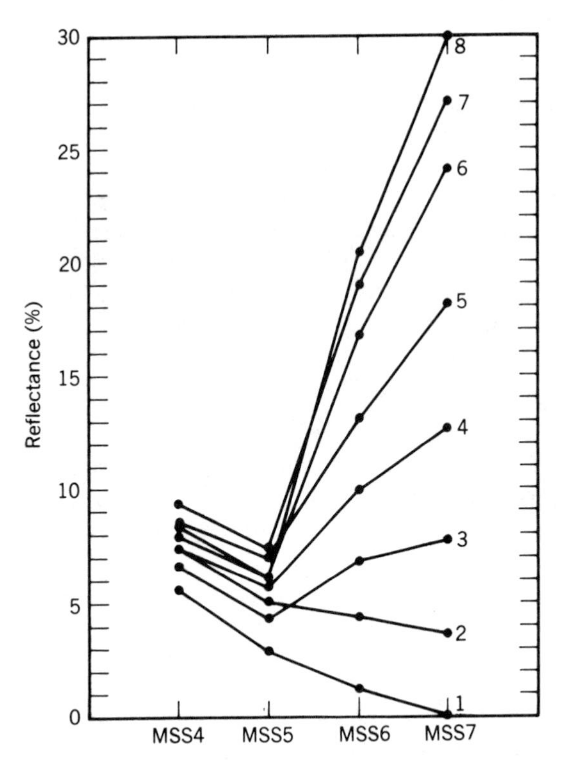

Figure 4-5-2. *Landsat* reflectance signatures of wet and dry marsh grasses. Transition from water (1) through wet marsh (2–5) to dry marsh grass (6–8), Galveston study site, October 3, 1972. After Weisblatt (1974).

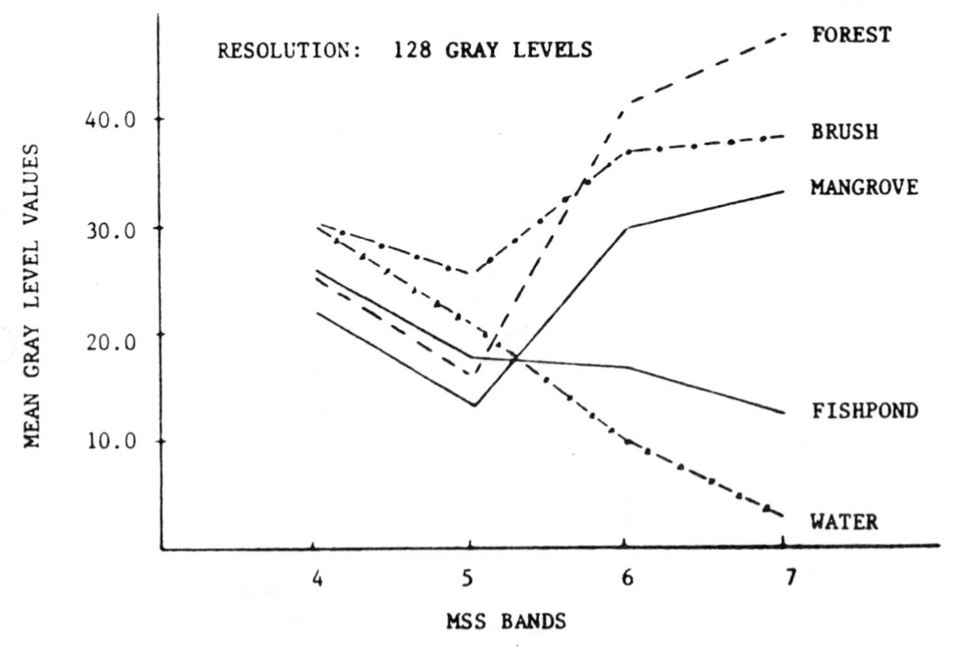

Figure 4-5-3. Mean spectral signature of *Landsat* classification units. After Capehart et al. (1977). Reproduced with permission from *Photogrammetric Engineering & Remote Sensing*, copyright by the American Society for Photogrammetry and Remote Sensing.

A series of reflectance signatures of water, and wet and dry marsh grasses are shown in Figure 4-5-2. The data follow the general trends in the reflectance spectra of vegetation and water, especially in the near IR part of the spectrum.

The variability of land cover features within mangrove areas is reflected in the very wide range of *Landsat* reflectance values found in three study sites and reported by Biña et al. (1978b) (Fig. 4-5-3).

5

CONCEPTS IN DATA PROCESSING AND INTERPRETATION

5-1 GENERAL CONSIDERATIONS

With the introduction of machine processing of remotely sensed data, the actual boundaries between the processing of data and the interpretation of that data have become more diffused, because in recent image processing systems the processing of data from a tape to the image product is already part of the interpretation whereby resources specialists extract specific features from the original data.

To be interpreted by human analysts, the image products must allow the detection and correct identification of features of interest. Detection requires at least the simple recognition or awareness that a feature is present. Identification of a detected feature requires further synthesis of spectral, spatial, textural, and associative characteristics (Hay, 1982).

Due to the complex nature of processing and interpretation and the difficulty in differentiating between them, specialists have defined certain boundaries between interpreted and processed data:

"Primary data" is raw data acquired by remote sensors borne by a space object and that are transmitted or delivered to the ground from space by telemetry in the form of electromagnetic signals, photographic film, magnetic tape, or any other means.

"Processed data" are the products resulting from the processing of the primary data to make such data usable.

"Analyzed information" is the information resulting from the interpretation of processed data, inputs of data, and knowledge from other sources (United Nations General Assembly Resolution, 1986).

In reality, such strict boundaries do not exist, and the interpretation is, in digital mode, already a conceptual input into the processing. This chapter touches on the concepts and basic understanding of the fundamentals that enter into the final interpretation of image material. Although data sets similar to *Landsat* imagery use similar techniques and principles as in photogrammetry, the interpretation of radar data is more complex and has led to the term *radargrammetry*.

Photointerpretation of satellite data is coupled with the processes involved from the sensor design, data acquisition, and its processing, with input from the interpreter.

5-2 SPECTRAL BANDS AND GROUND RESOLUTION

To identify targets, the satellite need not always record continuous spectra; instead, selected bands of the spectrum can be used for distinguishing specific features in a recorded scene. For instance, if two different targets have in different spectral bands different spectral responses, they may be separated on a spectral basis alone. Bands are normally selected for specific detection of spectral properties related to the composition of the target. Over unknown targets, on the other hand, the recorded energy in different bands can be used to define the criteria for data processing and extraction of information in the data. Under the *Landsat* program many spectral classes using the spectral bands of the MSS have been developed, a few of which will be briefly discussed here. For instance, to identify conifer species groupings from *Landsat* digital data, Mayer (1981) reported several band patterns, of which a few selected ones are shown in Figure 5-2-1.

Saline soils, river sand, ravines, and barren lands, having a more or less similar pattern of spectral signature, have higher reflectance in bands 5 and 6 as compared to bands 4 and 7 in *Landsat* data (Kachhwaha, 1983). Saline soils show the highest reflectance in all four *Landsat* bands, next to which is river sand followed by barren land and ravines. Although black soil fallow shows similar patterns, it has very low reflectance in all four bands compared with the categories mentioned above and therefore can be well separated. Spectral reflectance from water continuously declines from bands 4 to 7, being lowest in band 7. Hill shadow also reflects much less energy in bands 4 and 5, but it does not mix with water since it shows a rising trend toward bands 6 and 7.

Field spectral measurements of homogeneous wetland classes illustrate typical vegetation and water responses. The chlorophyll absorption of red light causes classes containing vegetation to exhibit a slight decrease in the percentage of reflectance within band 5 (0.6 μm–0.7 μm). These same classes show a marked rise in the percentage of IR reflectance, bands 6 and 7 (0.7 μm–1.1 μm), which is typical of healthy green vegetation. The range of near IR reflectance values for the wetland types is much greater than that of visible reflectance values.

According to the spectral bands and the optical properties of a recorded scene, each band shows distinctive features in the pictorial display. As shown in Figure 5-2-2 for narrow spectral bands from 0.5–0.6 μm, vegetation has low reflectance and only in the channel 0.7–0.8 μm can a significant response be seen. Sediment patterns in the water show up clearly in the nearshore area in the spectral range of 0.6–0.7 μm, and sandy beach regions can be distinguished from vegetated areas

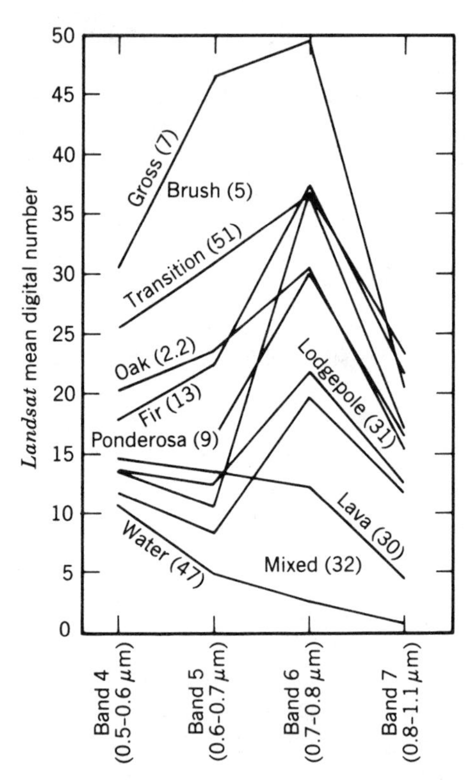

Figure 5-2-1. The four digital number patterns for the representative spectral classes of the major resource categories defined. Note: Band 7 digital number values are scaled from 0 to 63; bands 4, 5, and 6 digital number values are scaled from 0 to 127. After Mayer (1981). Reproduced with permission from *Photogrammetric Engineering & Remote Sensing,* copyright by the American Society for Photogrammetry and Remote Sensing.

mainly in the spectral region 0.6–0.7 μm. These examples show that an understanding of spectral signatures and a knowledge of a sensor's spectral response are essential for proper image analysis and interpretation.

In addition to the spectral understanding of a scene, the ground resolution of a recorded image must be known. Generally speaking, resolution can be defined as a parameter to distinguish between signals that are spatially near or spectrally similar. Therefore, spatial resolution can be defined as a measure of the smallest angular or linear separation between two objects, whereas spectral resolution is a measure of the discreteness of the bandwidths and the sensitivity of the sensor to distinguish between spectral intensity levels. The spatial resolution of a remote sensing system is also a function of the spectral contrast between the objects in the scene and their background, of the shape of the objects, and of the signal-to-noise ratio of the sensing system.

Definition of the ground resolution also takes into account the nadir angle of the sensor. With increasing nadir angle, the size of the ground resolution element increases. With the *Nimbus* satellite's HRIR for instance, at a 50° angle, the ground resolution of the HRIR was 35.2 km long (east and west) and 15.3 km wide

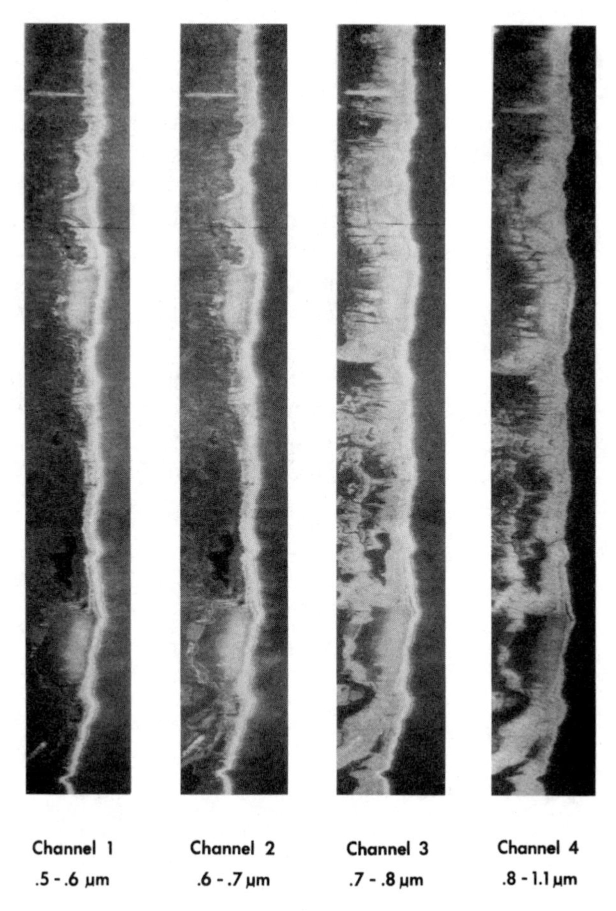

Figure 5-2-2. Images from the four-channel data obtained by the multispectral channel simulator during a flight over the Assateague Island offshore bar. The top half of each image corresponds to the lower portion of the island. After Short and MacLeod (1972).

(north and south) compared with a resolution of about 8 km at nadir. Since more atmosphere is probed at high nadir angles, it is obvious that the signal to be detected also interacts more with the atmospheric constituents; data should therefore be collected at low nadir angles, or corrections as a function of nadir angles must be considered.

There is a relationship between radiometric resolution, spatial resolution, and spectral resolution from the point of view of data applications, but there is a strong relationship between these three parameters from the point of view of instrument design. Generally, improving one parameter comes at the expense of some deterioration of the other two. This is a direct result of the fact that radiometric fidelity is better when the energy received is higher. The energy an instrument registers from the area of a pixel is higher when the pixel area is larger and the wavelength band is broader.

In the technical approach to the definition of resolution, one of the most useful concepts is that of the modulation transfer function (MTF). Modulation describes the variation of the intensity of a signal. Consider, as in the lower part of Figure 5-2-3, hypothetical ground surfaces with alternating dark and light stripes. The intensity of the light reflected at the ground is indicated by the curves on the surfaces. The amplitude (peak intensity) is the same in both cases, but the spatial frequency of change is greater for the surface at the lower right. The relationships in Figure 5-2-3 indicate that the MTF decreases as the spatial frequency of the variation in the ground scene becomes greater.

In praxis, qualitative resolution tests can be conducted by identifying known structures in the imagery materials, which has been done by estimation of the size of the resolvable smallest structure, as shown in Table 5-2-1. Figure 5-2-4 compare two remote sensing systems observing the same scene but with different ground resolutions. Figure 5-2-5 compares data obtained with the MSS on *Landsat* and an

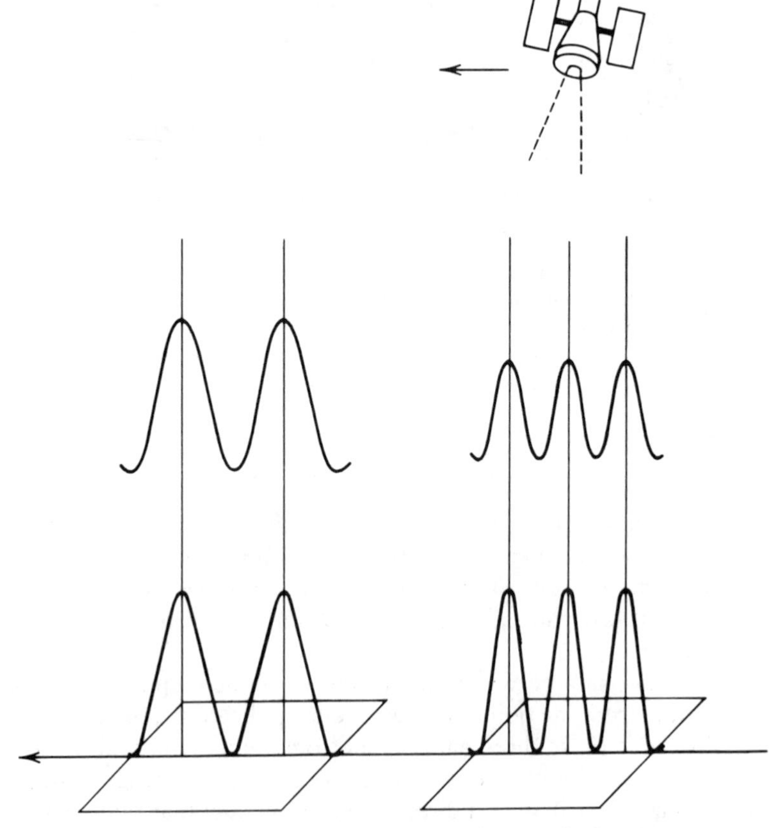

Figure 5-2-3. Concept of the modulation transfer function.

TABLE 5-2-1. Experimental Resolution Tests for the RBV and MSS on *Landsat*

		Performance	
Target	Separation (m)	RBV	MSS
Parallel roadways running into Dulles airport	60	Just resolvable	Not resolvable
Two parallel taxiways	96	Easily resolvable	Not resolvable
Divided highway	120	Easily resolvable	Just resolvable
Runway with parallel taxiway, Dulles airport	216	Easily resolvable	Easily resolvable

NOAA image recorded with the very high resolution IR radiometer having a ground resolution of about 900 m.

The improvement of higher ground resolution compared with a system with lower resolution but with almost identical spectral characteristics is shown with data from the *Landsat* TM and *Landsat* MSS data. Although at the same scale, the MSS shows the single picture elements with a distortion of features, such as streets and boundaries between the fields, whereas the TM shows more details (Fig. 5-2-6).

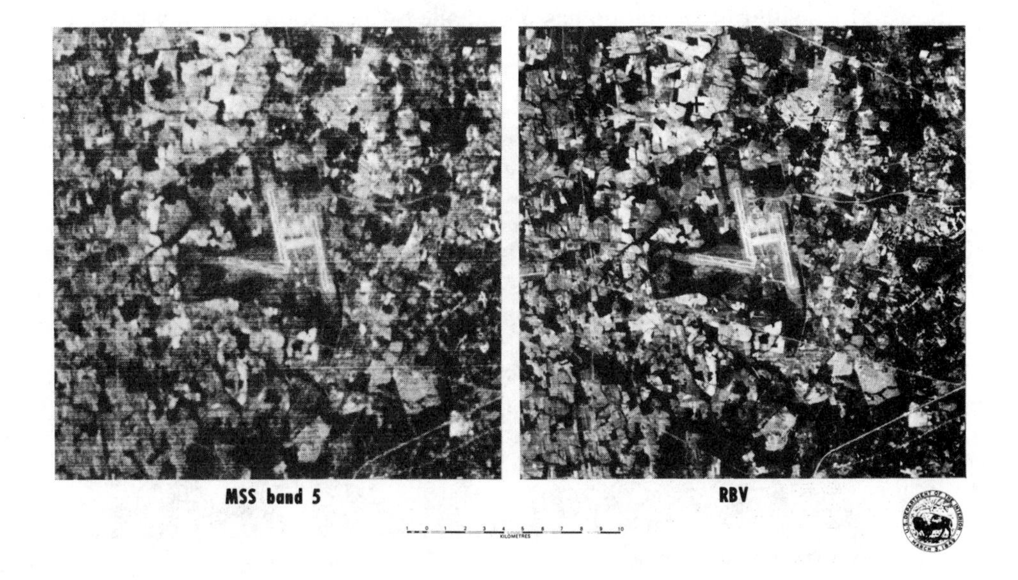

Figure 5-2-4. Comparison of MSS band 5 and the RBV of the Dulles/Reston (Virginia) area as recorded by *Landsat 3* on June 30, 1978. The enlargements were made to 1 : 1,000,000 scale, 34 times (enlargement) for the MSS and 17 times for the RBV. Source: USGS report 78-1107.

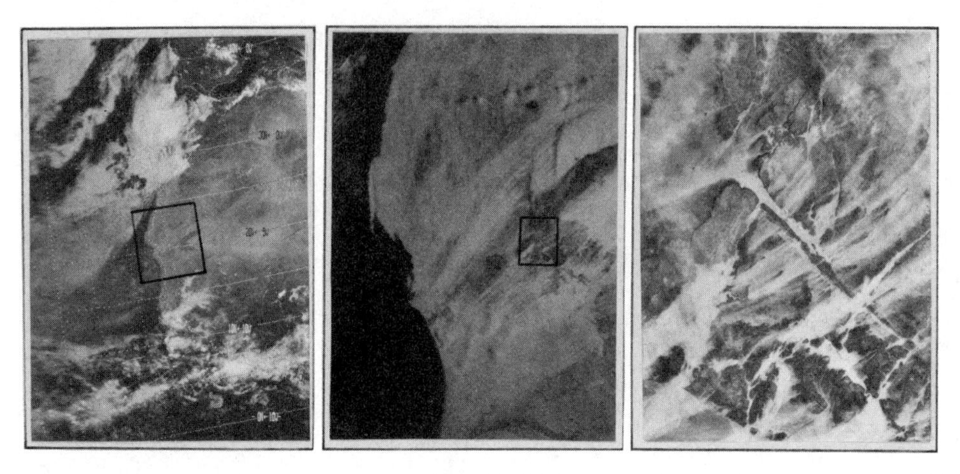

Figure 5-2-5. Comparison of images recorded with different ground resolution over northwest Africa. *Left:* NOAA image recorded with 900 m resolution. *Middle:* Image as indicated in quadrangle on left side. *Right: Landsat* image recorded with 80-m resolution; area covered is outlined in the quadrangle of middle image.

5-3 PATTERN RECOGNITION

In a broader sense, pattern recognition is considered as a technique that identifies objects within an image that can hardly be identified by the human eye but can be made visible through techniques exemplifying spectral, spatial, or additional signatures. Most pattern recognition has been done by a computer, mainly based on

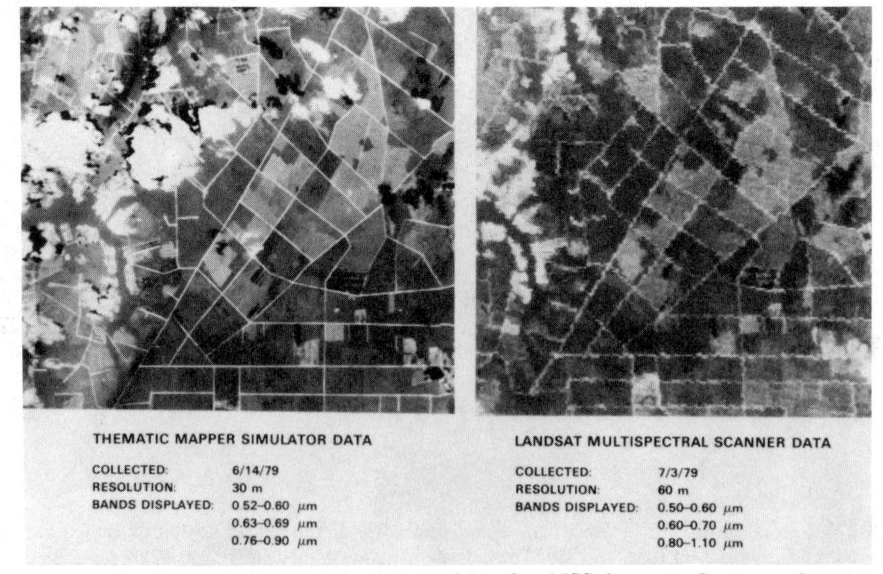

Figure 5-2-6. Comparison of TM simulator and *Landsat* MSS data sets. Courtesy of NASA.

digital data processing and clustering techniques for a variety of classification techniques. However, one should not forget that simple image enhancement through photographic means, such as color coding, analog density slicing, and so on, are still important processes to recognize pattern in photographic material. Some concepts will be briefly described in order to introduce the principles of pattern recognition as an additional tool to interpret data and extract information from images obtained from space missions.

In the use of multispectral data collected by remote sensing techniques, pattern recognition is the act of classifying an unknown object or data set as belonging to a specific category or class of data. The graphical display of different reflectance signatures for *Landsat* data, for instance, shows that different targets build clusters that can be mathematically described in order to set boundaries for assigning to a set of data a quantitative description. These patterns of known classification constitute training patterns against which patterns from unknown fields can be compared or identified.

Clustering is a data analysis technique by which one exploits the "natural" or "inherent" relationships in a set of data points. Clustering is used extensively in the study of statistical data, in numerical taxonomy in the investigation of social and biological data, and in the analysis of remotely sensed data, where it is used to check the homogeneity of data, detect class boundaries, and develop nonsupervised classification.

Clustering of reflectance data over a target, as shown in Figure 5-3-1, can be assigned to spectral features that can be separated while quantitatively a decision

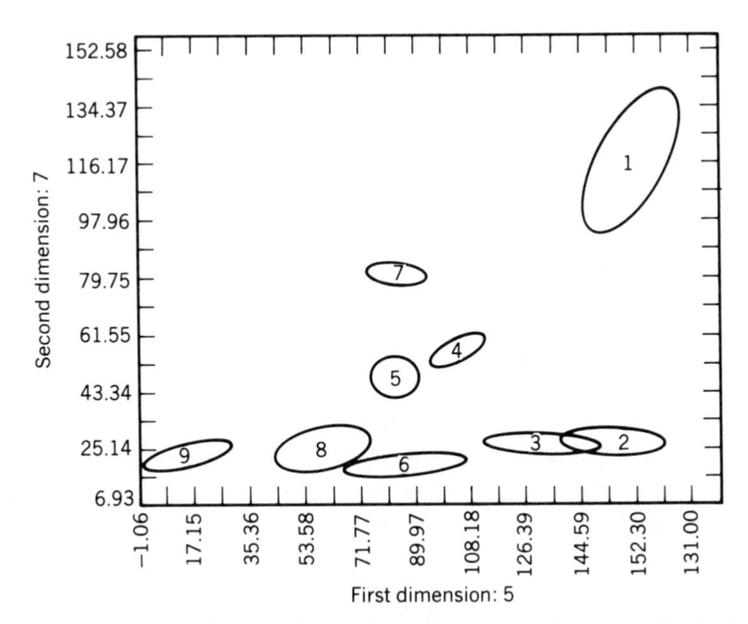

Figure 5-3-1. Clustering of reflectance data over a target in a two-dimensional distribution of different classes using *Landsat* channels 5 and 7: 1. Calcerous rocks; 2. pure beech; 3. mixed deciduous forest; 4. grassland; 5. agricultural area; 6. mixed deciduous (shadow); 7. village; 8. back pine; 9. water. After Reichert (1985).

boundary separating the clusters is established. This example uses two bands to identify clusters that can be attributed to reflectance signatures. By the addition of n additional bands, the signatures can be further separated in n-dimensional space.

In order to separate clusters into regions, decision boundaries have to be defined through which all points lying on one side of a boundary are classified into one class, while all points lying on the other side of a boundary are classified into another class. For instance, decision boundaries can be placed through the sets of known data vectors, and any unknown features that lie in the boundary can be classified accordingly.

If two classes look spectrally alike; that is, if the components of the data vectors represent the responses in different spectral bands and are alike, they will not be separable and may build a common cluster.

Unsupervised, or "clustering," algorithms examine a large number of unknown data vectors and divide them into clusters based on properties inherent in the data. The resulting clusters stem from differences observed in the data. In particular, data vectors within a class should be in some sense close together in the measurement space, whereas data vectors in different classes or clusters should be comparatively well separated. The basis of this algorithm is that data points are assigned to the nearest cluster center, the cluster centers are modified either by combining or splitting the existing clusters, and each class or cluster can be considered as a union of an appropriate number of subclasses of nominal size. A cluster is considered to be of nominal size when the standard deviation in every channel is less than a predefined threshold.

5-4 ANALOG ENHANCEMENT

Much of the material obtained from satellites is delivered to the end user in the form of images and photographs. Because most of this material is black-and-white, different processes are used to enhance the information contained in the images. For instance, color density slicing is used to enhance pattern recognition. Color density slicing converts black-and-white imagery into a color television display in which each color represents a particular gray level in the original image.

In a color presentation, the intensities of the light collected by the imaging device are represented as different colors so that subtle changes in scene contrast are enhanced. Discrete density levels in the film material are assigned a unique color, and the scene is regenerated in these colors.

Density analysis and contrast stretching are used to accurately measure or subdivide film densities in order to detect and delineate small variations in the density of an image. Density analysis can be accomplished through photographic processing, photometric measurement, video processing, or digital processing.

For the composition of data reproduced from different spectral channels of the same scene, the additive process for color composition may be used. By this process, light at different wavelengths is employed for each spectral band. With the primary colors red, green, and blue, all other colors may be produced. For instance, when all three primaries are combined in a definite proportion, white is the resulting effect for the human eye. Combining red and green gives yellow, red and blue produces magenta, and blue and green make cyan. Yellow, magenta, and cyan are thus called secondary colors, and are the complements of blue, green, and red, respectively. When a secondary color is combined with its complementary primary, white is produced. When added together, the complementary colors produce white. Specific proportions of these colors must be used in order to produce white. We summarize this as follows (see also Fig. 5-4-1):

$$\begin{array}{llllll} \text{red} & + & \text{green} & = & \text{yellow} \\ \text{red} & + & \text{blue} & = & \text{magenta} \\ \text{blue} & + & \text{green} & = & \text{cyan} \\ \text{yellow} & + & \text{blue} & = & \text{white} \\ \text{cyan} & + & \text{red} & = & \text{white} \\ \text{magenta} & + & \text{green} & = & \text{white} \end{array}$$

Yellow + blue = white, and red + green = yellow; thus,

$$\text{red} + \text{green} + \text{blue} = \text{white}$$

Since the addition of the three primaries can produce white, the addition of the correct proportions of the three complementaries can also produce white:

$$\text{cyan} + \text{magenta} + \text{yellow} = \text{white}$$

Many satellite sensors provide for each scene different images, which can be simultaneously viewed on a color-additive viewer. The viewer system is able to use either positive or negative transparency material in such a way that, on the screen, a false-color image can be adjusted through filtering and brightening each frame independently.

Color additive systems can be employed in both image masking, positive/positive and positive/negative. Positive/negative masking is accomplished by placing a negative of an image over the positive of another image of the same area for a different season or by placing a negative of one spectral band over a positive transparency of another band.

Color (or false-color composite) is a useful method of presenting data from a

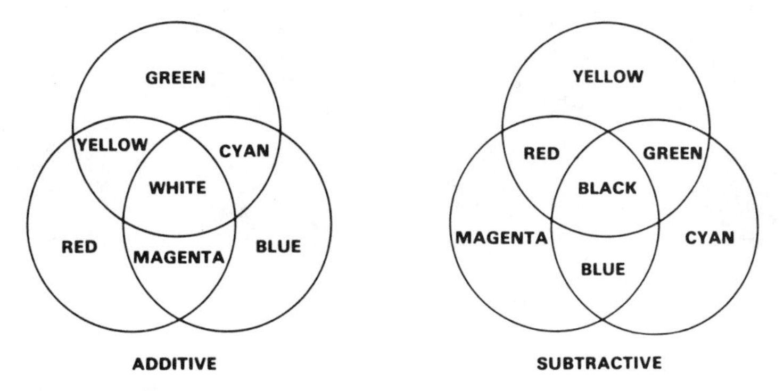

Figure 5-4-1. Color production by additive and subtractive processes.

TABLE 5-4-1. Positive Transparencies of Each Channel of the Assateague Image

Strip	Channel 1	Channel 2	Channel 3	Channel 4
A	Blue	Green	Red	
B	Blue	Green		Red
C	Green	Red	Blue	
D		Red	Green	

After Short and MacLeod (1972).

MSS; it takes several wavelength bands from the MSS and represents each as a specific color whose intensity is proportional to the energy in that band. The rendition, either as a video or as a photographic image, is then available for interpretation.

False-color presentation of *Landsat* data can be simulated in a very simple process with diazo color film. The properties of diazo color film are very similar to those of ammonia copying ("blueprint") paper used in engineering and architecture for the copying of plans. It consists of an inert transparent base supporting a photosensitive "emulsion" layer that is colorless or pale yellow in the undeveloped condition.

In order to produce a color composite of a *Landsat* MSS image, one need only reproduce two or three MSS bands of the same scene, each in a different color diazo material, and superimpose these colored diazo copies in register. For more versatility in processing, 9-in negative transparencies and diazo film in the additional colors black, blue, green, red, and so on, are required (Malan, 1976).

The application of additive color principles can be used also for multitemporal color composites. Eyton (1983) showed that the use of three images of different

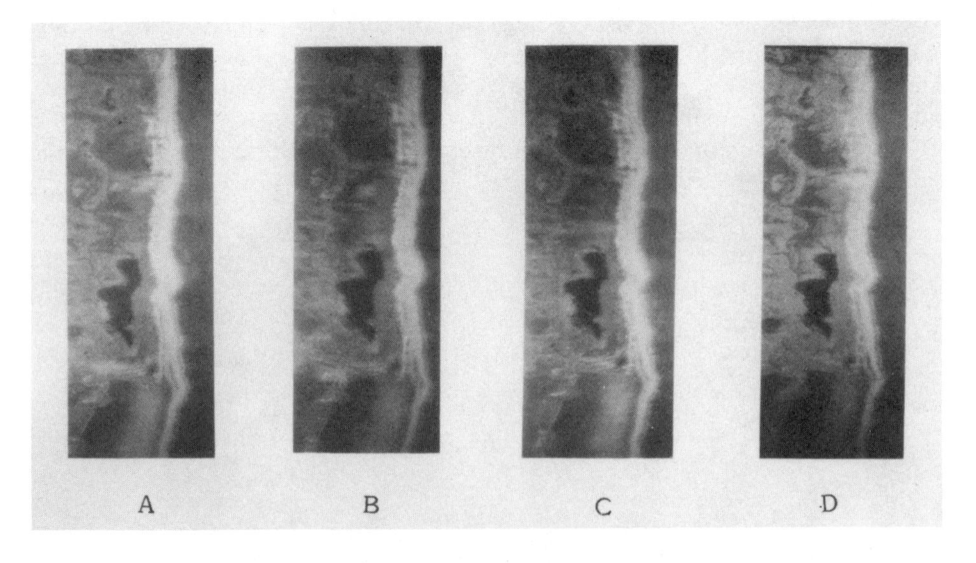

Figure 5-4-2. Color-additive composites of lower part of Assateague Island as recorded on different channels. The images were projected through viewer filters given in Table 5-4-1. After Short and MacLeod (1972). See insert for color representation.

dates in the construction of a color composite produces an image in which features changing through time are displayed as various colors, and areas of no change are displayed as gray tones. Because the same scene is recorded in different channels, many alternatives exist to enhance the features through additive color processing (Short and MacLeod, 1972); see Table 5-4-1 and Fig. 5-4-2.

Landsat MSS and RBV images can also be photographically color composited with scenes of the same spectral band obtained at three different dates. The resulting composite exhibits temporal information only, with areas of change displayed as various hues, and areas of no change displayed as black-and-white tones. For instance, anniversary date images from different years and images acquired at different periods within the same year can be used to construct several examples of a multitemporal color composite.

5-5 DIGITAL IMAGE PROCESSING

5-5-1 Digital Display Enhancement

The term "display enhancement" is used in the application of computers and photographic techniques to modify and display data from remote sensors. Inputs for the display enhancement are data in digital form that are eventually transferred into two-dimensional color plots. This is mainly done to expand the range of reception for patterns by the human eye. Digital plots can either be displayed in different colors, that is, a color is assigned to a given gray level, or a new symbol in black and white can be assigned. An example of both procedures is given in Figures 5-5-1 and 5-5-2.

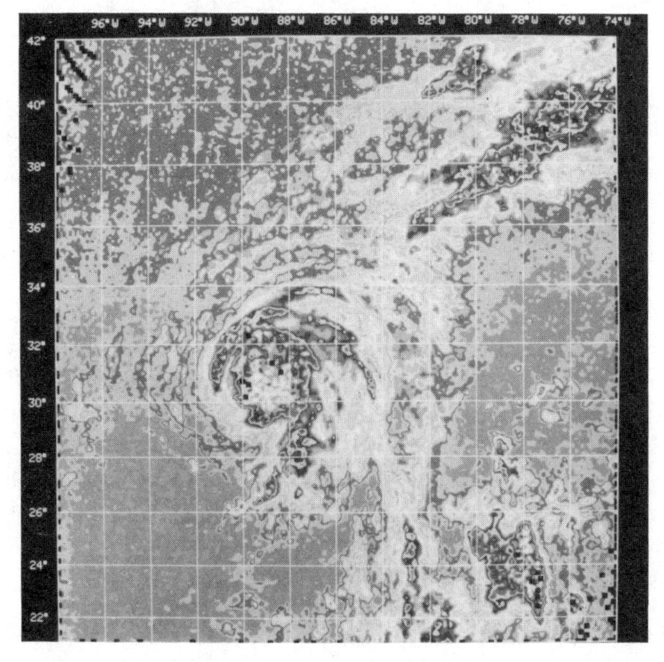

Figure 5-5-1. Digital image of Hurricane Camille on August 18, 1969. Courtesy of NASA.

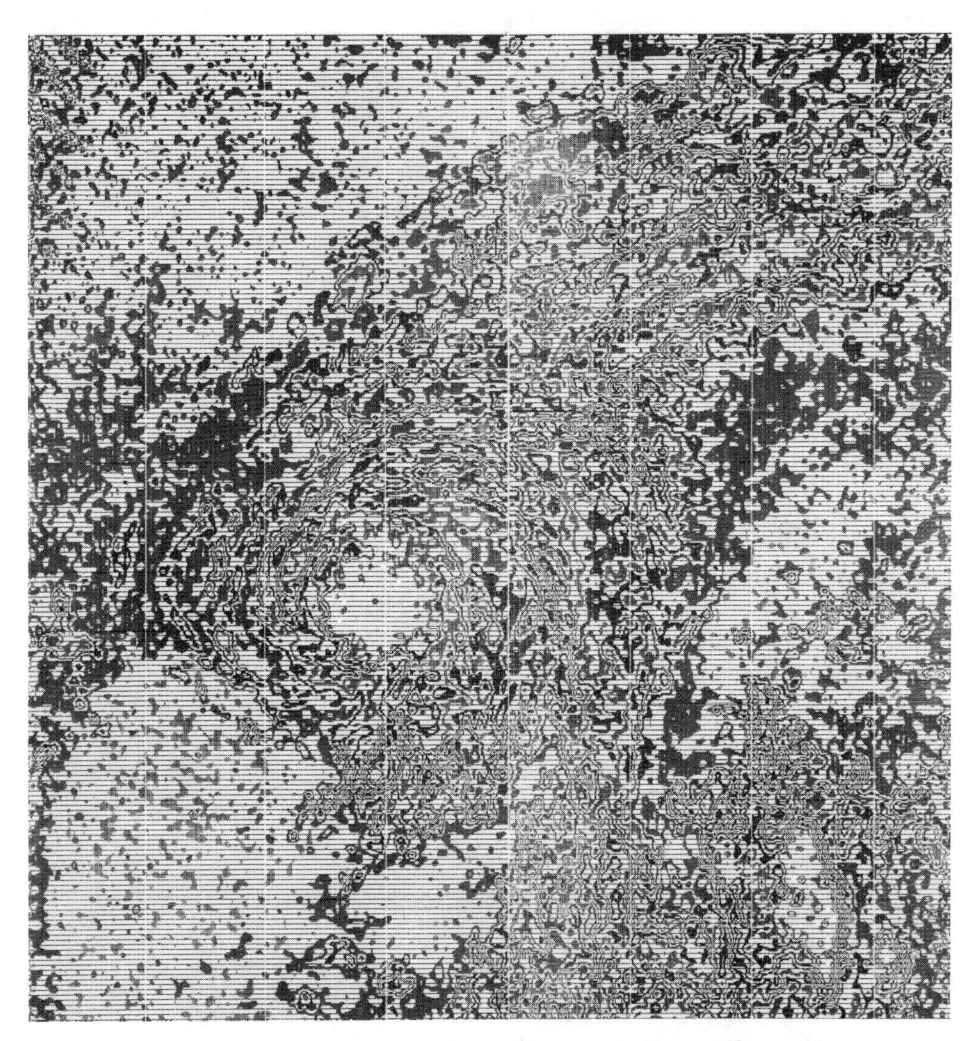

Figure 5-5-2. Black-and-white digital display of Hurricane Camille as recorded on August 18, 1969. Courtesy of NASA.

The selection of symbols with different gray-level information in coding an area is very essential. It has been indicated that only five gray tones are differentiable by standard printing techniques. However, to preserve identity within similar categories, one can map these tones with different symbols of similar intensity. A sample of digital display of gray level through symbols is given in Figure 5-5-3.

In the approach of contrast stretching and data enhancement, enhancement curves are used to separate certain regions of a gray scale and enhance them over a larger portion of the gray scale. An example of annotations to enhance meteorological features is described in Table 5-5-1 (see also Figure 5-5-4).

A hurricane curve annotation provides a broader range of temperatures to depict the cirrus canopy to include warmer storms and a better presentation of

```
   44444444445555555555666666666677777777778888888888
   01234567890123456789012345678901234567890123456789

30 ,*,,**,:x0xx0bxxx???xxx0H0b000HHbx/::??xQbxxb0xxxx
31 '',,,'/0b0bb0H0xx?/?xbb0QH00bb0Hb/?x?xbQH??bx?xxbb
32 :::':xHHbbH0HH0bx????HH00HHb?bb0x:?HQQHHHHHbxx0Q%%
33 :::/?xxxxbH000xxxxx?xb00HQQbxbbxxxxH%NQ/x%HQNNNNNQ
34 /:/?xbxxxxbb0xx?xx0x?xbbb00bbbx?/xb0QQH?:xQ%NNNNNN
35 /?b00b0bx0Hbx?/xbxbHH0bbbbb0b/::::/bQQQHbbQNNNN%NN
36 ?0Hxbbxxxxb???xxxxxbxxxxxxx/:''''/xHQQQHx?0M%NNN%b
37 HHb000bbx0x//???///??????/:':/?/':xbHH0QQQHQ%NN%Q0
38 QHb?00xxbb?/?bb??//?/?/:''':??x:':xb0bHHHQHQNNNN%Q
39 H0HHQHxxxxxxxxxxbb?///:'**,:/??:''?xxxxbHHH00H%%%NN
40 H0bbHQHxxx?xbx??bx??/'',,*,:x/:'':xx?/xbb00bb0HHHH
41 QH0bxxb0bx//000xb?//''',***:/:'''/?/::?bHHHHHHHHHH
42 Q00bx??:/xx??xb/':::',,,,**,:::/:'':/bbH000HH00H0
43 0b00x?/:/??//:''':::'',,,,':::/:'':::?00000H0HH00
44 ?0HHH0b?xb/:'**':::':::'''':::/:':::/0H0HHH0H0000b
45 ?xbHQHbxx/',**:/:'*,,'''''':///?xxxxb0bbb0H00bxb0b
46 ///?bx?:''',*,:??:*,*,*,''''/x?/xxxbbbbbb0b0b0xxx/
47 ?x/?b/:'**,''/b?/'**,,,,,',':?//?xxxb00bbH00b00xxx
48 x?///:'*,*,'??':?/',',,,,,,,''/????xxxb000bb0bbx?/?
49 ??::',*,*,,*,,*,':/',**,:'',,:????/?xbbbbx000bb/?b
50 x:''':/:',,,,***,:/'**':'**':xxxx//?bbb0H00bbbxxx
51 x'*,'//:'',,,,***'?:,*,,*,,':/0bx?xxx?xxb0b??b0bxx
52 :?'*,'/?:'****,,''/:,*,,,*''':??//xbxxxxbbbxxxxbxx
53 :?/:'**'''***''''/:'*,,,,'///x/:::/00xxx?xbx?xx??x
54 :xx/***,,'**'/::?:',,,,,,:???xx::::/bb?????//?bHbx/
55 ':/:*****'',:?/x/?'*,*,:???/::/::/bbxx/::::/H%Q/:
56 *'::**,*,'':?/???',*'':?xbx/''xH0?bbxx?/:'::?HQQ/:
57 ,,,'***,,,://///:*,,'/xxxx??//?b0?xbx??/::///H%%%Qb
58 ,,,,,***,'':///?:,**,/xxHx????/?xx?xx???:':/?bHQQQ
59 '',****':::::/:/:*,*':://0b//?/////?/??/?////x0x/:/H
60 ,,***,::'''xx??/::,/??/??/??//?/?????/???//??///?x
61 ,***,'//'':??x:::xx?/////??????/????x??///://:/?x/
62 ,,,,'/:'':?//?::/?xbxb??/??//?xxbx???/':/?::':xbbx
63 ,**,::,,''''''':/?xxbbx??/?x/://??/:''''/?:/x???xbx
64 *':?:''*,*'''''///?xbbxx??xx?/:::''''''':?//x0bx?xH
65 ''''',,,,,,,*'/////?bb?//::::::'''''''''''''':0?/0HH0??b
66 ''','',,',*,:??????/????:'',,',''''':??/'/0H0xbb/://
67 ,''::::'::///////////////?/''''''''''''''//:'/QQQ0xx/::/
68 ''':///:/??????//::/::::::''''''''''''''':/:'xQHb?:/::''
69 ',,'/?//?xxxxbxx??x/:/?/''''''''''''''/x?xQQ0x?//:::'
70 ,''/??xxxx?xxxxxxxbxxxxx:'''''''''''':x0xbQQ0xxx???:
71 '':?xbbb00b?bbbbb0Hbbbx?'''''''''''':x?/xH///?x/:'/
72 ??x?xb00H0H0xxbbbH0b0bx?'''''''''''''?x:'/?/xb?/:,'?
73 xxbb?xb000HQHx//x0bbbbb/'''''''''''':::?/',,':??::/:':
74 :/?/::bHHHHHHH?/x0x??bx:''':'''''':/?/:,*,::''''''::
75 ,,''::/b0bHHH00xbxxxxx/'''''''''''':xbQQb/,,*,,,*':?
76 ',*':'::?xxH00x??????x?:''':'''''':?bHQQQ?,,,,,,,'/
77 :''':::::/00b?//////?x/::'':'''''':xH0HHH:,,,'/??//
78 '::::':::://xxxx?xbxxxx/'':''''::':/xxxQHxx/:/bb/:/
79 ::::::::::::::://///?//:/:::///??/:'':/xQH00H0HQ0?x?
```

Figure 5-5-3. Digital display of gray level through symbols.

TABLE 5-5-1. Annotations to Enhancement Curve

Segment Number	Temperature (°C)	Reason for Segment Enhancement
1	+56.8–+35.3	Black
2	+35.3–+15.3	Land-sea contrast for registration
3	+15.3–−15.3	Depict cumulus cloud bands
4	−15.3–−44.2	Buffer zone (black)
5	−44.2–−68.2	Depict cirrus canopy, 2° increments
6	−68.2–−70.2	Marker band (black)
7	−70.2–−90.2	Cirrus canopy, 2° increments
8	−90.2–−110.2	White (above hurricane tops)

After Corbell et al. (1980).

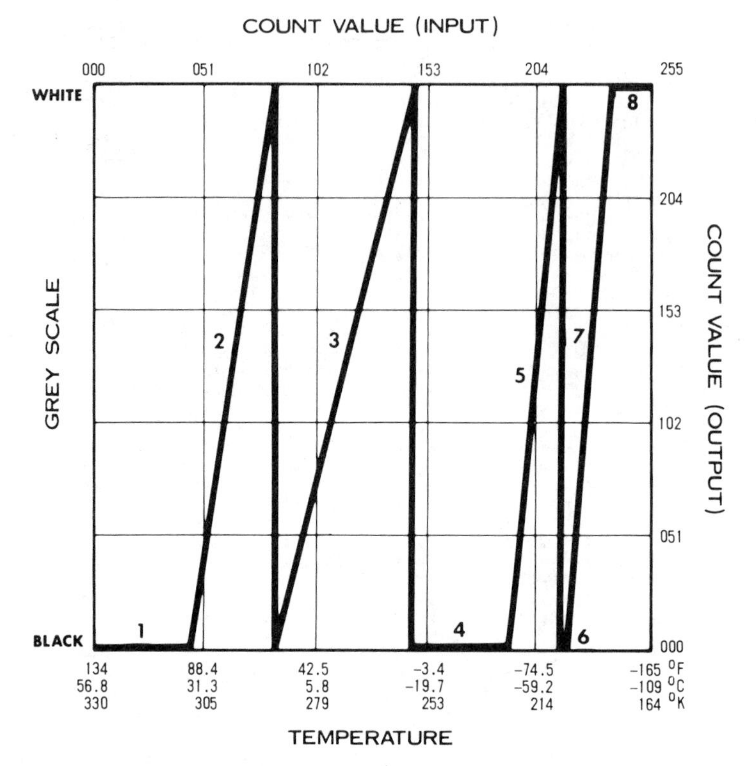

Figure 5-5-4. Annotations to enhance meteorological features. After Corbell et al. (1980).

cumulus cloud bands. It retains the buffer between the cloud bands and cirrus canopy for ease of use. In addition, it provides enhancement of 2°C steps in the cirrus canopy with a marker band in black to allow determination of the gray shade representing the storm top. The primary use of this curve is to assist the analyst in locating tropical cyclones at night, by using IR data, and to depict the structure of the cirrus canopy with underlying heavy convection and the surrounding cumulus banding.

5-5-2 Digital Image Processing of Multispectral Data

Digital image processing has developed rapidly because of the need to process digital MSS data from *Landsat*. In principle, digital image processing involves data processing, image enhancement, and image classification, the latter two at times already part of image interpretation.

Computers are used in various phases of data manipulation and analysis and in more sophisticated information extraction methods. They are also instrumental in integrating remote sensing results into other data sources in models for status assessments and interpretations, decision making, and operational control.

There are two basic approaches to computer-aided processing of *Landsat* data: (1) the batch mode, and (2) the interactive mode (usually with graphics or pictorial TV console display). In either processing mode, users draw upon their knowledge of the areas being investigated and their experience in making interpretations and assessments according to the types of problems or data analysis. Ancillary information, such as maps, aerial photos, previous field trips, or general familiarity with the discipline-related results expected, are commonly integrated into the processing as real inputs into the computer or as conceptual models in the user's mind (Short, 1982).

Several programs for data processing include the adjustment of system-introduced errors, adjustment for atmospheric and solar illumination variations, and the registration of data. Image enhancement is performed to make spatial patterns, displayed in tones and color, more apparent on imagery, whereas image classification is performed to delineate multispectral patterns in image data. The steps involved for processing the data include evaluation of histograms and adjustment of brightness values and errors in the remote sensing system. Stripping results from differences in the response characteristics of the detectors in an MSS, replacing bad-data lines, and adjusting for line length are also a corrective part of data processing.

The rotation of the earth and movements of the satellite itself introduce geometric distortions in the data. During data processing, each scene is therefore "deskewed" by an algorithm that shifts scan lines to the right as a function of latitude of the line. This type of geometric adjustment ensures that landscape features are in relative position with respect to each other throughout the scene.

Adjusting for solar illumination involves the analysis of adjacent scenes of *Landsat* data in order to compare spectral properties of cover types in scenes acquired under different conditions of solar illumination. Sun elevation angle adjustments can be made by multiplying all brightness values in a scene by a constant that is a function of the sun angle, but these adjustments do not correct for differences in solar illumination caused by changes in the azimuth of illumination.

Linear contrast enhancement can be done through assignment of new brightness values to each pixel in a scene according to the following equation (Rohde et al., 1978):

$$BV_0 = \frac{BV_I - MIN}{MAX - MIN} \times 255$$

where BV_0 = enhanced brightness value of pixel in output image
BV_I = brightness value of pixel on computer-compatible tapes (CCT, input value)
MIN = minimum brightness value parameter
MAX = maximum brightness value parameter

When the expanded range of brightness values is recorded on film, the result is an expanded density range. Thus, features in the scene are more easily distinguished because scene contrast is higher. In nonlinear contrast-enhancement techniques, an algorithm that redistributes data values is applied to the original data in such a manner that increments of scene brightness are unequally distributed over a range of 0 to 255.

Nonlinear contrast enhancements can be extremely useful in the analysis and interpretation of imagery for geologic applications. In areas dominated by rock and soil cover, for instance, the pdf stretch can make subtle differences in rock and soil brightness more apparent.

The edge-enhancement technique has the effect of producing a sharper image, but it may introduce artifacts into the data by producing shadows adjacent to features that have abrupt changes in brightness values. Enhancement of drainage and landforms for geologic applications works best in areas of uniform cover. Ratios of two *Landsat* bands are obtained by dividing the brightness value in one band by the brightness value in another band for each pixel within the scene. The ratioed values are usually multiplied by a factor from a table so that all values will lie between 0 to 255. If *Landsat* data are not corrected for atmospheric effects, the most practical band ratios are obtained by dividing the brightness values of band 4 by 2 times the brightness values of band 5 ($4 \div 2 \times 5$), the brightness values of band 5 divided by 2 times those of band 6 ($5 \div 2 \times 6$), and the brightness values of band 6 divided by 2 times those of band 7 ($6 \div 2 \times 7$).

Another useful technique for determining changes in landscape cover conditions with time is to form a temporal ratio of the same *Landsat* band.

Because *Landsat* MSS does not record a blue visible band, standard color products are printed such that the green visible band (band 4) is blue, the red visible band (band 5) is green, and the invisible IR band (band 7) is red. Simulated natural color images, however, can then be developed from a computer-produced blue band. This blue band is displayed as visible blue, the existing *Landsat* band 4 (visible green) is displayed as visible green, and the existing *Landsat* band 5 (visible red) is displayed as visible red in order to simulate natural color images. Raw *Landsat* data are corrected for atmospheric scattering. Then a 5/6 band ratio is used to identify pixels as belonging to soil-rock (ratio value of 1.5), vegetation (ratio value of 0.45), or water classes (ratio value of 1.45). An algorithm is then applied to the data in all spectral bands to determine the brightness for pixels in the new "blue band." This technique is explained in detail by Chavez et al. (1977).

Image classification is the process in which a set of rules is used to assign each ground resolution element to one of several classes by machine or human methods.

Computer classification eliminates the repetitious judgment of a human operator and increases the detail of the classification. Extensive interaction with a human interpreter is still required because the "ground truth" requirements are very similar to those of photo interpretation. Known examples of the classes selected for classification must be identified for the computer as training sets. The computer abstracts the mean and standard deviation of each of the spectral measurements for all elements in each training set (Lawrence and Herzog, 1975).

For maximum error at the 95% confidence level, the minimum sample size per field must vary between 1 to 58 for ground radiometric and airborne MSS measurements (Curran and Williamson, 1986).

For each class ($1, 2, \ldots, i$) the computer constructs a prototype vector

$$F_x = \begin{bmatrix} f_{ia} \\ f_{ib} \\ f_{ic} \\ f_{id} \end{bmatrix}$$

where a, b, c, and d indicate four spectral bands. Each element of this vector is the mean of the corresponding training set for that class. Classification of the unknown vector is based on the Euclidean distance between the unknown vector and each of the prototype vectors. Each unknown vector F_x is constructed from the four data of a given resolution element such that

$$F_x = \begin{bmatrix} f_a \\ f_b \\ f_c \\ f_d \end{bmatrix}$$

and a set of D_i are calculated against the prototype vectors as

$$D_i = (f_{ia} - f_a)^2 + (f_{ib} - f_b)^2 + (f_{ic} - f_c)^2 + (f_{id} - f_d)^2$$

The threshold value T_i for each class is defined by the human operator. This value is included to allow a "none of these" classification if the distance is larger than expected for a given class. Now D is the minimum value of the set of D_i and of $D \leq T$.

Image classification of *Landsat* data involves both machine-assisted delineation of multispectral patterns in up to four-dimensional spectral space and identification of machine-delineated multispectral patterns that represent landscape cover patterns. The first process is called multispectral analysis, and the second process is called multispectral classification. However, it should be kept in mind that image data are not classified until an interpreter determines which spectral classes are representative of particular cover conditions.

If an algorithm is employed to determine "natural" groupings of multispectral data in four-dimensional spectral space, then the image is said to have been analyzed by an unsupervised approach. If the analyst "trains" the analytical processor by selecting samples of classes to be recognized, then the image is said to have been analyzed by a supervised approach.

The concept of the parallelepiped classification algorithm is shown in Figure 5-5-5. For purposes of illustration, only three spectral band axes are shown in Figure 5-5-5 (Taranik, 1978).

The digital processing of remote sensing data and enhancement techniques and their effects on the final image display can be demonstrated with a multispectral image in Figure 5-5-6. This false-color composite of multispectral data represents bilinear interpolation to resample the 20-m data to 10-m resolution and uses a high-frequency convolution filter to sharpen the image. In inset 1 a filter extracted vertical edges such as roads and docks. The red, green, and blue lines represent edges in the corresponding channels (3,2,1). White lines are edges detected in all three channels. Using the red and green bands, inset 2 illustrates the first two components of a principal-components analysis. Image channels were processed through a color-space transformation, and new orthogonal color axes were assigned. The red band represents the greatest amount of variance in the original data, with supplemental features shown in the green band. Inset 3 is a composite image of SPOT panchromatic data overlain by digitized roads and urban features.

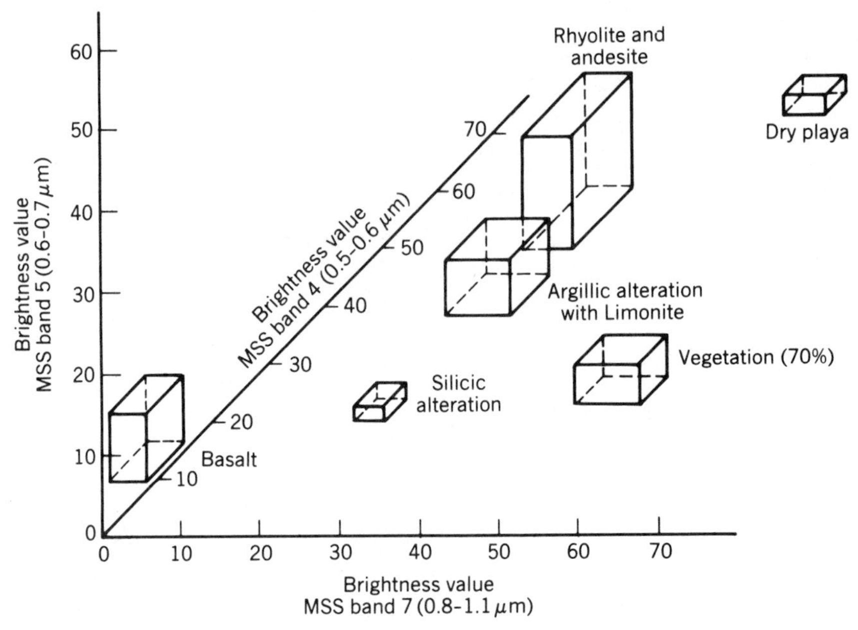

Figure 5-5-5. Concept of parallepiped classification algorithm. After Taranik (1978).

The 10-m panchromatic channel was rectified to Universal Transvers Mercator coordinates by using bilinear interpolation, and was enhanced by a high-frequency filter.

5-5-3 Principles of Digital Image Processing Systems

Image processing is, in principle, a special form of two-dimensional, and sometimes three-dimensional, signal processing of scenes collected by sensors in cameras or sensing devices. Digital data of those scenes are stored in computers and image processors in bits, which are the smallest units of information. One bit of information is either on (1) or off (0). If an image had 1 bit of information in it, there would be two gray levels in it: white and black. As more bits of data are added, the number of gray levels in the picture increases. With 6 bits of data (this stands for 2^6), there are 64 gray levels, ranging in value from 0 (black) up to 63 (white). In this last example, 32 would be a medium gray since it is halfway between 0 and 63.

The number of bits in a given pixel determines the number of unique gray values or colors available. Eight-bit pixels provide 256 different gray values in black to white shades, or 256 unique colors in a pseudocolor mode. This is where an arbitrary color is assigned to each different gray level. Color image processing normally involves assigning one of three colors to three different black-and-white images. Each image has a certain number of gray levels, which are then represented by a color. Color image processing systems have 24-bit pixels, 8 bits each for red, green, and blue, which translates into over 16 million unique colors, which are more than can be displayed on a monitor and more than one can distinguish visually.

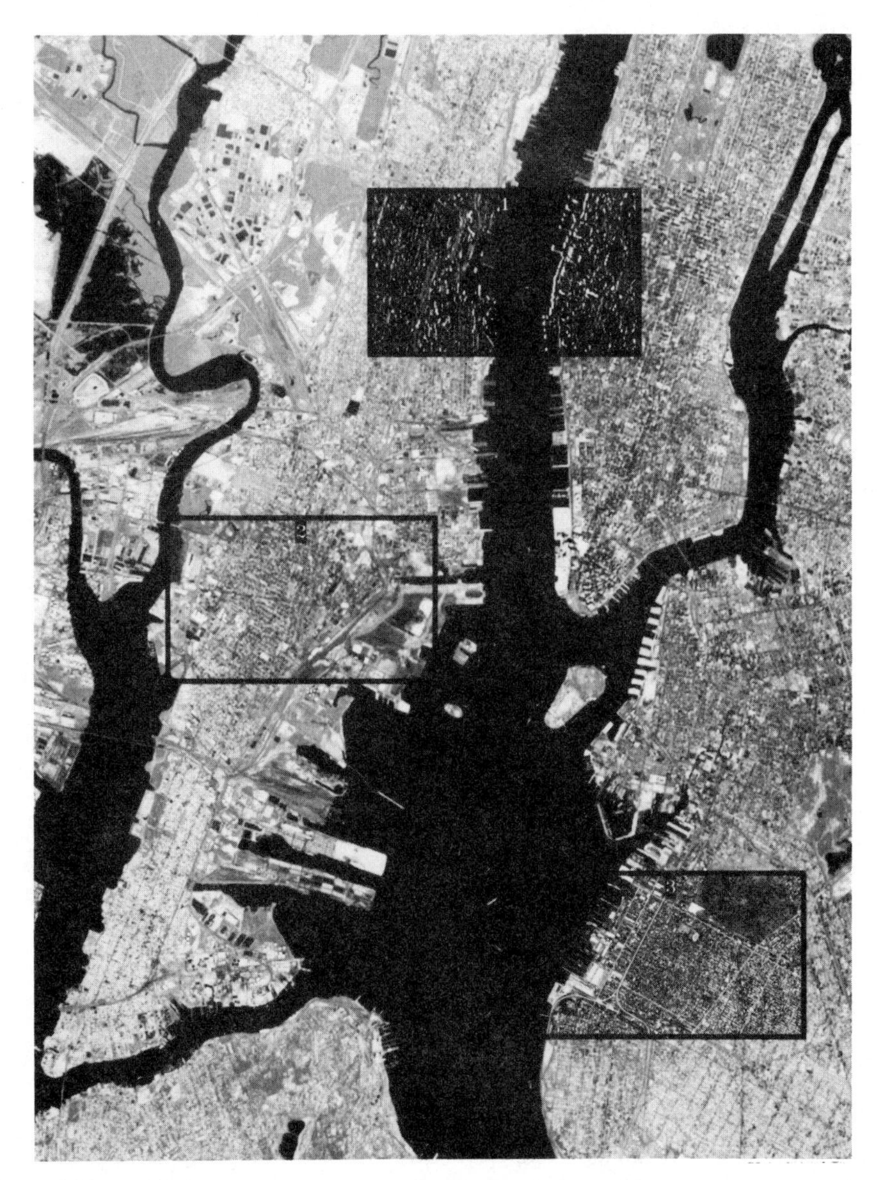

Figure 5-5-6. Digital processing and enhancement techniques as applied on multispectral image of Manhattan. See text for explanation of insets. Courtesy of ERDAS. See insert for color representation.

Computer systems for image processing range over all of the computer field, from Apples and IBM personal computers (PCs), through small minicomputers, to mainframe installations. The standard has changed from the small PDP-11 to the Motorola 68000 microprocessor and digital VAX systems. The large mainframe is often too expensive and usually not necessary, but small PCs cannot handle all of the necessary operational processing tasks.

Dedicated image processing systems include display memory, a video processor, a parallel interface to a computer, a human-machine interface, digital-to-analog (D/A) converters, and a comprehensive software subroutine library. The basic subroutine library should contain all the necessary software for manipulating the internal parts of the image processor. Source code should be provided to allow sophisticated users to do their own programming to solve unique requirements. Upgrades include high-level, user-friendly software, an embedded 32-bit minicomputer with Winchester disk drive and nine-track magnetic tape storage. As the needs of the individual image processing lab increase, so should the power of the computer, the number of image processing stations, and the number of input/output peripheral devices that interface to the image processing system.

A frame buffer is the key to any image processing system. This bank of memory stores the image data. Most medium-sized systems are several banks of 512 × 512 elements. The rows of the frame buffer matrix are the lines of the image, and the columns along each line are the samples. A typical choice for a color system is to have four memory banks or channels, one each for red, green, and blue, and a fourth for intermediate calculations or superimposition of graphics and annotation.

A D/A converter transforms the contents of the image memory into a form compatible with the monitor. The number of different intensity levels that a D/A converter can output is related to the number of bits it is designed to handle; the more bits, the more distinct colors or gray levels it can produce. Few systems use D/A converters with more than 8 bits of resolution. For a full color system, this arrangement translates into 8 bits for each of three channels (red, green, and blue), a total of 24 bits of color information per pixel, or over 16 million unique colors. The outputs of the D/A converts are generally formatted to either a standard composite video or, in higher-resolution systems, sent to the display via separate red, green, and blue cables.

An important part of an image processing system is a look-up table, which is a table of stored data for reference purposes. The look-up table performs a transformation or mapping between each unique input data value and some predefined output value. Applications include color or density mapping and calculations that must be performed rapidly.

Many low-cost systems for data processing, especially *Landsat* and SPOT data, have been marketed in recent years. They are stand-alone, turn-key, image processing systems capable of serving as an intelligent workstation to a larger computer. Such systems are represented, for instance, by ERDAS (Earth Resources Data Analysis System) (Fig. 5-5-7) with a high-resolution display, programs, and menus for image enhancement, classification, geometric corrections, GIS (Geographic Information System) merger and analysis, and scale hard-copy output. This system can be interfaced to and exchange data with larger systems to expand to a more powerful workstation. This allows the inclusion of ground truth data for large-scale processing and full scene classification.

A general schematic of an interactive system used at NASA/GSFC (Goddard Space Flight Center) for NOAA AVHRR data is shown in Figure 5-5-8. Data may be displayed on this system with grid overlays and mapped to several different projections; gray shade and color enhancement capability is provided. A color bar of up to 64 levels can be selected from 2084 different colors. Linear modification of image data can be performed on-line in the form of the contour of the image, difference of two images, ratio of two images, differentiation of an image, and combinations of these capabilities. A selected portion of an image may be enlarged

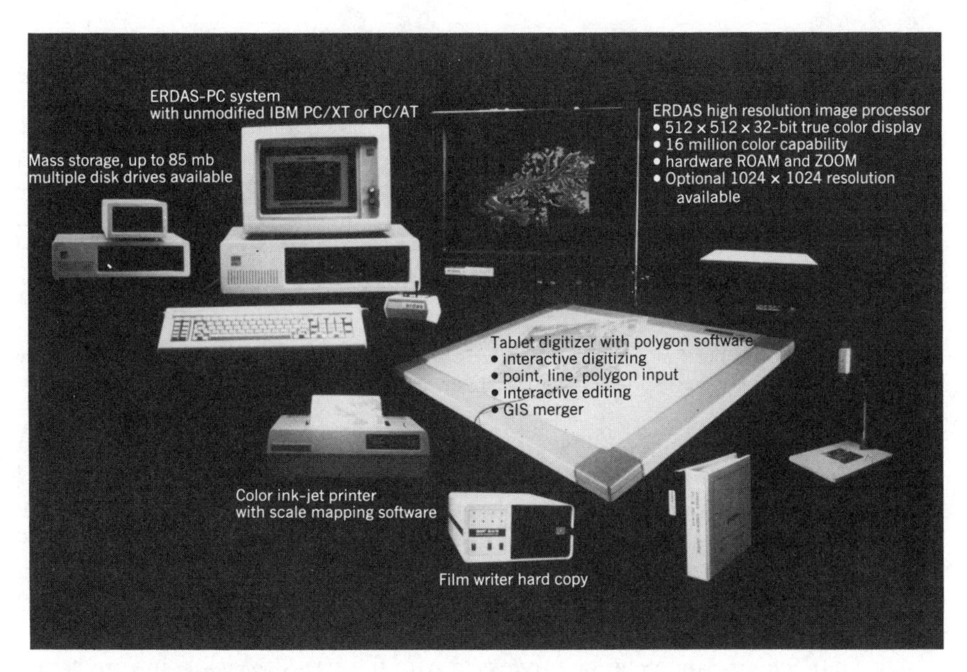

Figure 5-5-7. ERDAS processing system. Courtesy of ERDAS.

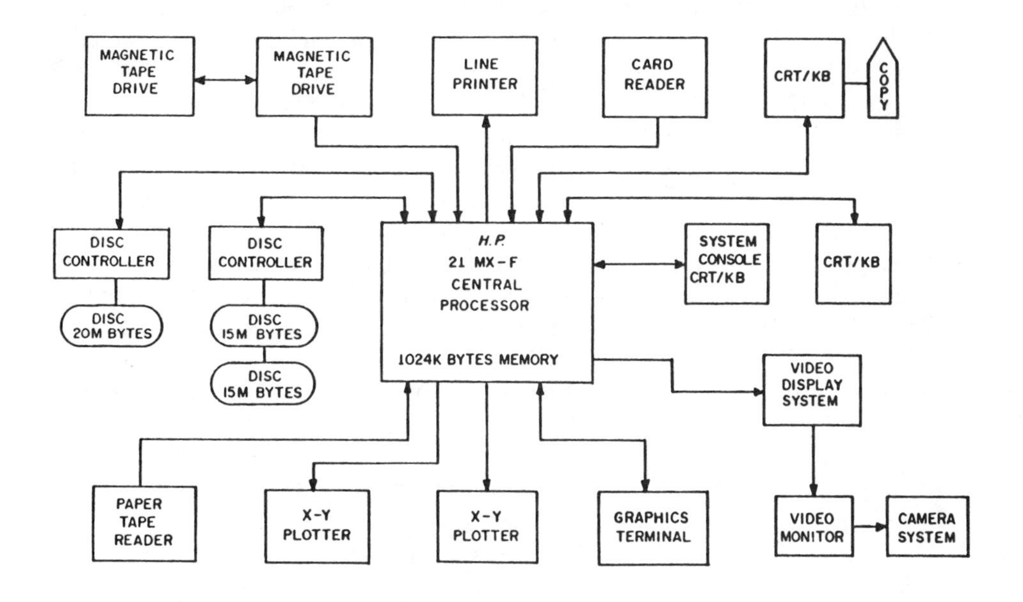

Figure 5-5-8. General schematic of the HP 1000 computer interactive system used for analyzing AVHRR data.

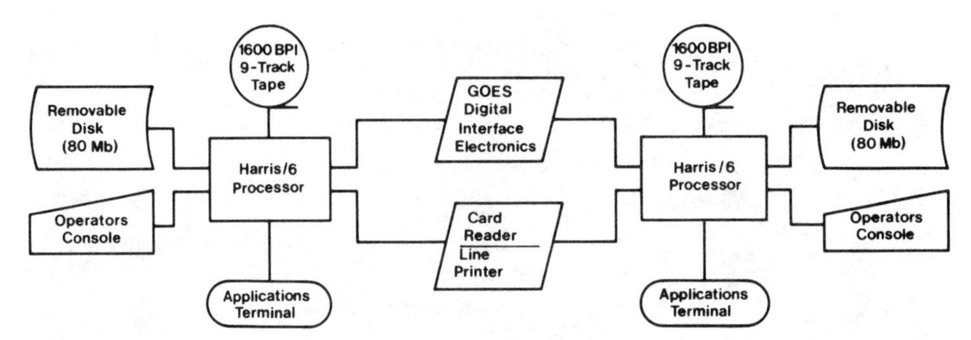

Figure 5-5-9. VIRGS hardware configuration. GOES data are received through the digital interface electronics to the removable disk and the applications terminal.

or compressed by a floating-point factor such as 2.2 or 3.0. On-line filtering and enhancement techniques are selectable. Profiles of the image data may be obtained by horizontal, vertical, radial, or circular sampling of the pixels. Hard-copy output is available by means of a plotting device and a video camera system that uses Polaroid 8 × 10 color film, which can be exposed and developed in less than 2 min for any selected image.

The GOES digital data can be processed with the VISSR (visible infrared spin-scan radiometer) image registration and gridding system (VIRGS). Data output from VIRGS is in two formats: color-annotated photographs of GOES imagery and numerical arrays of the GOES digital data. The VIRGS is an interactive system designed primarily for providing specific data sets and information from GOES digital data. The VIRGS (Fig. 5-5-9) comprises standard computer industry components. The interaction between analyst and computer takes place mainly at the applications terminal (Fig. 5-5-10), which comprises a 40-cm color video CRT monitor, a smaller alphanumeric monitor, a pair of joysticks, and a line printer. The color monitor has 500 scan lines, each line containing 640 pixels, and is used to display GOES imagery. The monitor can store eight image frames and a graphics frame. The graphics frame, with red, green, and yellow colors, displays a variety of map backgrounds and lettering fields. The joysticks are used to position a cursor on the TV monitors. Data are output to the printer and controlled by the applications programs once the cursor is positioned over the specific location.

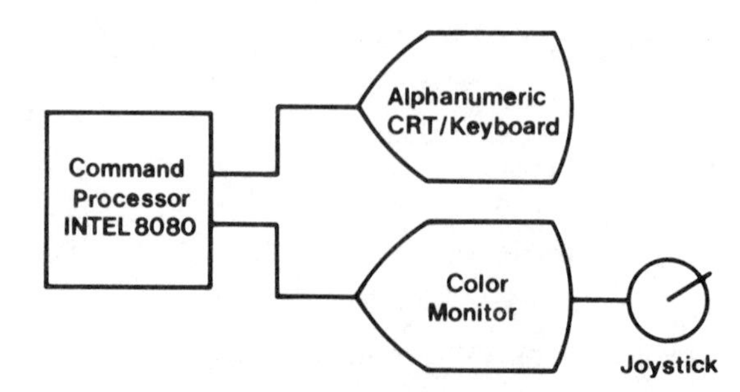

Figure 5-5-10. Interactive applications terminal on the VIRGS. The color monitor displays the image and graphic frames. The joystick controls the position of the cursor on the video monitor.

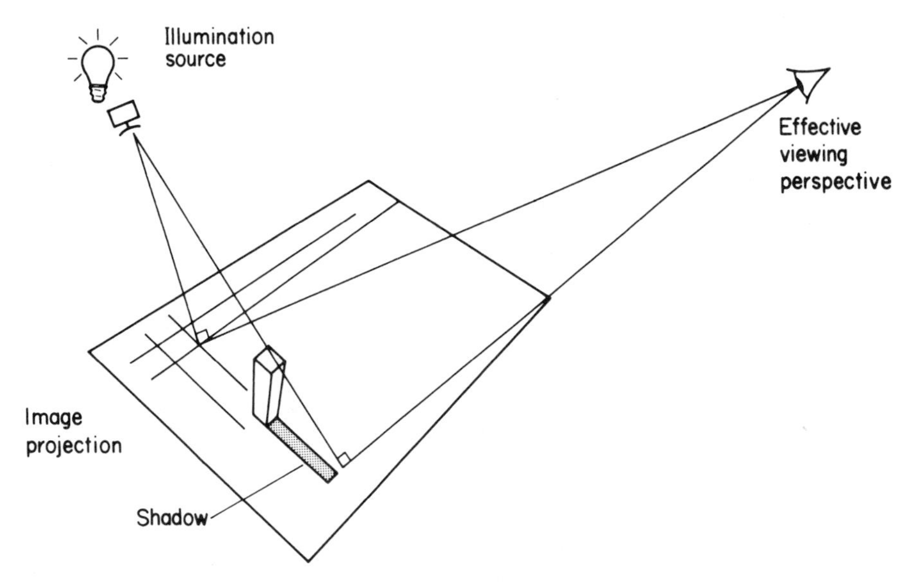

Figure 5-6-1. Interpretation aid for interpreting radar images. After Raney (1986). Reproduced with permission from Graham & Trotman, Ltd.: Szekielda, K.-H., 1986. *Satellite Remote Sensing for Resources Development.*

5-6 INTERPRETATION OF RADAR IMAGES

Interpretation of radar images is similar to photo interpretation, although the illumination coming from the side of the scene creates a viewing direction from above the image at right angles to the transmitted microwave energy.

As an interpretation aid for radar imagery, Figure 5-6-1 can be used to establish the effective viewing perspective (Raney, 1986). Through this perspective, geometric effects, such as elevation displacement, become equivalent between radar

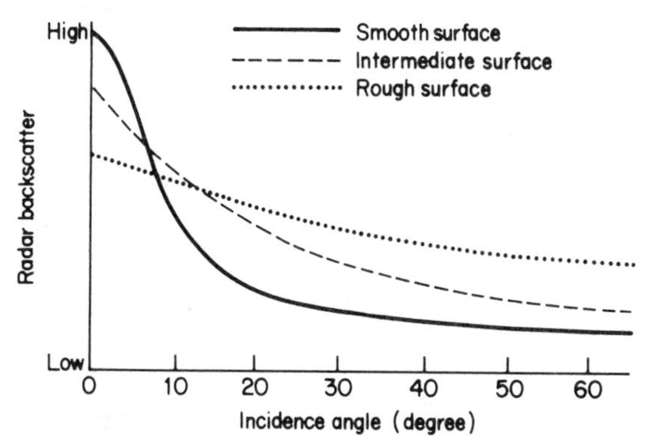

Figure 5-6-2. Schematic radar backscatter curves showing relative echo strength as a function of incidence angle. Note the relative separation of the various surfaces at the various angles. After Blom and Dixon (1986). Reproduced with permission from Graham & Trotman, Ltd.: Szekielda, K.-H., 1986. *Satellite Remote Sensing for Resources Development.*

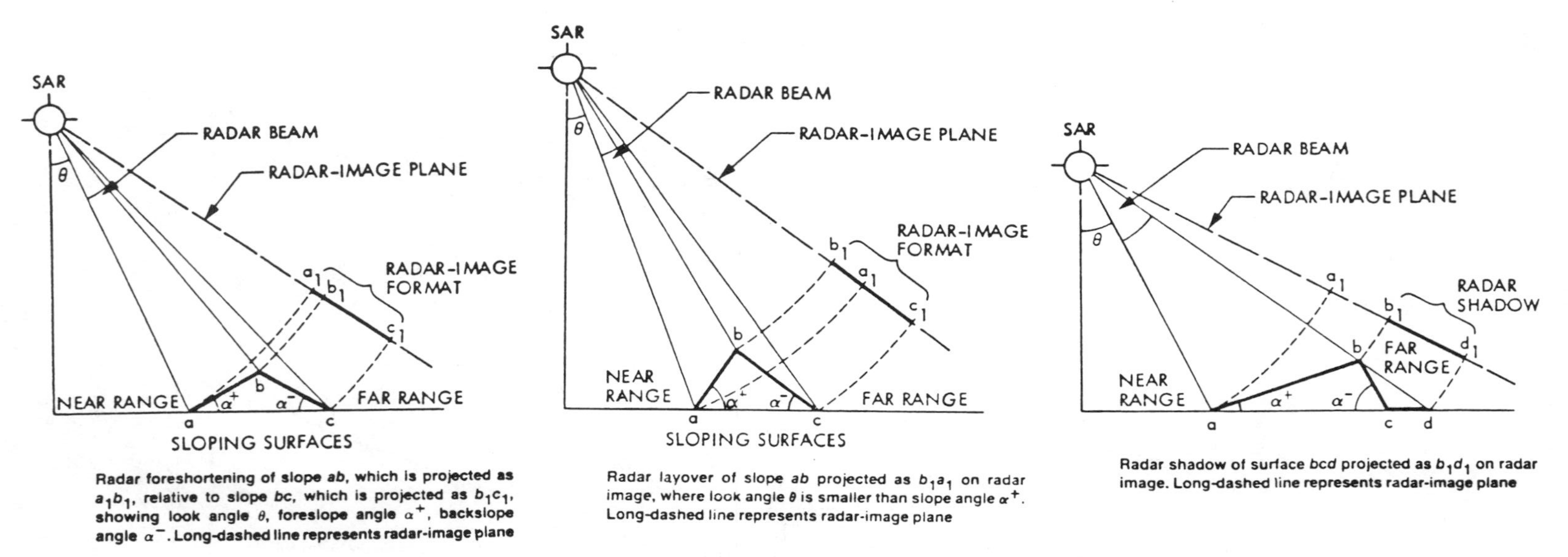

Radar foreshortening of slope *ab*, which is projected as a_1b_1, relative to slope *bc*, which is projected as b_1c_1, showing look angle θ, foreslope angle α^+, backslope angle α^-. Long-dashed line represents radar-image plane

Radar layover of slope *ab* projected as b_1a_1 on radar image, where look angle θ is smaller than slope angle α^+. Long-dashed line represents radar-image plane

Radar shadow of surface *bcd* projected as b_1d_1 on radar image. Long-dashed line represents radar-image plane

Figure 5-6-3. Factors to be taken into account for SAR interpretation. Courtesy of Dr. C. Elachi, J.P.L.

and aerial photography, as reflections from taller objects are shifted away from the viewpoint nadir toward the radar in the side-looking geometry.

The primary variables that control the brightness in radar image are, in order of decreasing effect, slope, roughness, and dielectric constant (including the effect of soil moisture). Each variable gives valuable information, especially to the geologist.

Because orbital radar images are acquired at a great distance from the terrain, the angle subtended by even a 100-km swath is quite small. With *Seasat,* for example, the incidence angle varied only 6° from the near- to far-range portions of the 100-km image swath. The SIR-A had a similar angular variation over its 50-km image swath, although the altitude was much lower. Thus, in contrast to radar images from aircraft, the imaging geometry is quite constant across any potential study area. The SIR-B could vary the incidence angle, with the goal of studying the additional information provided by observing the same area at different angles.

To illustrate the general behavior of different surfaces as a function of incidence angle, we give the schematic backscatter curves in Figure 5-6-2, which show the difference in relative backscatter values for rough, intermediate, and smooth surfaces. It is also evident that the greatest differences between the surfaces are at relatively small incidence angles. The ability to image the same surfaces at different incidence angles, therefore, gives information on the roughness characteristics of the surface (Blom and Dixon, 1986).

Figure 5-6-3 summarizes certain factors considered for image interpretation of *Seasat* SAR. Because SAR was a single, right-side-looking system orbiting in an inclined plane (108°), images of certain scenes were acquired from two different look directions. As pointed out before, the direction of the radar illumination has a pronounced effect on perception, especially of linear features, whereby the extent to which linear features are modified on the imagery is a function of their orientation relative to the direction of illumination. Consequently, some linear features that appear dark or are imperceptible in one look direction may appear bright in the other look direction.

Slopes inclined toward the radar appear compressed relative to slopes inclined away from the radar. The foreshortening factor F_f is approximately

$$F_f = \sin(\theta - \alpha)$$

where the look angle θ is the angle between the vertical plane and a line that links the imaging radar antenna to a feature on the ground, and α is the slope angle of the surface. This angle is positive (α^+) where the slope is inclined toward the radar (foreslope), and negative (α^-) where the slope is inclined away from the radar (backslope). Layover is an extreme case of foreshortening that occurs when the look angle θ is smaller than the foreslope α^+ ($\theta < \alpha^+$).

Slopes inclined away from the radar are in shadow when the look angle θ plus the backslope angle α^- is greater than 90°. Shadows are caused by ground features that obstruct the radar beam and prevent illumination of the area behind them. This effect occurs on *Seasat* SAR images whenever the backslope in the radar-viewing direction exceeds about 70° (Ford et al., 1980).

6

OBSERVATIONS OVER THE OCEANS

6-1 GENERAL CONSIDERATIONS

6-1-1 Temperature

If an area to be observed is cloud free, satellite IR recordings can directly be used to detect thermal gradients. For large-scale mapping, however, in order to eliminate atmospheric contributions to the measurements, statistical methods need to be applied. So far, the most advanced large-scale mapping systems are the former global ocean sea surface temperature computer program (GOSSTCOMP) and the more recent multichannel sea surface temperature (MCSST) method, which provide data on a large grid, but, compared with ship measurements, especially in southern oceans, have very dense monitoring grids.

With respect to SST measurements, one has to consider that radiation observed by a sensor is a function of radiation, with atmospheric components and radiation reflected from the surface. Due to the high emissivity of water, a value of unity can be assumed, which concludes that reflection is negligibly low.

Mathematically, the component sources of radiation in a clear atmosphere can be written as

$$N_s = \int_{\lambda_1}^{\lambda_2} \varepsilon_s(\lambda)\beta(\lambda, T_s)\, d\lambda$$

where N_s = total energy emitted by a radiating surface in a bandwidth (λ_1, λ_2) at T_s

β = Planck radiance

ε_s = surface emissivity

and

$$N_a = \int_{\lambda_1}^{\lambda_2} \int_{P_0}^{\infty} \beta(\lambda, T(P)) \frac{d\tau(\lambda, P)}{dP} \, dP \, d\lambda$$

where N_a = positive energy contribution of the atmosphere in the bandpass (λ_1, λ_2)
$T(P)$ = temperature of the atmosphere at pressure level P
$T(\lambda, P)$ = atmospheric transmittance from pressure level P to the top of the atmosphere
P_0 = atmospheric pressure at the surface

The total energy $N(T)$ over the bandpass measured by the sensor viewing an ocean surface is then

$$N(T) = N_s\tau_s + N_a = N_s - (N_s\alpha_s - N_a)$$

where α_s and τ_s are the absorptivity and transmittance, respectively, of the total atmosphere. Expressed in terms of equivalent temperatures, this equation becomes

$$T_{bb} = T_s - \Delta T$$

where T_{bb} = sensor observed brightness temperature
T_s = surface temperature
ΔT = atmospheric correction

The value of ΔT is obtained from the vertical temperature profile radiometer (VTPR) soundings.

As shown in Table 6-1-1, attenuation by water vapor far exceeds that by any of the other absorbers. Other residual absorption, such as that produced by salt and dust particles, has little effect unless there is high particle concentration. Ice crystals near the tropopause are assumed to be opaque.

Brightness temperatures associated with SST measurements are corrected for atmospheric attenuation, with coefficients derived from VTPR processing of a coincident temperature and moisture profile sounding. The coefficients are derived by integrating the effect of absorption and emission of radiance by each layer of the atmosphere. This integration is based on the VTPR sounding and theoretical transmittance functions. The computed coefficients are used to correct for the

TABLE 6-1-1. Atmospheric Attenuation Corrections in the Range 10.5 to 12.5 μm

Absorber	Range (°C)
H_2O	0–9.0
CO_2	0.1–0.2
O_3	0.1
Aerosols	0.1–0.95

straight-down single atmosphere absorption and the absorption through a slanted atmospheric path.

The ability to eliminate cloud-contaminated scenes is the most significant factor for determining the eventual accuracy of surface temperature values. Clouds are themselves radiating surfaces. They absorb the radiation emitted from the ocean surface and emit radiation as specified by the Planck function. In a single-channel model, no effective means are available to infer a surface radiance value in cloud-contaminated scenes. For this reason the viewed scene must be limited to cloud-free areas. Considerable effort has gone into the design of the SST model, both in the retrieval process and in postanalysis, to eliminate any cloud-contaminated brightness temperatures.

Original research on SST retrieval techniques using scanning radiometer infrared (SRIR) data was done by Smith et al. (1970). A modification of this original technique, first suggested by Smith and Koffler (1970), was later developed into the present technique by Leese et al. (1971). The current method of retrieving SSTs from satellite data employs a statistical technique to extract SSTs from blocks of raw data organized into histograms.

For a uniform scene temperature field, with only random noise affecting the sensor output, a frequency distribution of the data elements from this field is gaussian. Simply calculating the arithmetic mean or locating the modal class of this histogram gives the actual scene temperature. Since most scenes have some clouds present, it is necessary to find the true surface scene temperature when cloud-contaminated samples are present in the histogram. In an uncontaminated histogram whose cold side is a mirror image of the warm side, the mean would be the surface scene temperature.

Because there is a certain degree of nonrandom noise in the data, two operations are performed to ensure the accuracy of the calculated mean:

1. The histogram is smoothed before attempting to retrieve a mean.
2. Since a single calculation of the mean from a slightly nongaussian histogram could be in error, repeated calculations of the mean are made from all possible combinations, taken three at a time, of the classes (each class being one IR response count wide) constituting the warm side of the histogram.

These estimates of the mean are distributed into a second histogram called the mean estimate histogram (MEH). The uncontaminated mean of the data histogram is then calculated from the MEH by finding the mean value of all mean estimates that fall within two classes of the mode of the MEH.

In addition to comparisons between ship observations and satellite SSTs, the quality of SST data for an area of investigation without any additional ship data can be evaluated. For instance, if a chosen area has rather low seasonal changes, variations around the monthly mean are also expected to be low. Therefore, any variation greater than the expected natural changes could indicate an error introduced by the data and the processing (see Fig. 6-1-1).

High-quality multispectral measurements from satellites and the recent development of multiple-window techniques to correct IR brightness temperatures for atmospheric attenuation have enabled marked improvements in global mapping of SSTs. Four-kilometer resolution data have been used in two visual bands and

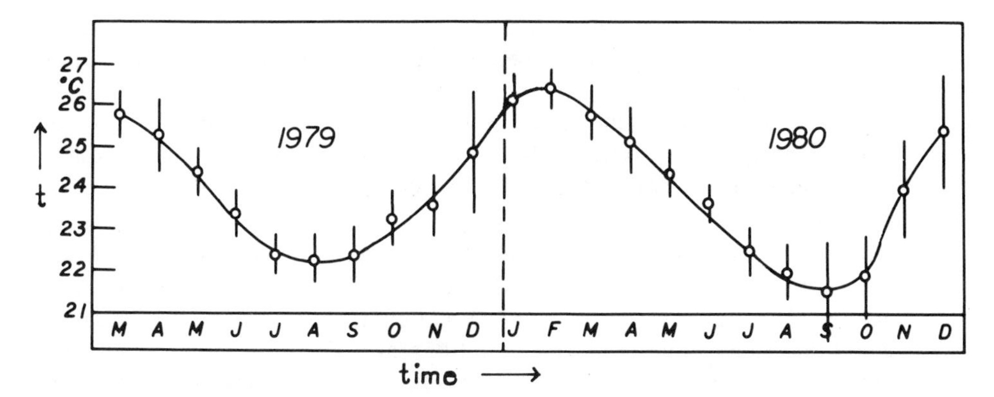

Figure 6-1-1. Monthly average SST obtained at 55°E and 25°S for 1979 and 1980. Included is the standard deviation of data around the monthly average. Reproduced with permission from Deutsches Hydrographisches Institut, Hamburg.

three atmospheric windows in the thermal IR band from the AVHRR on NOAA's operational polar satellites. Various threshold or spatial homogeneity tests are applied on small arrays to discriminate nominally cloud-free samples for subsequent processing. Multiwindow atmospheric correction equations were derived by simulation using temperature/humidity profiles and atmospheric transmittance models. Temperature-dependent bias corrections were obtained for equations by comparison of results with buoy temperatures. Tests of the bias-corrected equations against independent buoy data gave biases equal to 0.42°C and scatter equal to 0.62°C. Global statistical comparisons with ships, compiled daily, indicate significant improvements in accuracy and coverage over previously available satellite-derived surface temperatures.

The AVHRR on the NOAA satellites provides high-quality digital measurements that have a basic spatial resolution of 1.1 km at nadir in the visible (0.58–0.68 μm) and reflected IR (0.725–1.1 μm) bands and in two or three emitted IR window channels (3.55–3.93, 10.3–11.3, and 11.5–12.5 μm). *Tiros-N, NOAA 6,* and *NOAA 8* carried AVHRRs equipped with the first two window channels mentioned, whereas *NOAA 7* and *NOAA 9* have AVHRRs with all three. The full-resolution measurements are available locally by direct readout (high-resolution picture transmission, HRPT) and by means of limited temporary onboard tape storage (LAC). The GAC data is provided twice daily at a nominal resolution of 4 km (four of every five samples along the scan line are used to compute one average value, and the data from only every third scan line are processed) by means of onboard data reduction and tape recording. There is provision for onboard calibration of the emitted IR channels (McClain et al., 1985).

Various combinations of the AVHRR visible and IR channels are used to detect the presence of clouds in small array. Only cloud-free data arrays are processed for MCSSTs. The various cloud tests can be grouped into three classes (McClain et al., 1985):

1. *Visible or IR Reflectance Tests.* The bidirectional reflectance of the cloud-free ocean, as measured at a satellite, is generally less than 10%, whereas

the reflectance of most clouds is greater than 50%. Thus, thresholds can be established for the maximum expected AVHRR-measured reflectance in the absence of clouds. These thresholds are a function of solar zenith angle, satellite zenith angle, and the azimuthal angle of the viewed spot. No attempt is being made to process SSTs in regions of direct specular sun reflection, where the ocean reflectance can approach or exceed cloud reflectances.

2. *Uniform Tests*. Thresholds of the expected variation of measurement values from adjacent cloud-free fields of view, in either visible band or IR channels, are set to be slightly in excess of instrumental noise. With partially cloud-filled fields of view, the variations are generally larger than these thresholds. This test is particularly useful at night, because instrumental noise in the 11- or 12-μm channels is extremely small ($NE\Delta T \leq 0.1°C$), and visible-band data cannot be employed.
3. *Channel Intercomparison Tests*. At night, under cloud-free conditions, two or more independent measures of SST can be obtained from the three AVHRR window channels with equations of the form SST $= A + B(T_i - T_j) + T_i$. Equating two such equations, one finds that the ratio $(T_{3.7} - T_{11})/(T_{11} - T_{12})$ is invariant to changes in atmospheric conditions. Under cloudy conditions, however, this constancy does not apply. With transmissive or subresolution clouds, a measured channel blackbody radiance is a combination of the radiance of the cold cloud and the warmer sea surface. Because the radiance is more sensitive to temperature at 3.7 μm than at 11 or 12 μm, the indicated ratio increases under these conditions. With thick nontransmissive clouds, this radiance effect does not occur; but because the emissivity of such clouds at 3.7 μm is less than that at 11 or 12 μm, the ratio decreases. This characteristic is crucial for detecting low-level stratus clouds at night with AVHRR data. Such clouds often have uniform cloud top temperatures, and thus the IR uniformity tests do not detect them.

The theoretical basis of multiple-window SST techniques was reviewed by McMillin and Crosby (1984). In any spectral interval, thermal radiation emitted by the sea surface is absorbed by atmospheric constituents and reemitted at all levels in the atmosphere. The radiative transfer equation, which describes this process is simplified:

$$I_i = B_i(T_s)\tau_i + B_i(\overline{T}_i)(1 - \tau_i) \qquad (6\text{-}1)$$

where $\overline{T}_i$ = radiance at top of atmosphere
$B(T)$ = Planck function at temperature T
τ_i = atmospheric transmittance

Generally $\overline{T}_i$, which is the atmospheric temperature at some level, is a function of the spectral interval (i.e., the mean atmosphere altitude varies). The AVHRR on *NOAA 7* includes three IR window channels centered at 3.7, 11, and 12 μm. The atmospheric absorption in these channels occurs primarily at very low levels in the atmosphere. As a result, one may assume that the mean atmospheric temperature $\overline{T}$ is the same in these channels (McMillin and Crosby, 1984). Another result is that $\overline{T}$ is nearly equal to T_s, the SST. Furthermore, atmospheric absorption in these window channels is primarily by water vapor, and the transmittance can be approximated as $T_i = e^{-kX} = 1 - k_i X$, where k_i is the absorption coefficient and X is a function of the water vapor amount. With these approximations a Taylor series expansion of (6-1) in terms of temperature yields

$$T_i - T_s = k_i X(\overline{T} - T_s) \tag{6-2}$$

where T_i is the brightness temperature corresponding to I_i. Combining these results gives

$$T_s - T_i = \frac{k_i}{k_j - k_i}(T_i - T_j) \tag{6-3}$$

This equation states that the temperature deficit in one channel, $T_s - T_i$, which results from atmospheric absorption by water vapor, is a linear function of the brightness temperature difference of the two different window channels (Bernstein, 1982; Barton, 1983).

In comparison with in situ sensors, particularly those on ships, the AVHRR, with its nearly continuous onboard calibration, generates a uniform set of brightness temperature measurements. Except for unusual circumstances, the operationally derived MCSSTs generally constitute a spatially and temporally consistent data base. Changes made in the operational algorithms have affected the root-mean-square (rms) differences with respect to drifting-buoy temperatures only at the $\leq$0.25°C level. Recent drifting-buoy spot comparisons over a wide range of temperatures, geographic area, and seasons consistently indicate biases of $\leq$0.1°C and rms differences (or scatter) of 0.5–0.6°C (Strong and McClain, 1984).

In the MCSST method the use of a temperature-dependent bias correction derived from satellite/buoy matchup data presumably incorporates some sort of average skin temperature versus 1 m-depth temperature adjustment. The skin is almost always cool, and its temperature is usually 0.1–0.5°C (Robinson et al., 1984), although the temperature difference with depth is minimized during well-mixed conditions (McClain et al., 1985).

6-1-2 Chlorophyll

Gordon et al. (1980) used a regression analysis that relates the concentration of chlorophyll a and phaeophytin a to the radiance ratios:

$$R_1 = \frac{L_w^{443}}{L_w^{550}} \text{ and } R_2 = \frac{L_w^{520}}{L_w^{550}} \quad \text{through} \quad \log C_i = \log a + b \log R_i, \quad i = 1 \text{ or } 2 \tag{6-4}$$

where $\log a$ and b are coefficients, with values for R_1 of -0.297 and -1.269, respectively; for R_2, the corresponding values are -0.074 and -3.975.

The method to remove the aerosol effects is based on the approximation that the total radiance L^λ observed by the sensor at wavelength λ can be partitioned into Rayleigh scattering (L_R^λ), aerosol scattering (L_A^λ), and the radiance backscattered out of the ocean (L_w^λ) as transmitted through the atmosphere. This can be written as

$$L^\lambda = L_R^\lambda + L_A^\lambda + t^\lambda L_w^\lambda \tag{6-5}$$

where t^λ is the diffuse transmittance of the atmosphere.

As the aerosol-scattering phase function is assumed to be independent of λ, then L_A^λ at one wavelength can be considered to be approximately proportional to that at another wavelength. If Equation 6-5 is applied at wavelengths λ_1 and λ_2, it follows that

$$t^{\lambda_2} L_w^{\lambda_2} = L^{\lambda_2} - L_R^{\lambda_2} - \alpha(\lambda_1, \lambda_2)[L^{\lambda_1} - L_R^{\lambda_1} - t^{\lambda_1} L_w^{\lambda_1}] \tag{6-6}$$

where $\alpha(\lambda_1, \lambda_2)$ is a constant related to the optical properties of the aerosol through the single scattering approximation. If λ_1 is chosen such that $L_w^{\lambda_1} \sim 0$, then $L_w^{\lambda_2}$ can be determined for any λ_2, provided $\alpha(\lambda_1, \lambda_2)$ is known. The channel on the CZCS covering the 0.670-μm band is the most suitable, except for very high turbidity. In the analysis of CZCS data, reported by Gordon et al. (1980), $\alpha(\lambda_1, \lambda_2)$ was determined from Equation 6-6 by direct in situ measurement of L_w^λ at one location in recordings, and values have been used throughout the entire covered region for calibration.

From this, it has been derived that phytoplankton pigment concentrations extracted from the corrected CZCS radiances with surface measurements agree to within less than 0.5 log C, where C is the sum of the concentrations of chlorophyll *a* plus phaeopigments *a*.

Gordon et al. (1983) described similar algorithms that have been applied to the shelf and slope waters of the Middle Atlantic Bight and to Sargasso Sea waters. The resulting pigment concentrations, compared with continuous measurements made along ship tracks, suggest that over the 0.08–1.5 mg/m^3 range, the error in the retrieved pigment concentration is about 30 to 40% for a variety of atmospheric turbidities. In three direct comparisons between ship measured and satellite retrieved values of the water-leaving radiance, the atmospheric correction algorithm retrieved the water-leaving radiance with an average error of about 10%.

The CZCS-derived chlorophyll concentrations give the averaged pigment concentration to a depth of about 1 optical attenuation length, which corresponds to approximately the top 22% of the euphotic zone (Feldman, 1986).

Vertical distributions of photosynthetic pigments should be weighted by a function that takes into consideration that material near the surface is more important than material at greater depths. One determines this function by noting the exponential attenuation of irradiance with depth and assuming that the backscattered energy would be approximately attenuated in the same manner on its return to the surface. This means that in regions with high productivity, in particular, reliable information is provided on chlorophyll concentrations, as shown by Feldman (1984, 1986); also see Figure 6-1-2.

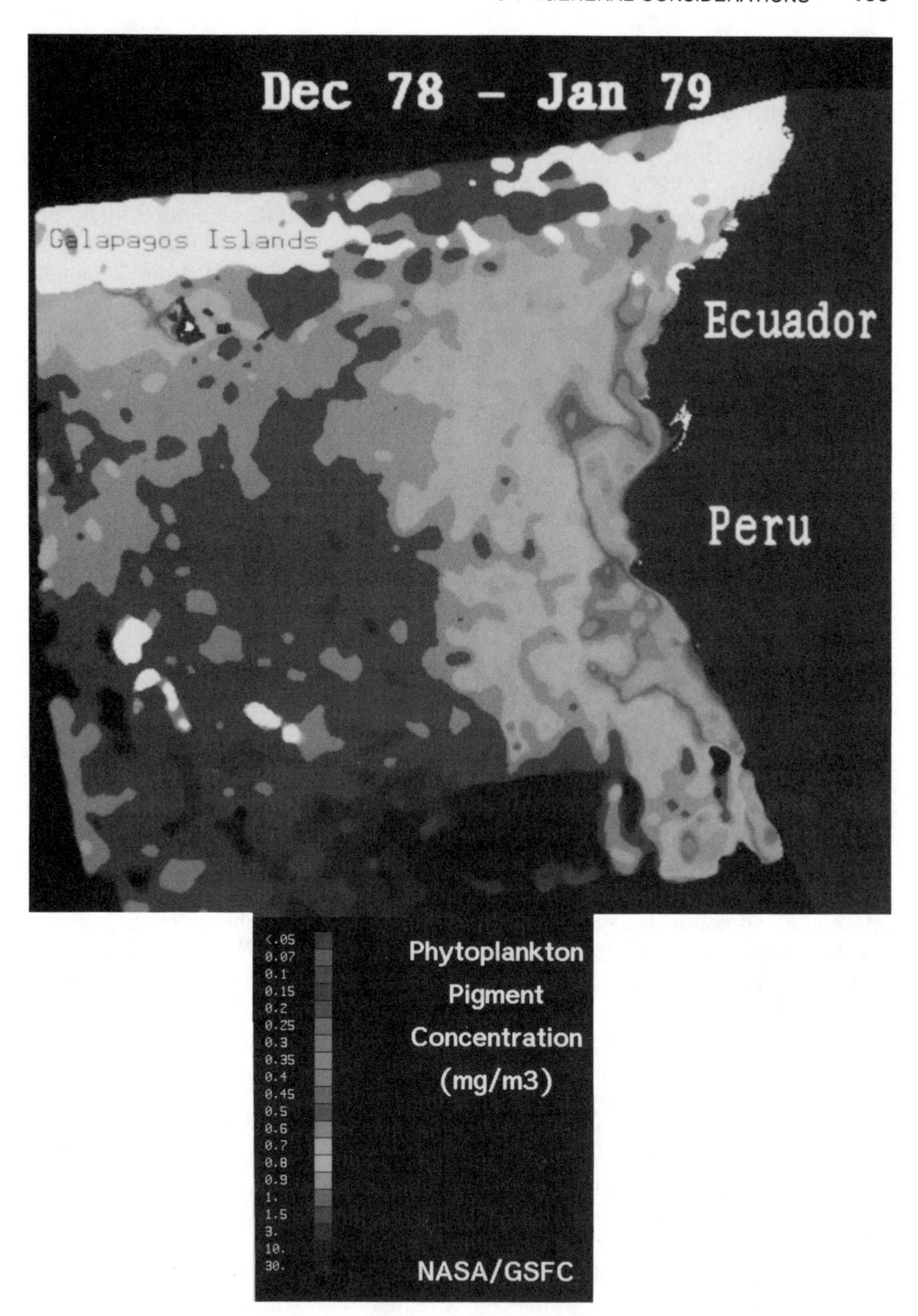

Figure 6-1-2. Chlorophyll concentrations along Peruvian upwelling region as recorded from December 1978 to January 1979. Courtesy of Dr. G. C. Feldman, NASA.

6-2 CURRENTS AND UPWELLING SYSTEMS

6-2-1 Gulf Stream

The Gulf Stream is the most energetic current in the North Atlantic, with its horizontal wave motions (or meanders), main current, and large eddies. The Gulf Stream has been observed from space since 1966, due to its high horizontal temperature gradient (referred to also as the North Wall) at the surface. Measurements with early satellites showed the large-scale monitoring of temporal changes that one can accomplish with IR measurements from space. As early as 1966, Warnecke et al. (1971) derived fluctuations of the Gulf Stream boundary (or the North Wall) over a five-month period. With increasing accuracy and better ground resolution, as well as repetition of data coverage, the recognition of more detailed features has been achieved. By combining IR data from different days, an analysis as outlined in Figure 6-2-1 can be achieved. With a high coverage rate of one to two per hour from the geostationary orbit, it is now possible to use cloud motion pictures to identify the location of temperature gradients, because ocean thermal features move at a rate of one order of magnitude slower than those of the atmosphere.

Under certain wind conditions, the boundaries between the Gulf Stream and its adjacent water have also been observed in the visible as a consequence of altering the sunglint pattern as a function of wind speed. An abrupt change in surface roughness has been observed (Strong and De Rycke, 1973) at the shoreward edge of the Gulf Stream current from Florida to Cape Hatteras, which results from the opposition of waves propagating against the flow of the Gulf Stream. Southward and eastward movement of the waves with the wind encounter the northerly direction of the Gulf Stream, and the waves are transformed into higher ones with shorter wavelengths and more white caps. These modifications of the water surface increase the diffusion of the sunglint and cause higher reflectance, thus increasing the contrast in the imageries. The strong horizontal temperature gradient connected with the North Wall makes it possible to monitor easily its fluctuations and the separation of eddies. Legeckis (1975), for instance, followed the successive displacement of an eddy as indicated by the deformation of surface isotherms.

A striking feature detected in IR data is the Gulf Stream eddies or Gulf Stream rings, which are cold, less-saline mass of slope water surrounded by rings of counterclockwise-flowing Gulf Stream water. The rings are formed from large Gulf Stream meanders that break off and become separated from the main Gulf Stream current. These eddies form for eight months in the region between 60° and 70° W just south of the Gulf Stream (Parker, 1971). They move in a complicated manner with speeds up to 6 mi$\cdot$day^{-1} (Fuglister, 1971), have a diameter of up to about 200 km, and a depth of few thousand meters (Evans et al., 1985).

The complex pattern in surface temperature, as created by eddy formation, is shown in Figure 6-2-2, where the Gulf Stream is shown as a band almost 100 km in diameter. These limits of the Gulf Stream, visible in IR data, represent the locations where the main thermocline (the transition zone between warm surface water and cold deep water) becomes horizontal and the current vanishes. The usual shape of rings is nearly circular, although variations are often found.

It has been hypothesized that strong westerly to northwesterly winds have

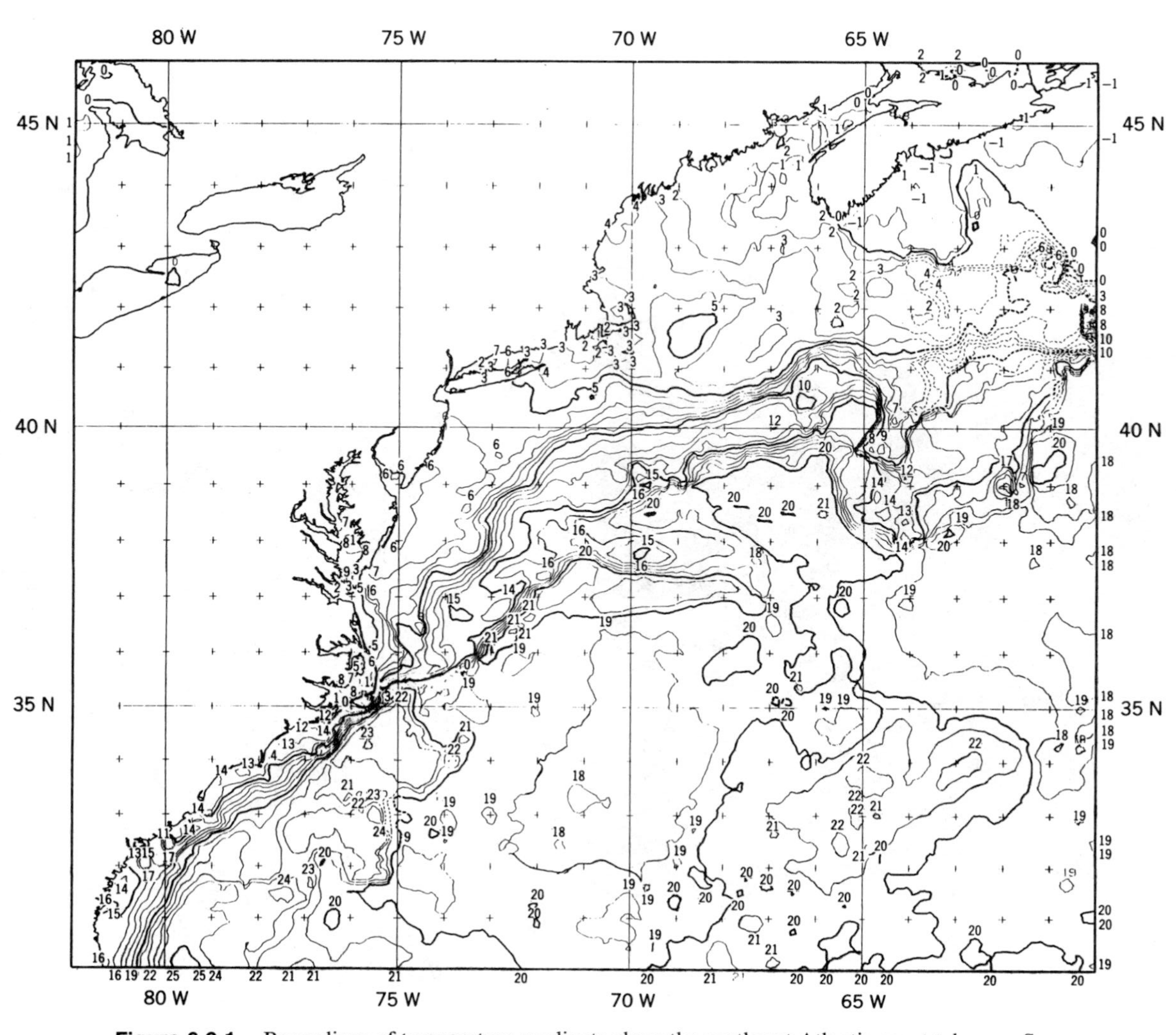

Figure 6-2-1. Recordings of temperature gradients along the northeast Atlantic coast taken on September 2, 1986. Courtesy of NOAA.

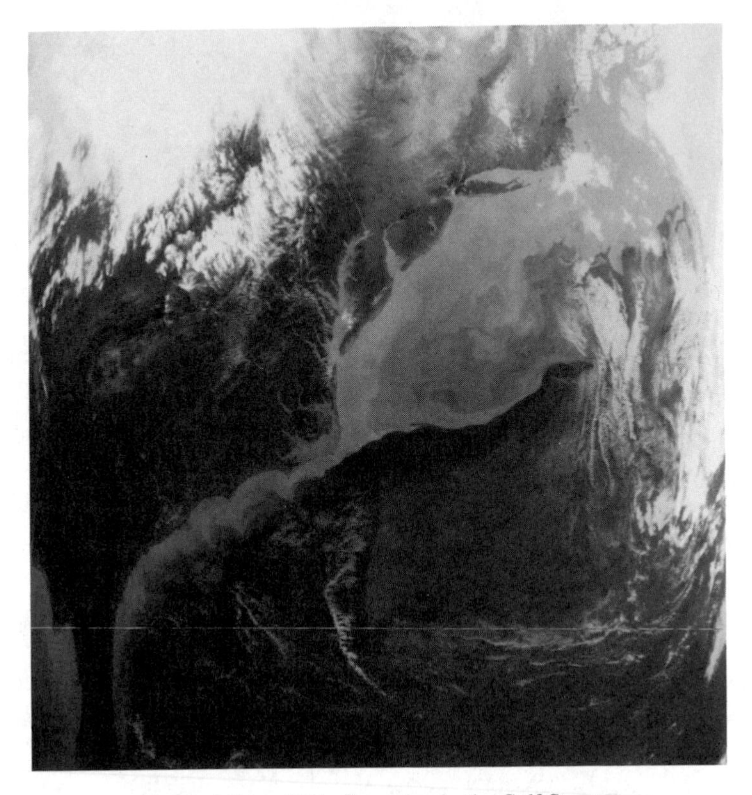

Figure 6-2-2. Eddy formation in the Gulf Stream.

been responsible for producing the meanders (De Rycke and Rao, 1973). It was believed that strong surface winds transport the cool shelf water toward the warm, northward-flowing Gulf Stream, and that the interaction between the two water masses moving in different directions sets up the development of meanders and eddies at the boundary.

Bottom topography and coastline geometry might also influence the generation of these large meanders, and there is a close relationship between the locations of the meanders and eddies and the projections of the capes along the coastline of North and South Carolinas (Stumpf and Rao, 1975). The observations show clearly that the eddies form from elongated and cutoff meanders.

Using a NOAA VHRIR (very high resolution infrared radiometer), Legeckis (1979) showed a persistent seaward deflection of the western boundary, which was found southeastward of Charleston, South Carolina, in the vicinity of a bulge in the continental slope. It was demonstrated that the maximum deflection angle was time-dependent and varied from 60 to 120° from true north. Downstream from the initial deflection, the western boundary was often wavelike, and there were several distinct and repeatable wave patterns. The separation between adjacent wave crests (wavelength) averaged 150 km, and the waves appeared to move northward at an average phase speed of 40 km$\cdot$day^{-1}. Monthly frequency distributions of wavelength showed that values range from 90 to 260 km downstream from the deflection, and wavelengths increased between February and April.

A cold eddy was observed with satellites from August 31, 1973 to April 1, 1974 (Vukovich, 1976). The genesis of the eddy took place at around 36°N, 68°W, and the dissipation of the eddy took place somewhere near 33°N, 74°W, suggesting that dissipation of the eddy was manifested through absorption by the Gulf Stream rather than through dispersion of heat mass. The average speed of the cold eddy was about 5.6 km·day^{-1}. The analysis of satellite data indicated that the cold eddy was elliptically shaped, with the major axis varying from 120 to 180 km and the minor axis varying from 100 to 120 km. The analysis also suggested that the circulation of the eddy was entraining warm Gulf Stream water, strengthening the warm ring around the eddy. Movement of a ring was also observed by Cheney et al. (1976) over a period of 11 months, using ships, floats, aircraft XBT (expendable bathythermo graphs) surveys, and satellite observations.

Cold eddies have been observed to move southwestward at approximately 1 to 2 mi·day^{-1}. These additional observations, taken in conjunction with other reported observations (such as Parker, 1971), suggest that perhaps the southwest movement of Gulf Stream eddies just offshore of the Gulf Stream is a common phenomenon, with two eddies per year following this path. The long lifetime of such eddies is demonstrated with satellite measurements in Figure 6-2-3, and

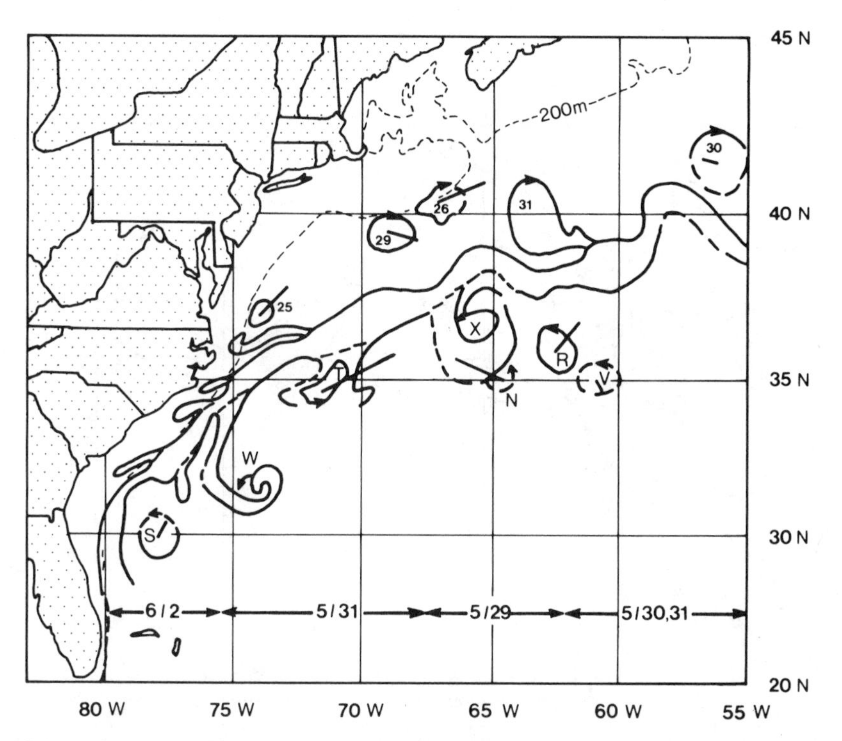

Figure 6-2-3. Sequence of eddies connected with the Gulf Stream observed over a time period of about four months. Eddy T used as an estimate for the life-span of an eddy, disappeared after June 28, 1983.

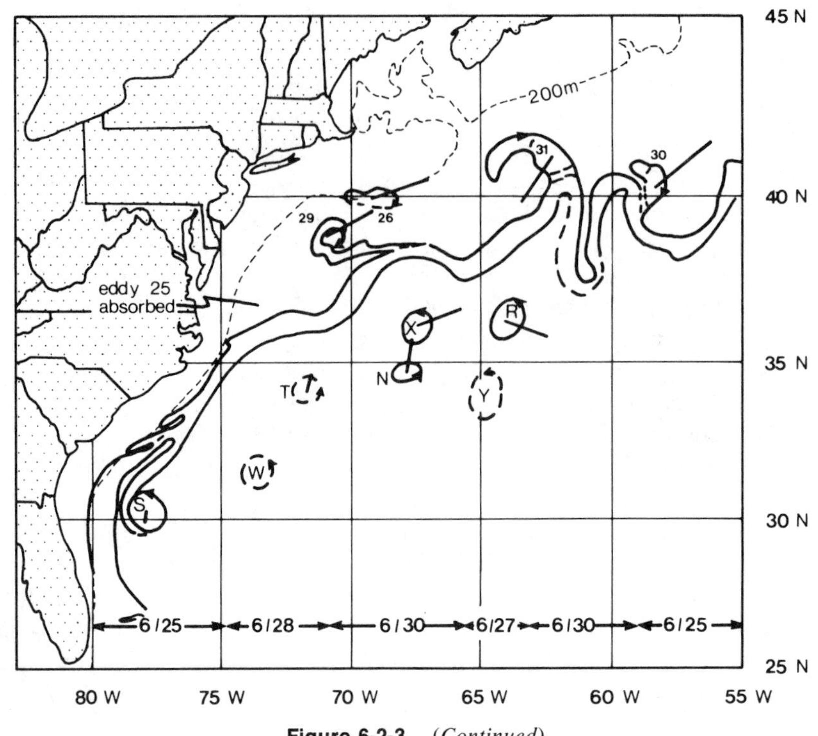

Figure 6-2-3. *(Continued)*

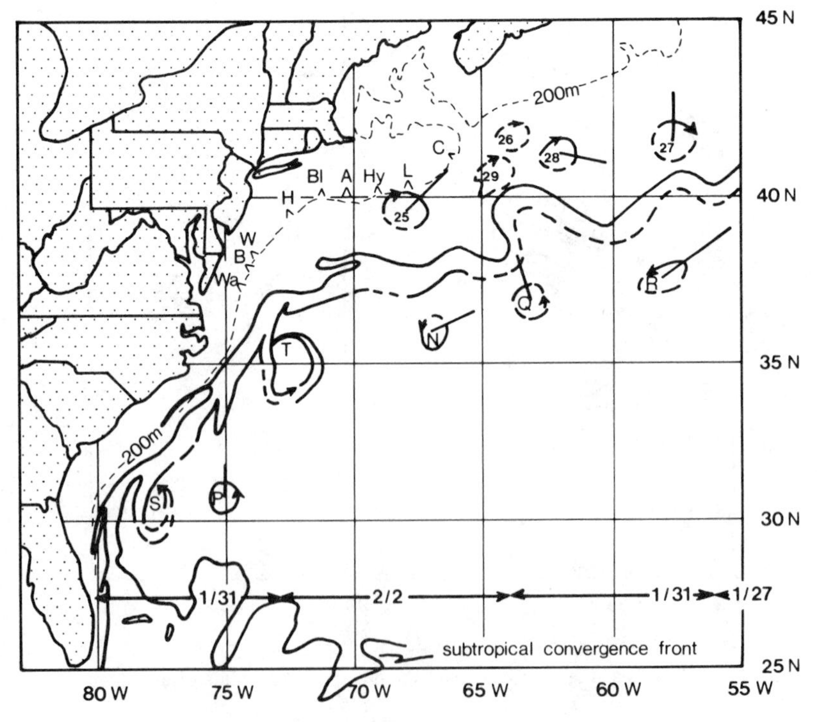

Figure 6-2-3. *(Continued)*

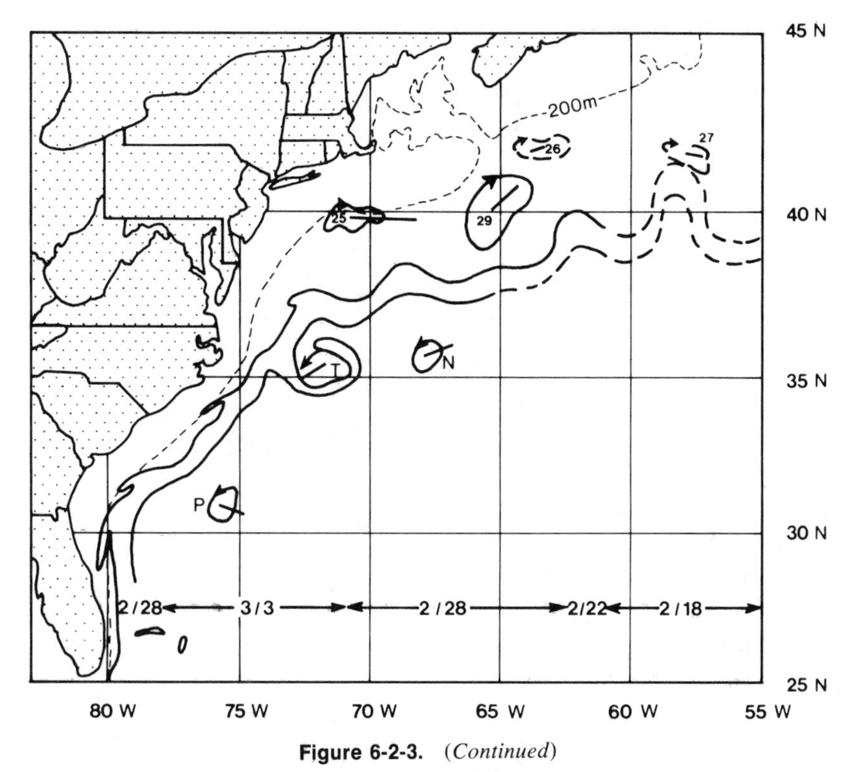

Figure 6-2-3. (*Continued*)

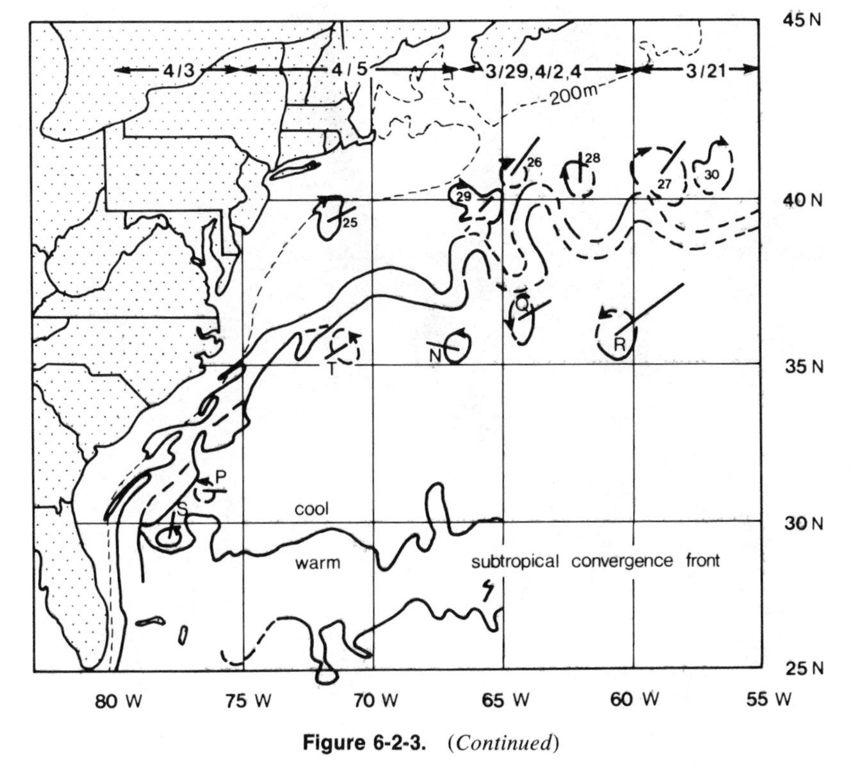

Figure 6-2-3. (*Continued*)

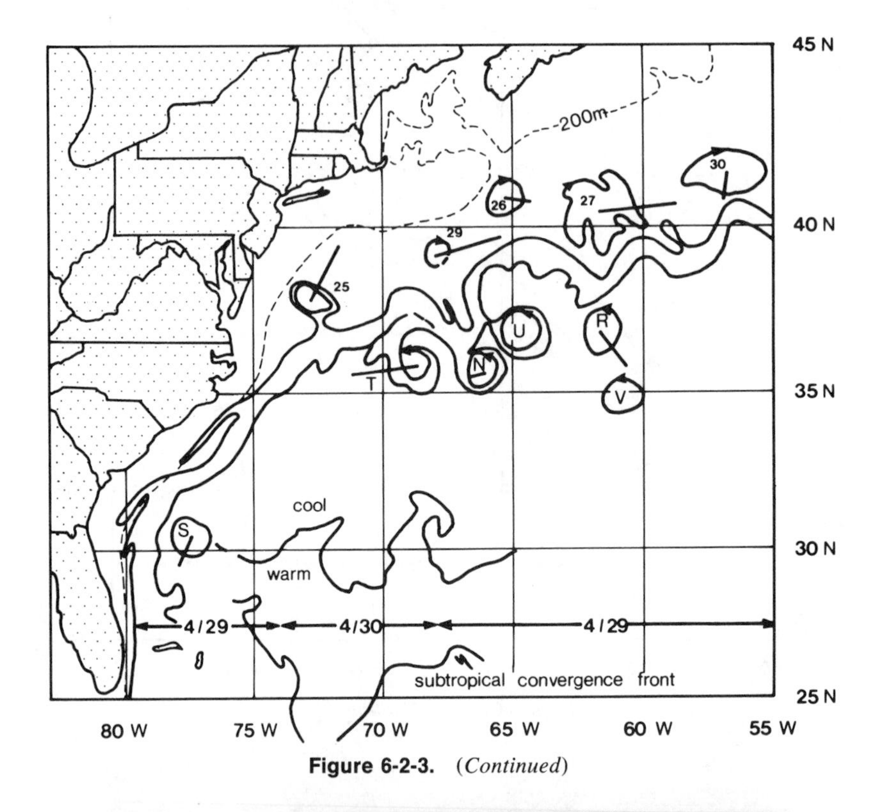

Figure 6-2-3. (*Continued*)

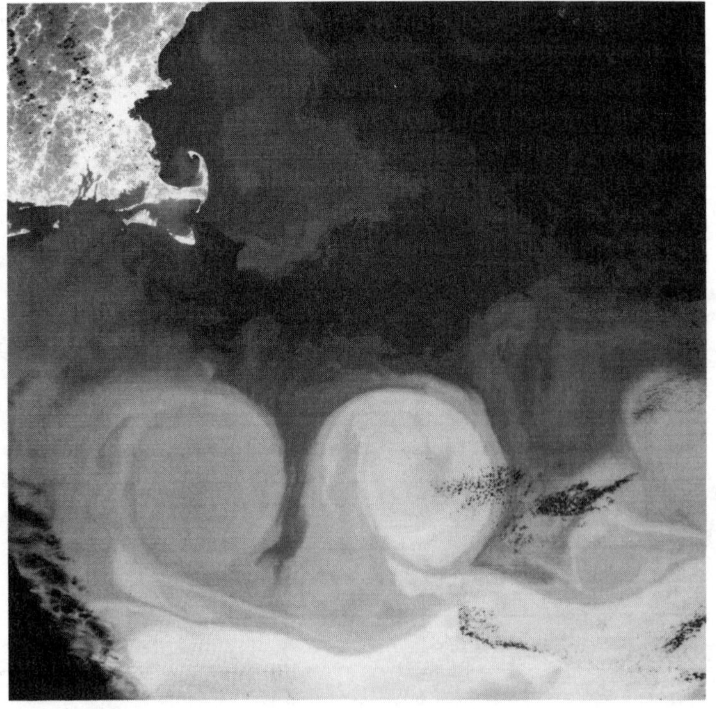

Figure 6-2-4. Infrared image of Gulf Steam eddy. Courtesy of NASA.

Figure 6-2-4 gives details in the thermal surface manifestations of a Gulf Stream eddy.

The generalized but complicated structure of the Gulf Stream system can be given with an analysis of the CZCS for chlorophyll estimates and SST (Fig. 6-2-5). It is particularly the warm Gulf Stream eddy, with its low pigment concentrations and the highly eutrophicated coastal region that can be recognized.

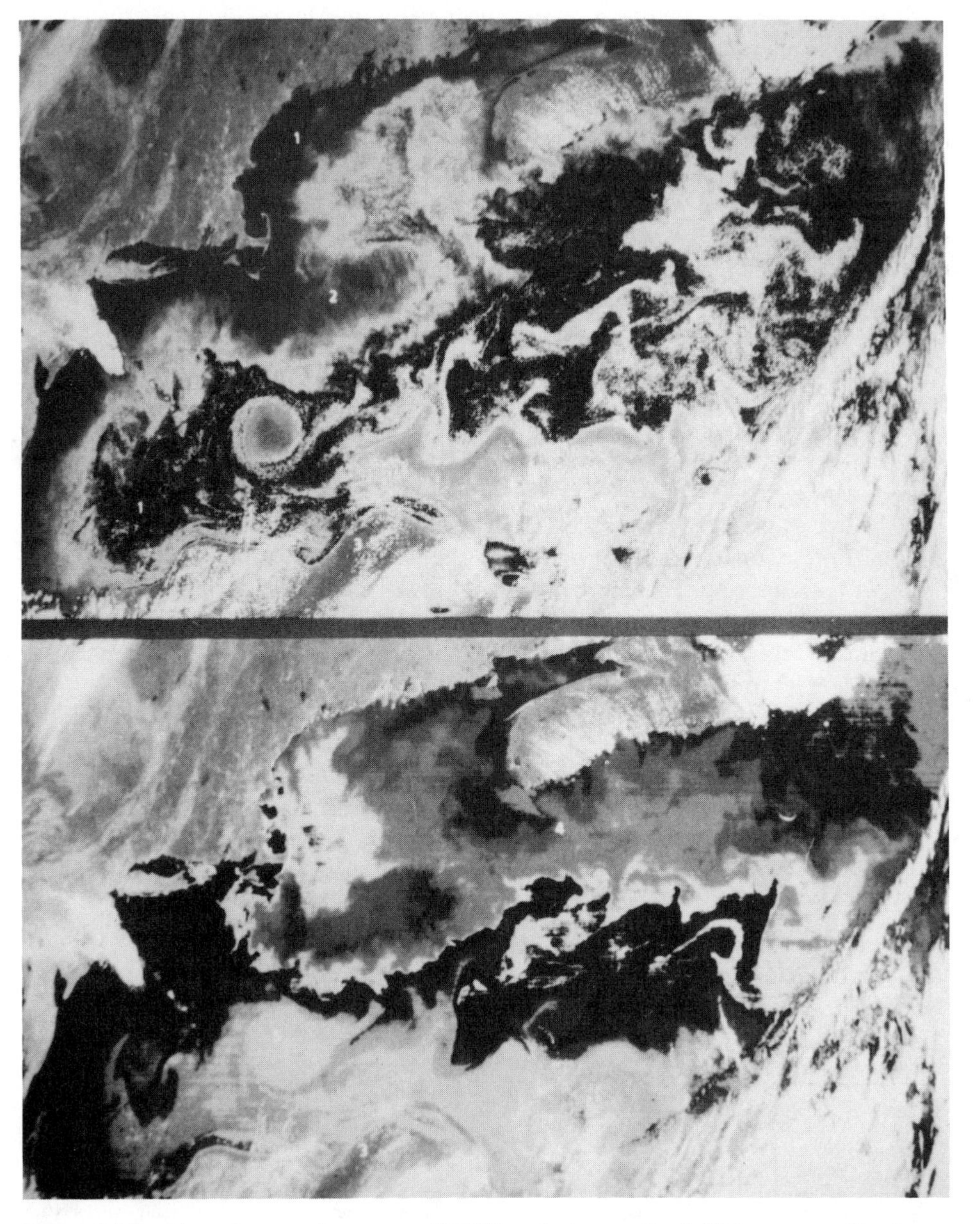

Figure 6-2-5. Coastal zone color scanner (CZCS) analysis over the Gulf Stream area. Top image shows chlorophyll concentration; bottom shows SST distribution. Courtesy of NASA.

The distribution of phytoplankton in connection with the Gulf Stream system shows the high diversity of phytoplankton distribution in the coastal and offshore region from Cape Hatteras to the Gulf of Maine. The region in Figure 6-2-6 can be subdivided into four areas, based on pigment content. Area 1 consists of shallow coastal waters (less than 25 m) that extend from Cape Hatteras to Cape Cod and over Nantucket Shoals and Georges Bank. In this area the pigment concentrations are generally greater than 2 $mg \cdot m^{-3}$. Area 2 consists of deep coastal waters extending from the edge of the continental shelf, circumnavigating Georges Bank, into the eastern regions of the Gulf of Maine. Pigment concentrations in this area range from 0.2 to 0.5 $mg \cdot m^{-3}$. These concentrations are generally lower than the values found on either side of this area. Area 3 encompasses all slope and basin waters greater than 100 m deep, with the seaward limit being the cold wall of the Gulf Stream. It is in this area where the largest diversity in pigment concentration occurs; ignoring the intrusive features, the pigment content ranges from 0.5 $mg \cdot m^{-3}$ to greater than 2 $mg \cdot m^{-3}$. Area 4 is characterized by the most prominent feature observed in the image, the frontal, cold North Wall of the Gulf Stream. This strong frontal feature extends from Cape Hatteras and meanders eastward to the waters of the open North Atlantic. Some of the Gulf Stream meanders form eddies, which are large rotating cells of water composed of Gulf Stream, Sargasso Sea, and slope water. An anticyclonic eddy consisting of a central core of Gulf Stram or Sargasso Sea water surrounded by slope water is seen in Area 3. A cyclonic eddy is also seen in the lower part of the image. This ring was spawned from an encirclement of coastal water by Sargasso Sea or Gulf Stream water.

Brown et al. (1985) used the CZCS to estimate the bloom magnitude during the onset of seasonal stratification due to the warming of surface water in spring for a Gulf Stream ring. Blooming occurred at two occasions, during an interval of about 10 days, showing pigment concentration of less than 0.6 $mg \cdot m^{-3}$ with increasing concentrations during the later part, while shelf and slope water showed pigment concentrations higher than 2 $mg \cdot m^{-3}$.

An areal estimate of total chlorophyll for the investigated region over the bloom period is given in Table 6-2-1, which also takes into account the different water masses and their productivity based on pigment concentration derived from satellite observations.

TABLE 6-2-1. Satellite-Derived Pigments (C_K and C_T) and Productivity (P_T) Values with Error Estimates[a]

Region	$\langle \bar{C}_K \rangle$ ($mg \cdot m^{-3}$)	$\langle \bar{C}_T \rangle$ ($mg \cdot m^{-2}$)	Area (10^{11} m^2)	$C^i = \iint C_T \, dt \, dA$ (10^5 metric tons)	$C^i/\Sigma C^i$ (%)	$\langle \bar{P}_T \rangle$ ($mg \cdot m^{-2} \cdot day^{-1}$)	$\iint P_T \, dt \, dA$ (10^6 metric tons)
Shelf	2.8 ± 1.7	65.7 ± 15.5	1.2	2.21 ± 0.52	46	924 ± 118	3.11 ± 0.40
Slope	3.0 ± 1.8	65.5 ± 14.4	1.2	2.20 ± 0.48	45	916 ± 103	2.82 ± 0.32
Warm-core ring	0.5 ± 0.3	28.9 ± 11.5	0.1	0.08 ± 0.03	2	529 ± 159	0.30 ± 0.09
Gulf Stream	0.3 ± 0.2	21.2 ± 9.0	0.6	0.35 ± 0.15	7	423 ± 138	0.71 ± 0.23

[a] Error estimates are based on the assumption that present algorithm inaccuracy is a factor of 2. Daily estimates are halved, doubled, and integrated, respectively, to produce a range for each parameter. This range is used to derive the error estimates.

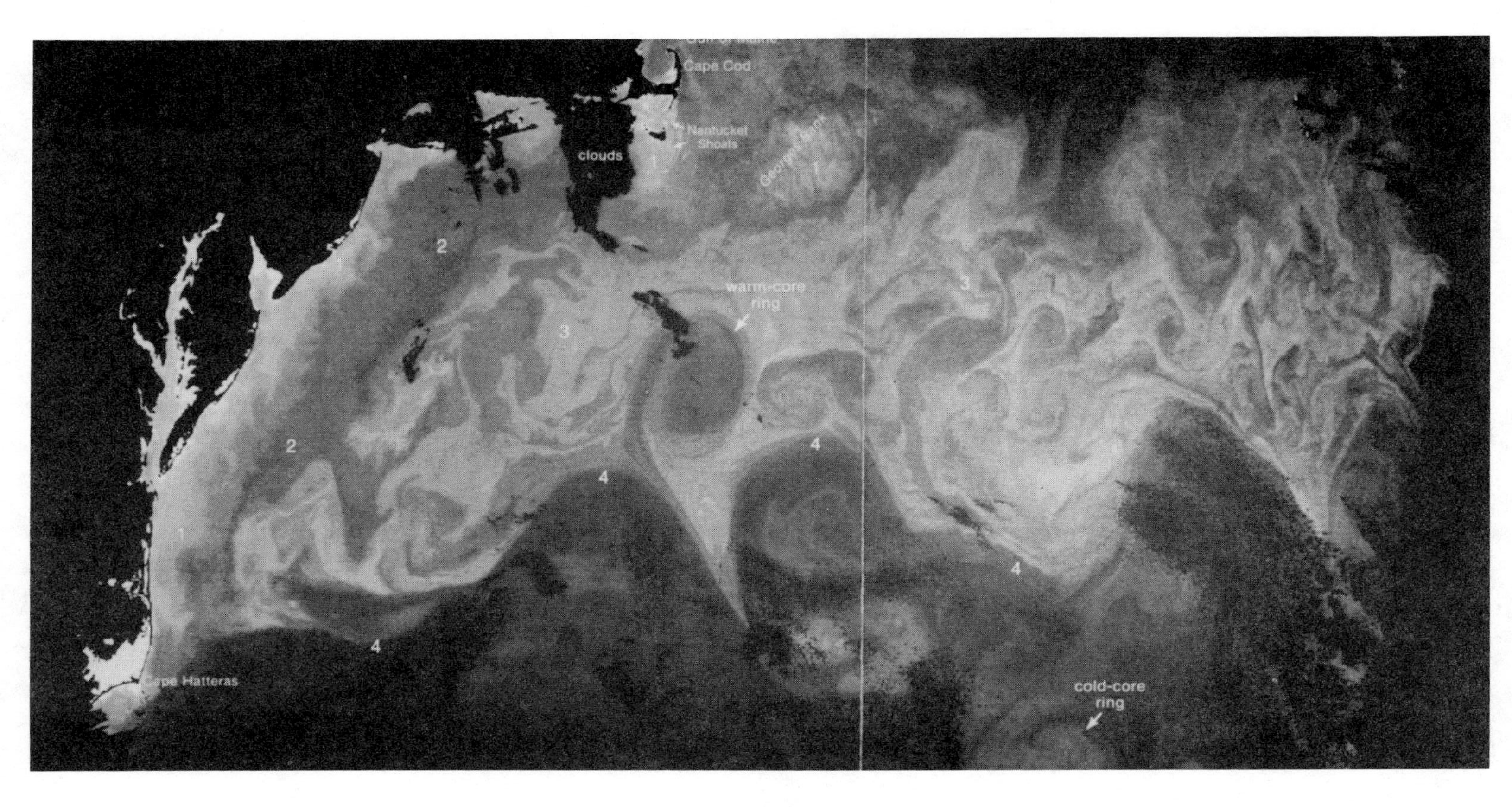

Figure 6-2-6. Pigment concentrations as recorded in coastal and offshore regions of Cape Hatteras to Gulf of Maine. For explanation on subdivided areas, see text. Courtesy of NASA.

6-2-2 Labrador Current

The Labrador Current is a density current with a southward direction and transports in part water of the north and westward-moving West Greenland Current. Originating at the David Strait, the joint current is then reinforced to the south by additional inflow through the Hudson Straits and the Belle Isle Straits. South of Newfoundland, the juncture with the northeast-moving branch of the Gulf Stream is at about 42°N latitude and 50°W longitude. An example of historical measurements made of the area, for the period of observations in November, shows that the mean SST difference between the Labrador Current and the North Atlantic Current is about 10 to 12°C (see Fig. 6-2-7).

Figure 6-2-8 shows an IR image depicting the main features of the Labrador–Gulf Stream system. Of particular interest is the very narrow cold band associated with the Labrador Current and its meandering before deviating into a southwestward direction. The IR recordings indicate that there seems to be no continuous inflow of the Labrador Current water or integration into the circulation scheme but rather discontinuities, and that there are injections of isolated Labrador water added to the North Atlantic Current. Also visible is the oscillating nature of the Gulf Stream boundary with eddies on both sides of the Gulf Stream axis. The hypothesis on the formation of cold Labrador rings generated at the edge of the junction of the Labrador and North Atlantic Currents can be supported with IR recordings from other dates.

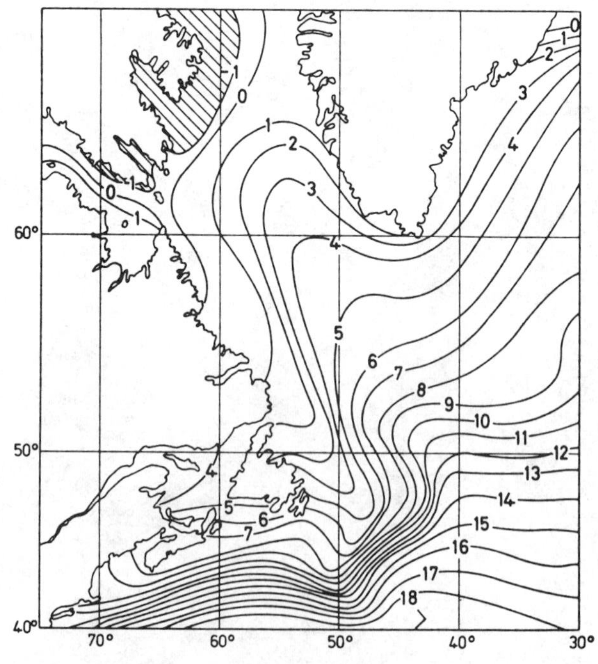

Figure 6-2-7. SST for the region under the influence of the Labrador Current and the Gulf Stream for the month of November.

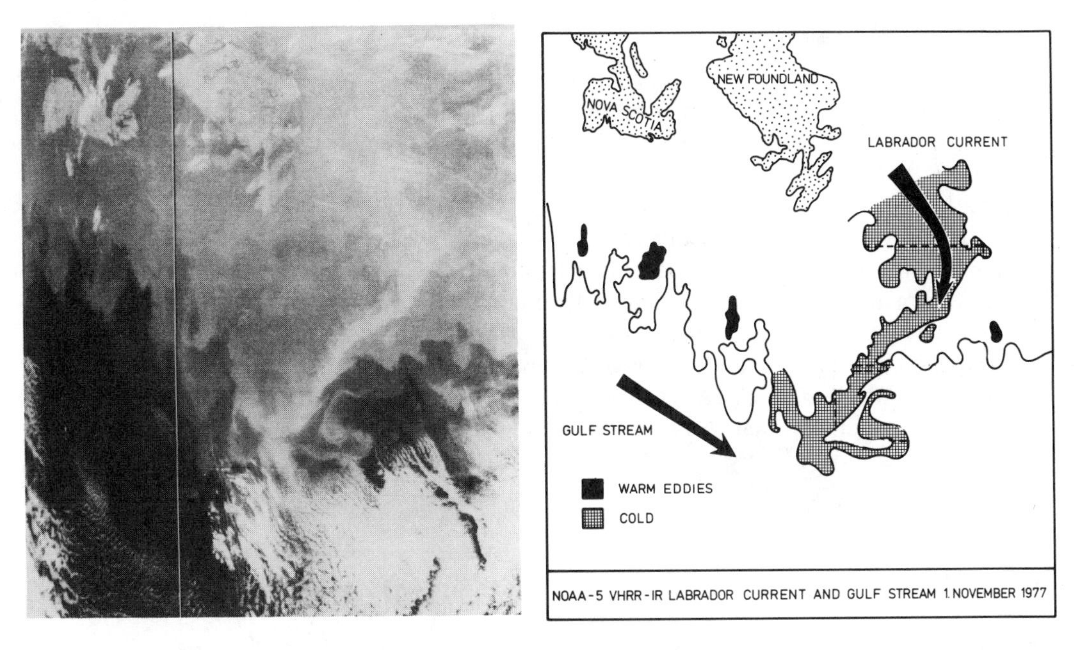

Figure 6-2-8. Infrared image of the Labrador–Gulf Stream System.

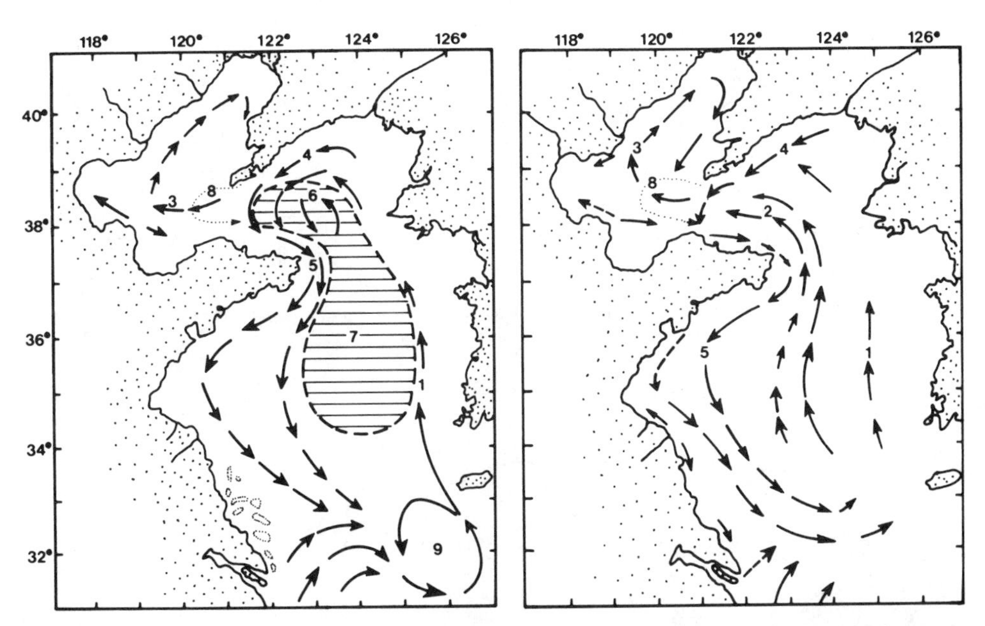

Figure 6-2-9. Model of current systems in the Huanghai Sea and Bohai Sea. *Left:* Model of current system in summer (June-August). 1. Warm current of the Huanghai Sea; 2. remains of warm current of the Huanghai Sea; 3. circulation of the Bohai Sea; 4. coastal current of Lianonan; 5. coastal current of the Huanghai Sea; 6. density circulation of the northern Huanghai Sea; 7. cold water mass of the Huanghai Sea; 8. density current of the Bohai Sea; 9. counterclockwise circulation with upwelling. *Right:* Model of current system in winter (December-March). After Szekielda and McGinnis (1985). Reproduced with permission from Prof. Dr. E. Degens, University of Hamburg.

6-2-3 Huanghai Sea

The Huanghai Sea and the oceanographic situation of the East China Sea are highly connected, and changes of this system are predominantly influenced by the seasons and prevailing wind conditions. The most important influence on the hydrography of the Huanghai Sea is the warm Kuroshio Current coming from the south.

The current patterns in the East China Sea may be roughly divided into two parts, with the division from the northernmost point of Taiwan to Chejudo Island. On the eastern side of this division is the main body of Kuroshio (entering the East China Sea through the passage northeast of Taiwan) and its branches, namely the Tsushima Warm Current (flowing northward and eventually entering the Japan Sea) and the Huanghai Sea Warm Current flowing northwestward and entering the southern Huanghai Sea. A part of the water of the Kuroshio entering Bashi Strait flows northward closely along the west coast of Taiwan. In the area between the

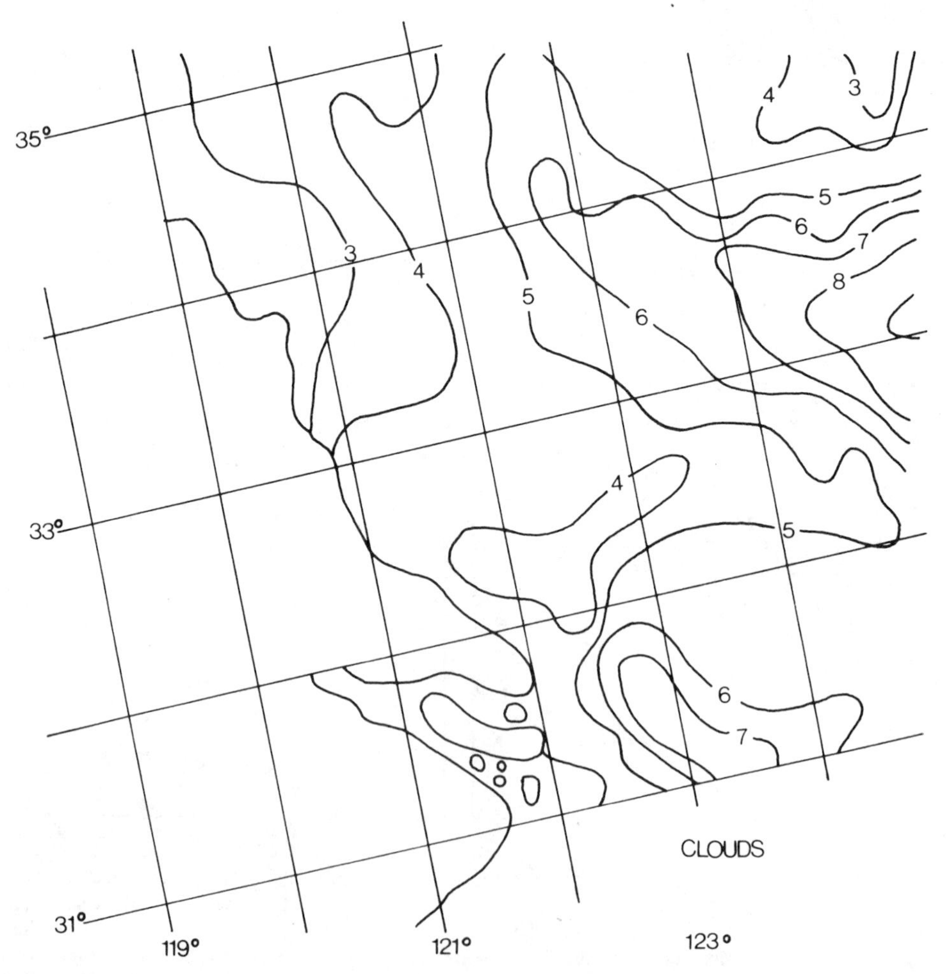

Figure 6-2-10. Temperature distribution on March 9, 1982 based on satellite IR measurements. Contours have been smoothed. After Szekielda and McGinnis (1985). Reproduced with permission from Prof. Dr. E. Degens, University of Hamburg.

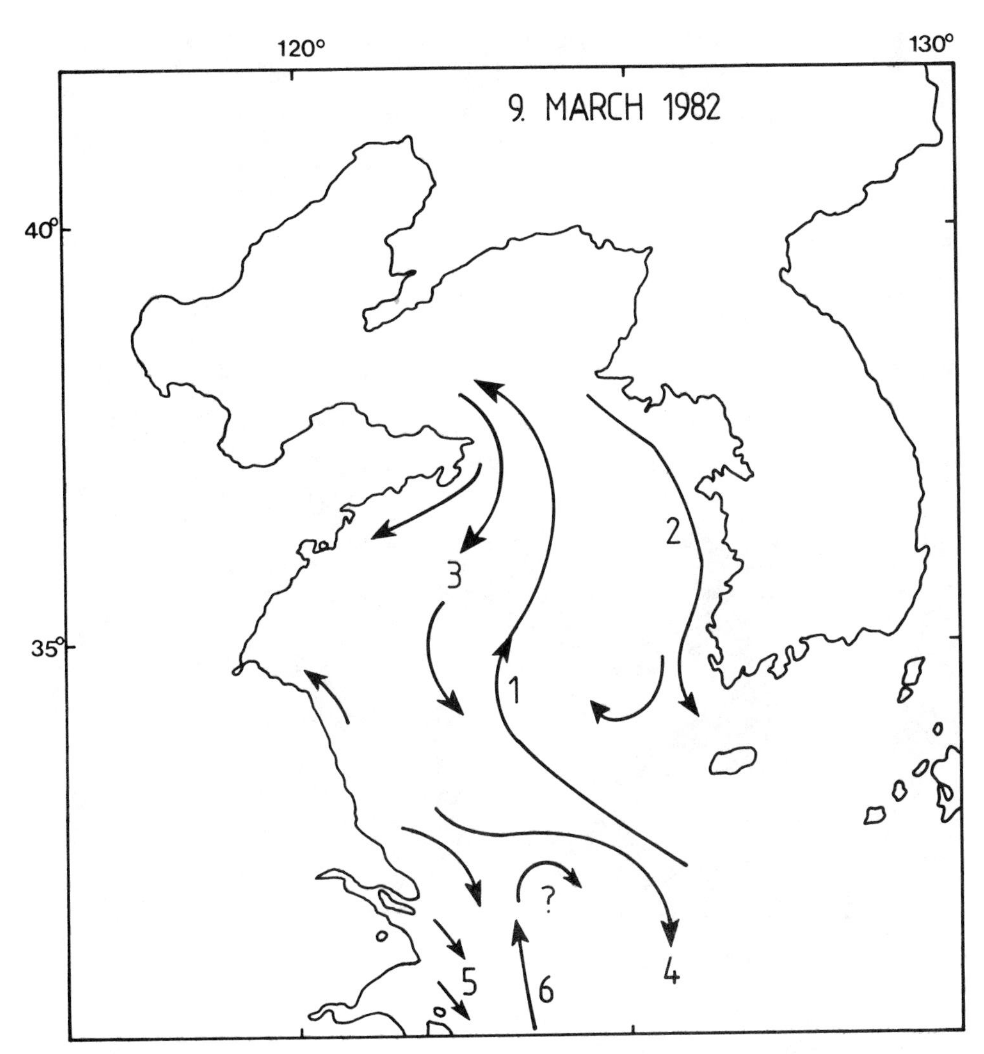

Figure 6-2-11. Current system based on observations in the visible and thermal IR on March 9, 1982: 1. Huanghai Sea warm current; 2. West Korean longshore current; 3. Huanghai Sea longshore current; 4. water mass with low temperature and high particle load; 5. river flow; 6. Taiwan warm current. After Szekielda and McGinnis (1985). Reproduced with permission from Prof. Dr. E. Degens, University of Hamburg.

main body of the Kuroshio and Ryukyu Island is what may be called the Kuroshio Countercurrent, which flows southwestward all year round. On the western side of the division, the current flows from southwest to northeast along the direction of Taiwan Strait and the coastline of Fujian and Zhejiang. The part of the current from north of the Strait to the mouth of the Changjiang is the Taiwan Warm Current. To the northeast of the mouth of the Changjiang, there exists in summer the "Changjiang diluted flow" at the surface layer, under which the Taiwan Warm Current enters (Fig. 6-2-9).

A temperature map for March 9, 1982 in Figure 6-2-10 shows the effect of the major current system with a branch of the Kuroshio carrying warm water of about 9°C to the north. Most probably under the influence of the northern winds, cold

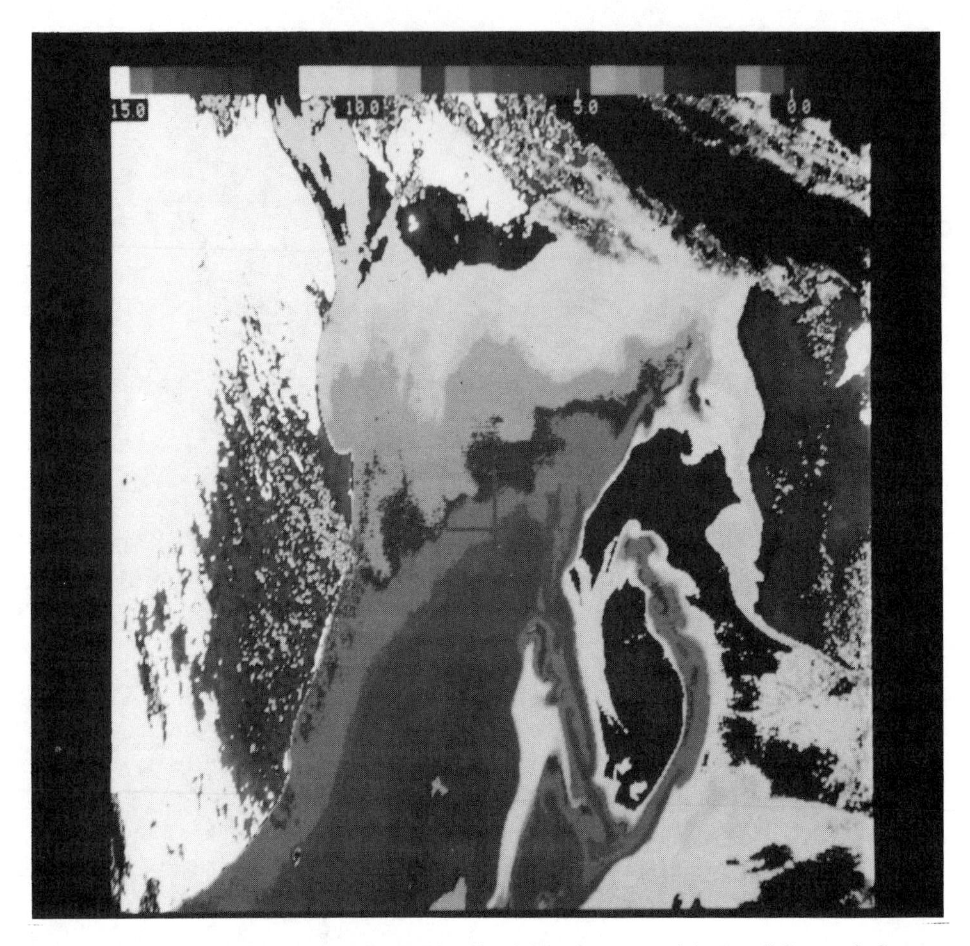

Figure 6-2-12. Infrared image over the Falkland/Malvinas Current and the Brazil Current in September 1983.

water is carried by the coastal current of the Huanghai Sea, with local low temperature less than 4°C when the current leaves the coast, creating upwelling. The outflowing water from the Changjiang is marked temperatures lower than the adjacent water. the current scheme can be derived from observations made in the visible and thermal IR, as given in Figure 6-2-11.

6-2-4 Falkland/Malvinas and Brazil Currents

The Brazil Current is known as a tongue of water with high temperature and high salinity. Close to the coast of Argentina, a branch of the Falkland/Malvinas Current carrying cold water forms a sharp temperature gradient where it meets the Brazil Current. The thermal gradient in the nearshore area, visible in Figure 6-2-12, reflects the coastline and the Rio de la Plata. Further offshore there is a second thermal discontinuity, which indicates the boundary between the southward-flowing warm Brazil Current and the Falkland/Malvinas Current, which transports cold water closer to the coast in a northerly direction.

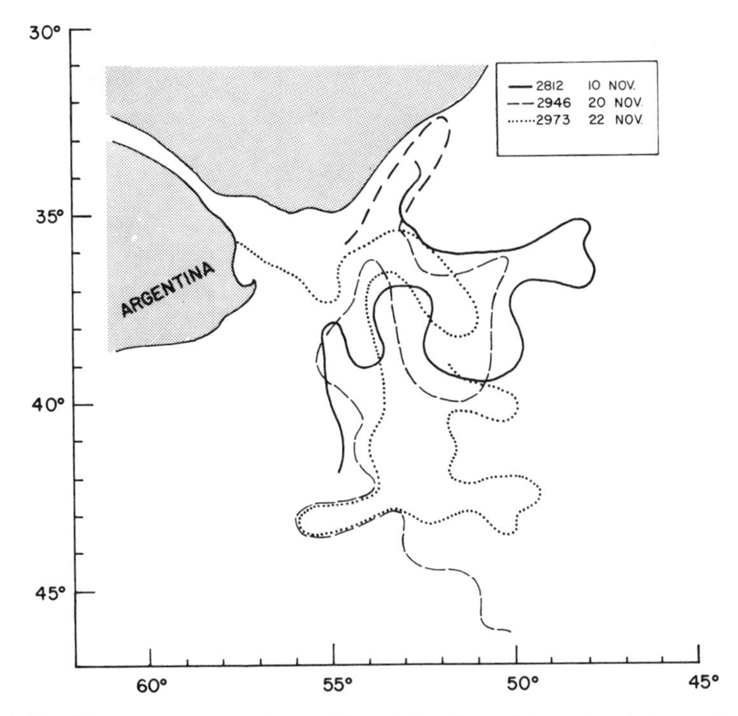

Figure 6-2-13. Time sequence of the position of the thermal boundary between the Falkland/Malvinas Current and the Brazil Current.

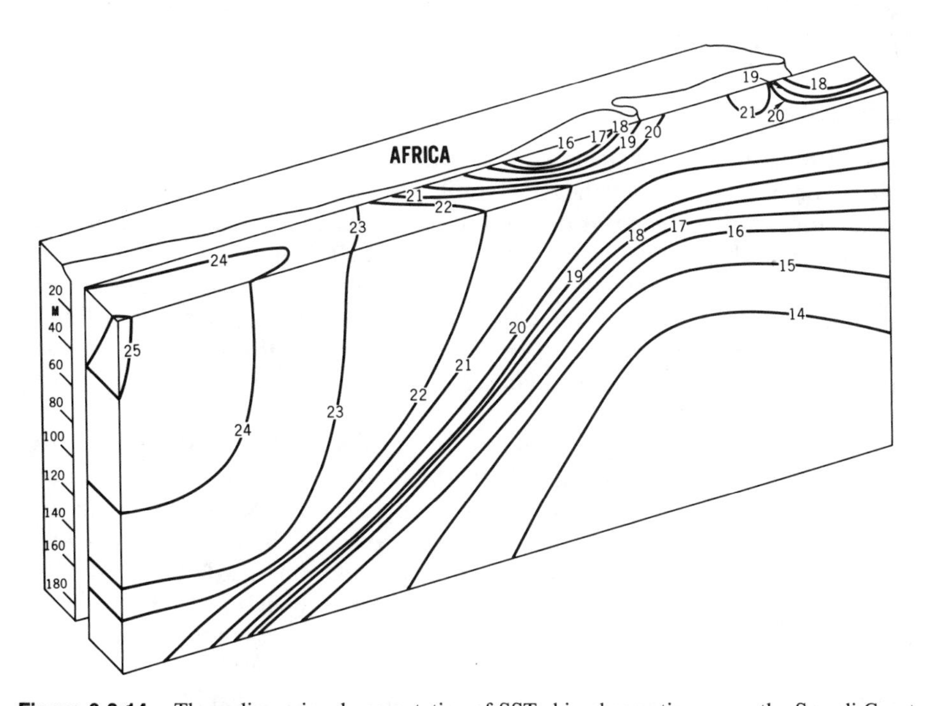

Figure 6-2-14. Three-dimensional presentation of SST ship observations over the Somali Coast.

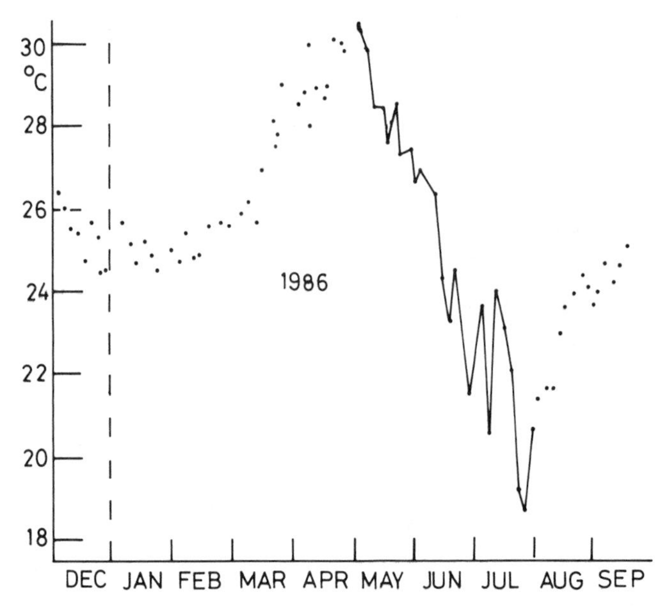

Figure 6-2-15. SSTs in upwelling region of Somali Coast.

Figure 6-2-13 shows the location of boundaries as estimated from IR imageries, and it is evident that the general structure of gradients change significantly over several days. Comparing the data from recordings obtained from different dates, one may estimate that the boundaries can fluctuate with a velocity of 5 $mi \cdot day^{-1}$ and that fluctuations in the wind system are the major causes for the observed oscillation of the current boundaries.

6-2-5 Somali Current

In response to the wind system over the Arabian Sea, the Somali Current changes its direction of flow from northeast to southwest along the east African coast. The changes appear after the onset of the southwest monsoon and the northeast monsoon, respectively. In response to the onset of the southwest monsoon, the current builds up from essentially no transport to approximately $60 \times 10^6\ m^3 \cdot sec^{-1}$. The zone of extremely high velocities greater than 300 $cm \cdot sec^{-1}$ is within the upper 50 m of the Somali Current. In response to the southwest monsoon, strong upwelling exists along the Somali coast, as Figure 6-2-14, based on conventional ship operations, shows.

Temperature changes in the center of the upwelling, close to Ras Hafun, were monitored with the HRIR and the MRIR (Medium Resolution Infrared Radiometer) in the *Nimbus* satellites. In order to compare the temperature readings from the radiometers with different ground resolutions in Figure 6-2-15, blackbody measurements were normalized to the ground resolution of MRIR, about 50 km, which also corresponds to the grid of MCSST data for the most recent observations. Figure 6-2-16 shows the recordings for different years in the same location.

The sharp temperature gradient associated with the upwelling area along the Somali coast may be easily monitored from space and can also be detected in the

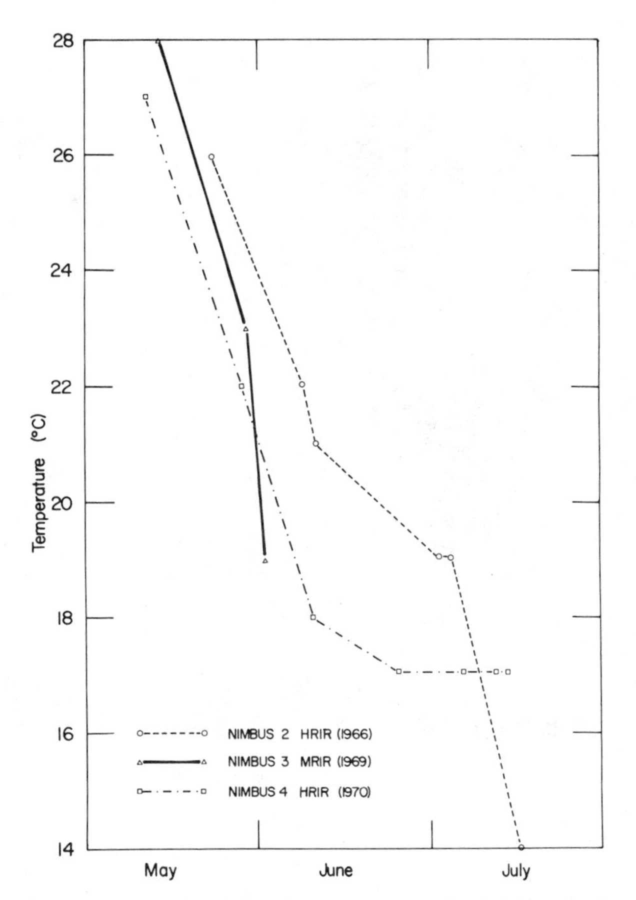

Figure 6-2-16. SST recordings taken in 1966, 1969, and 1970 along the Somali Coast.

imagery material of IR recordings. For instance, the development of the upwelling can be recognized with IR images, of which an example is shown in Figure 6-2-17. The time-dependent development of horizontal temperature gradients at the sea surface serves as an indicator for the formation of the baroclinic structure of the Somali Current. A comparison of SST gradients has been made with simultaneous developments of the southwest component of the monsoon winds. The results in Figure 6-2-18 show that the temperature gradients during the early formation stage are directly proportional to the wind speed. The phase lag between the development of wind and temperature gradient (i.e., upwelling) is surprisingly short during the development of the boundary current, ranging between 3 and 10 days. During the decay period in late summer and autumn, the temperature gradients lag 14 to 40 days behind the wind. The decreasing branch shows that the decay of the temperature gradients is considerably slower than their formation.

6-2-6 Cape Blanc

Between Cape Blanc and Cape Timiris, along the northwest coast of Africa, two main formations of water masses can be recognized. The first is within the upper

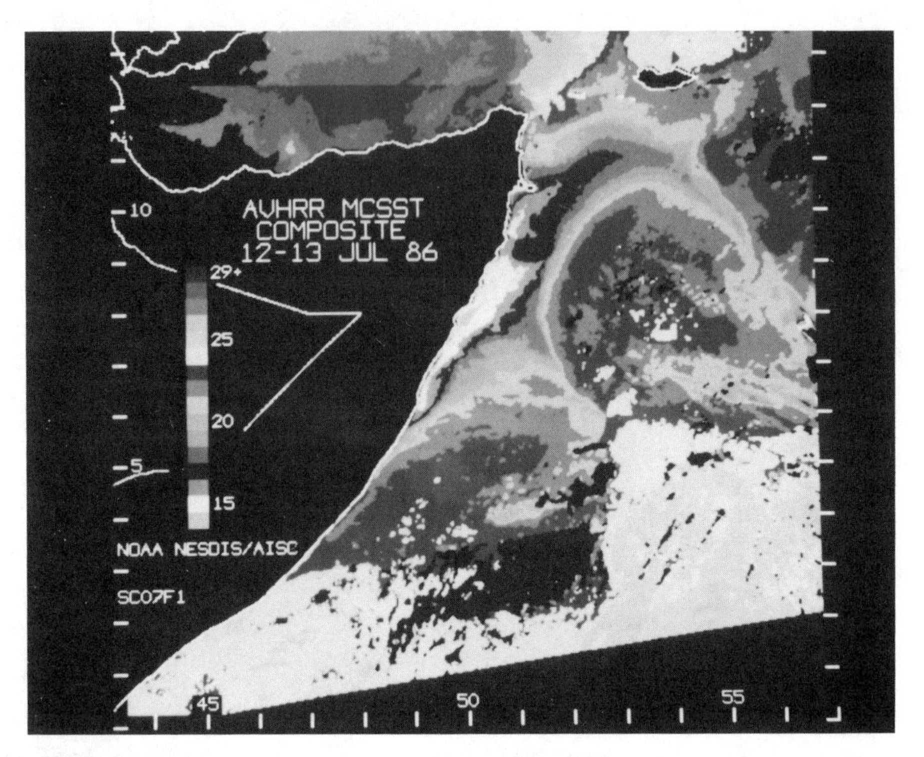

Figure 6-2-17. MCSST composite showing development of upwelling along the Somali Coast.

300 m and can be considered a mixture of surface water and Central South Atlantic water, with values of 12°C and 35.35‰ at 300 m. The layer between 400 and 800 m consists mainly of Central North Atlantic water from 12 to 8°C. The influence of Antarctic intermediate water is shown by water with salinity of 35.05‰ and a temperature of 7°C. Upwelling in this area is limited for water types with

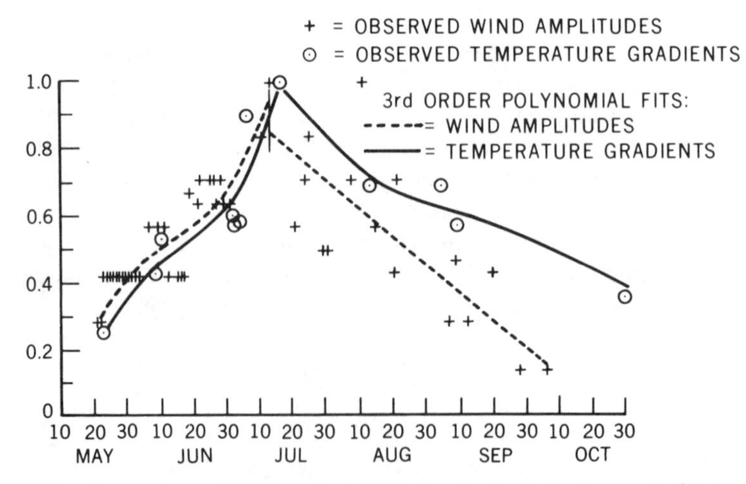

Figure 6-2-18. Comparison of SST gradients with simultaneous development of monsoon winds.

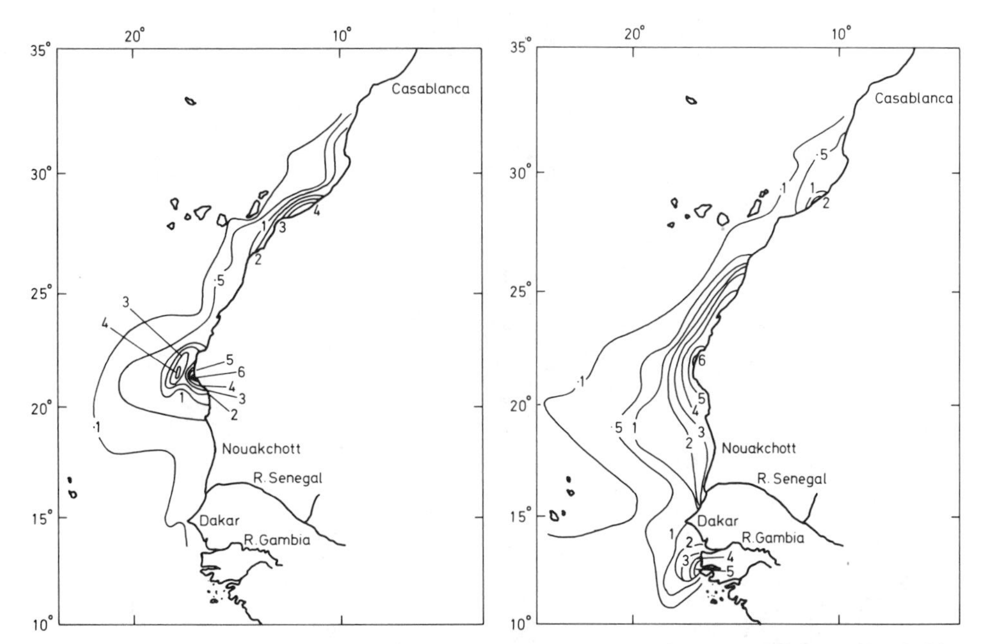

Figure 6-2-19. Composites of chlorophyll concentration in mg·m^{-3} as recorded from aboard ship. *Left:* January-June; *Right:* July-September. After Szekielda et al. (1977).

temperatures between 18.5 and 20.5°C and salinities between 35.8 and 35.95‰, which show that the upwelling has its origin in the upper 100 m. As a consequence of upwelling, high chlorophyll concentrations are observed throughout the year, as shown in Figure 6-2-19. The effect of increased particle concentrations is also reflected in *Landsat* data, and color gradients have been reported from the use of different sensors (see Fig. 6-2-20). The southern strong gradient in particulate material is connected with the converging water masses from the Banc d'Arguin and was detected during all four seasons. The gradient undergoes a fluctuation in space and time. Besides the larger fluctuations in the north, it is noticeable that the gradient shifts over only 10 mi.

The enclosed lines south of Cape Blanc are the relative positions of the suspended material patch observed from *Skylab 3* and *Skylab 4*. Between September 1973 and January 1974, the patch experience a broadening and shoreward movement, which may have been a result of seasonal changes in the circulation and in particle concentration.

Repetitive coverage with the recent NOAA satellites with LAC data at 2.5-km resolution allows the recording of more details on the structure and changes in the thermal bands (see Fig. 6-2-21). The CZCS has also given insight into the very dynamic features in temperature and pigment distribution (see Fig. 6-2-22).

6-2-7 California Coast

Strength and occurrence of upwelling in the California Current region appear closely related to the wind field. Strongest winds are observed off Baja California in April/May, off southern California in May/June, off northern California in

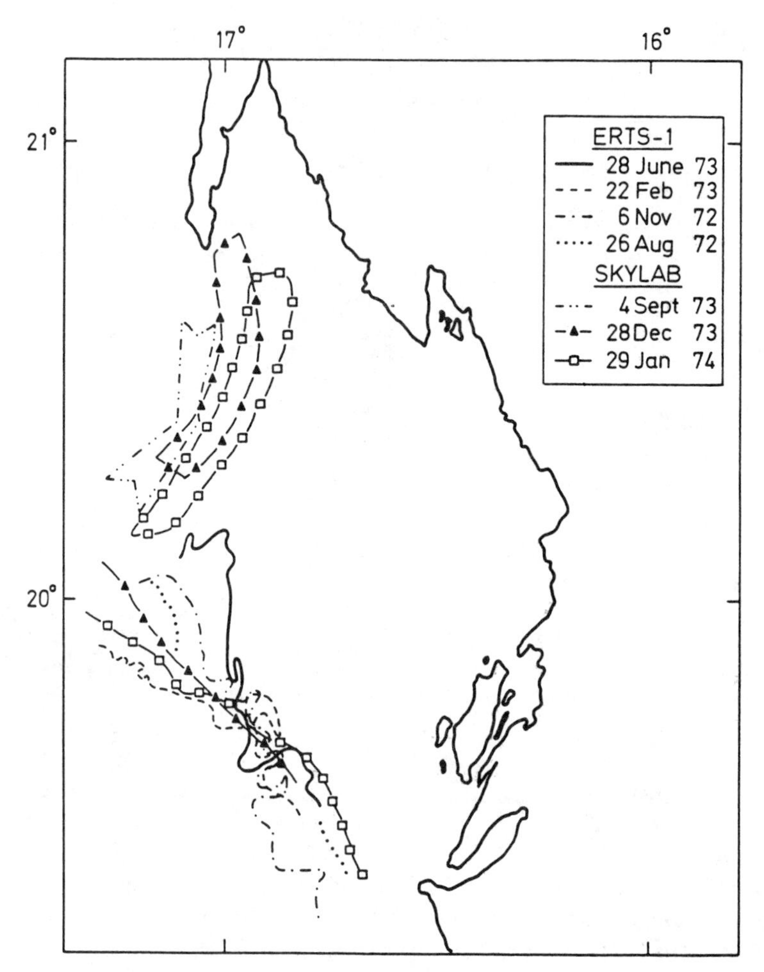

Figure 6-2-20. Offshore ocean color boundaries derived from *ERTS-1* and *Skylab*. After Szekielda et al. (1977).

June/July, and off Oregon and Washington in July/August. Consequently, high intense upwelling shifts in response to the northerly and northwesterly winds, which means a shift up the coast as spring and summer progress. In the fall and winter, the region north of Point Conception, California, experiences westerly or southwesterly winds under the dominance of the deepening Aleutian lows moving across the North Pacific and upwelling areas. In addition, nonseasonal occurrence of upwelling may appear, or the persistence of upwelling during the summer may be a result of variations in the wind field. Due to the wind field patterns, upwelling looks rather complicated, as evidenced in Figure 6-2-23. Satellite monitoring of ocean features is on an operational basis to identify boundaries as an input into fishery research and exploitation purposes. Figure 6-2-24 shows the thermal boundaries along the west coast of the United States, as determined from enhanced IR satellite imagery and ship and buoy observations.

Springtime warming, persistent positive SST anomalies, and the onset of significant coastal upwelling characterize the northern area of the region. K. Short and

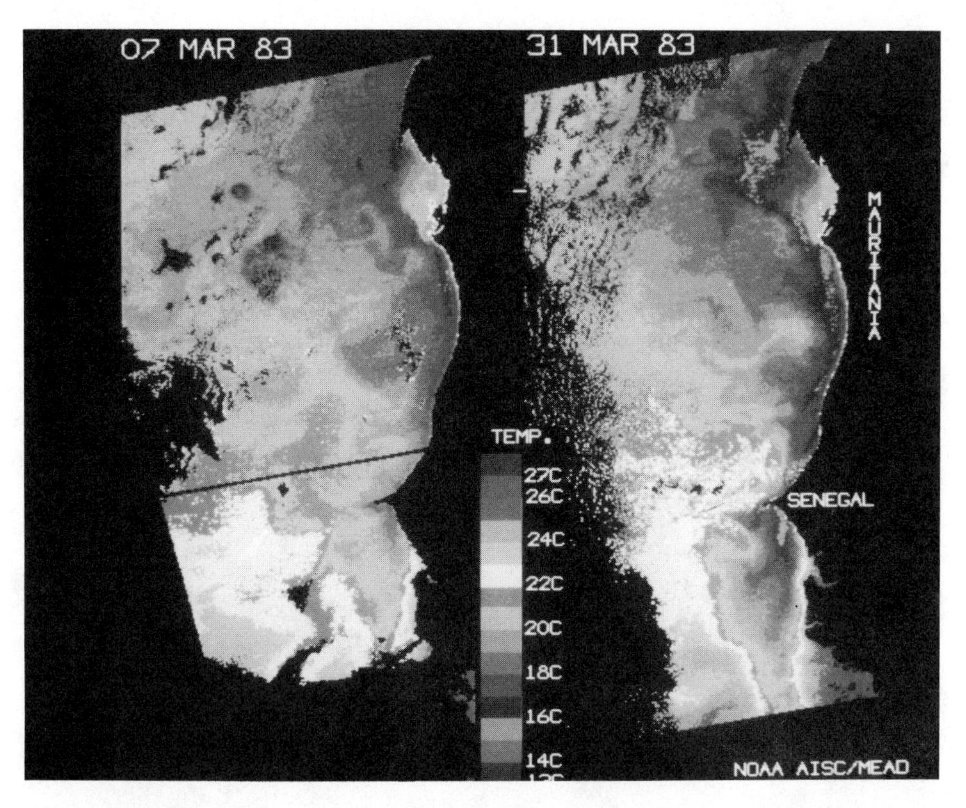

Figure 6-2-21. SST display for March 7 and March 31, 1983 LAC data at 2.5-km resolution. After Dennis et al. (1986).

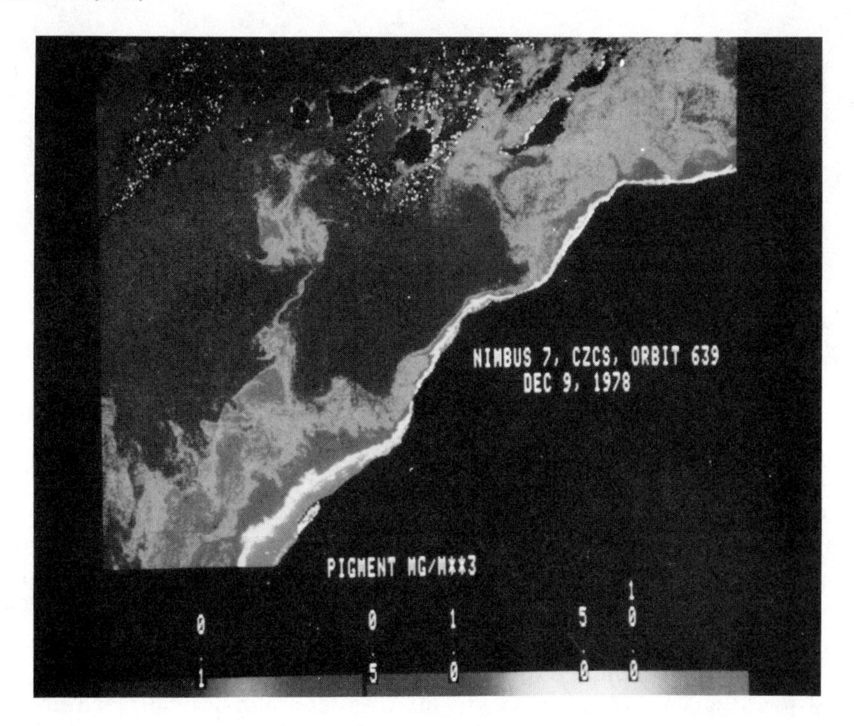

Figure 6-2-22. Estimates of photosynthetic pigments based on measurements with the CZCS on *Nimbus 7*.

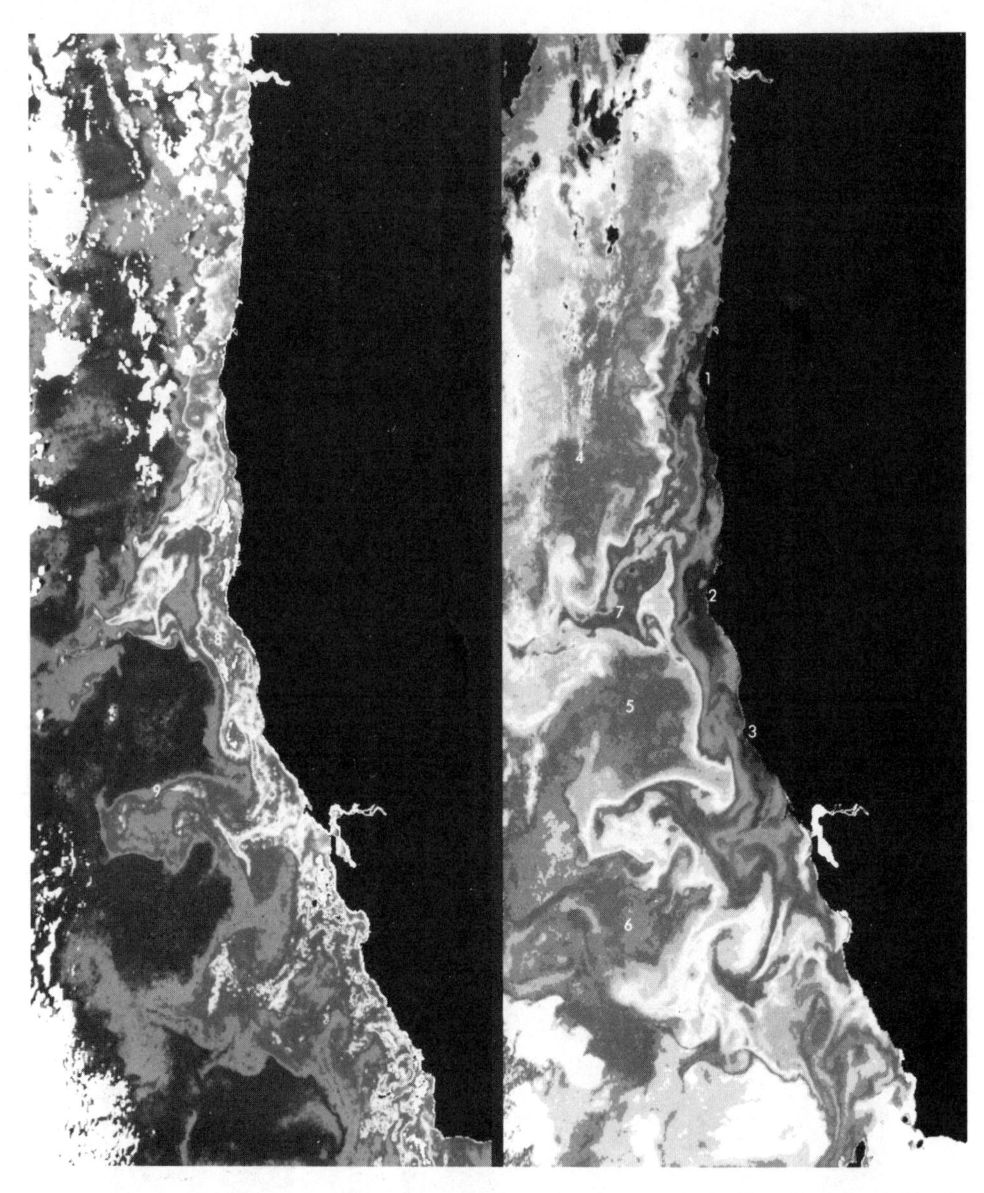

Figure 6-2-23. Temperature and pigment distributions in the coastal upwelling along the California coast, July 7, 1981. Courtesy of NASA. See insert for color representation.

Linder (1983) showed that SST increases over April values averaged about 1.5°C for both offshore and coastal waters. The overall coastal warming occurred in spite of a strong coastal upwelling event south of 45°N during May, during which a 1.1°C SST drop was recorded at 40.8°N, 124.5°W. The strongest upwelling-favorable winds were observed on May 20; and an associated ocean thermal boundary was seen on May 21 satellite imagery. Despite this event, upwelling was below normal for May for the entire coast, due to a less-than-normal mean northerly wind component.

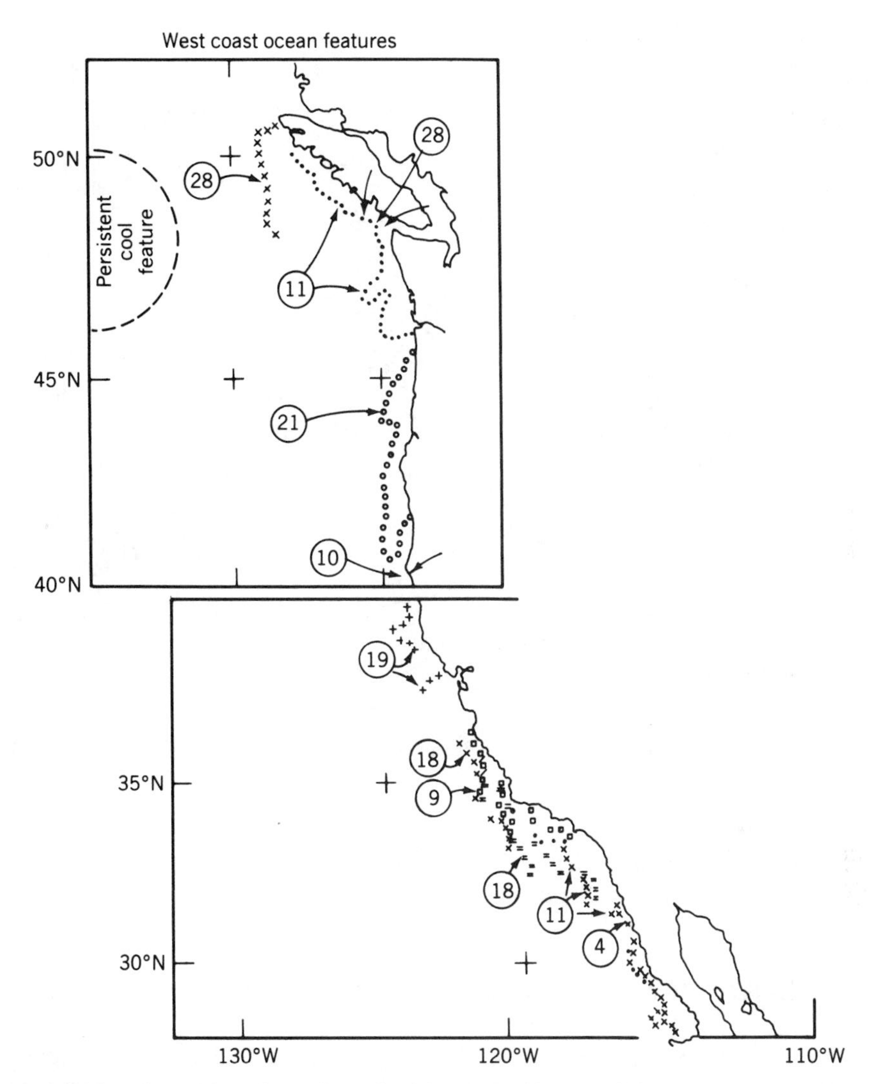

Figure 6-2-24. Thermal boundaries determined from a combination of enhanced IR satellite imagery and ship and buoy observations. Identification of the features by satellite is limited by cloud cover. Key: ·, ○, ×, ⊙, + = thermal boundaries; W = warm; C = cold; ③ = date of feature; ( ) = outline of water mass feature. After Short and Linder (1983).

In addition to the interpretation of IR data, ocean color boundaries have been used in the assistance of albacore tuna fishing activities along the west coast. Since pelagic tuna are associated with clear ocean water, but have the tendency to accumulate along fronts that separate the clear ocean water from the more turbid coastal or shelf water, the position of boundaries in the IR and visible data help identify potential fishing areas. Tuna, which are sight feeders, feed along these fronts as schools of coastal prey species move back and forth across the fronts. The NOAA National Marine Fisheries Service (NMFS) provides the tuna fishermen with ocean color boundary maps derived from the satellite data. The maps, of

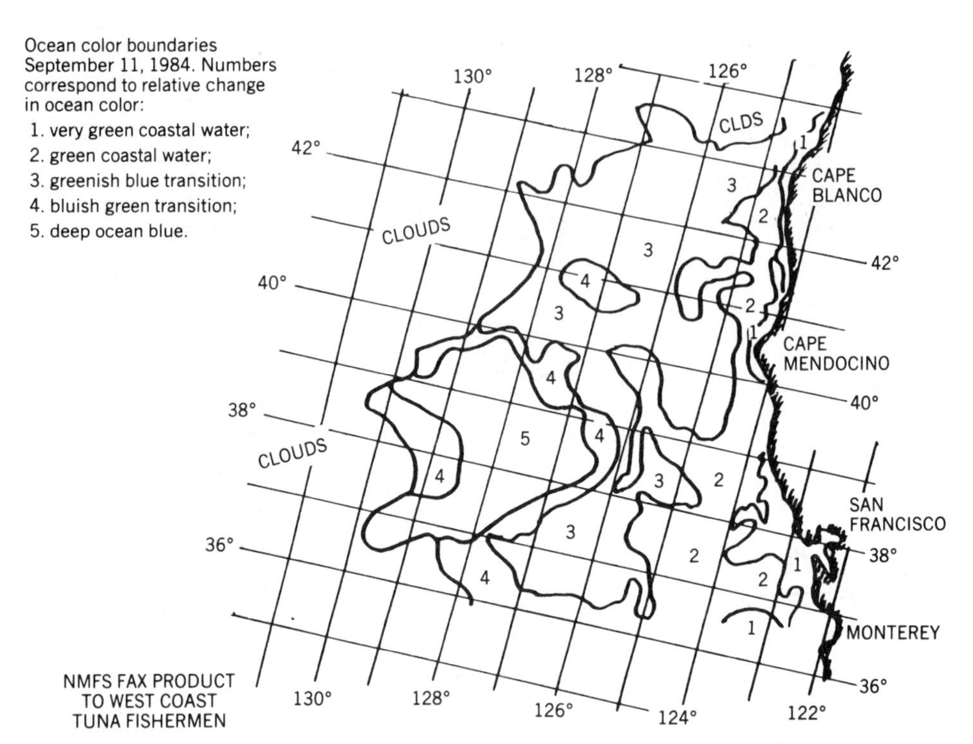

Figure 6-2-25. Satellite derived ocean color boundary map for waters off the California coast. After Turgeon (1986).

which a sample is given in Figure 6-2-25, can be obtained if the vessel is equipped with a facsimile machine (Turgeon, 1986).

6-3 SUBTROPICAL AND ANTARCTIC CONVERGENCE ZONES

The Antarctic Ocean can be divided into two separate regions: waters near the Antarctic Continent, and subantarctic waters. Both water masses can easily be recognized by their SSTs alone and have boundaries associated with the antarctic and subtropical convergence zones. The subantarctic water extends at the surface to the south of the subtropical convergence zone around 52° to 53°S, where antarctic convergence is located. Although the subantarctic waters exist throughout the whole year, significant temperature changes can be recognized.

The subtropical convergence with its position of 10°N of the antarctic convergence may show a temperature range of 10–14°C in winter and 14–18°C in summer, but its location may vary. For instance, the convergence is found further south on the western sides of the oceans where warm water is carried to the south. This transport of warm water through the eastbound currents was demonstrated as early as 1966 by Warnecke et al. (1971) with satellite data for the Agulhas Current.

Separated eddies from the Agulhas Current have been reported in the vicinity

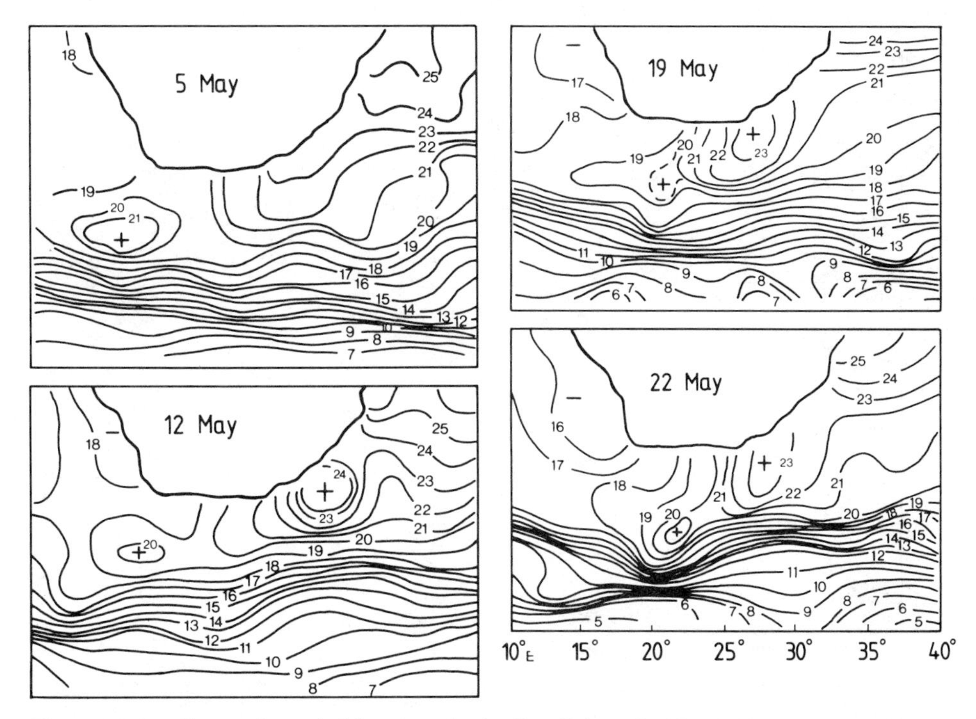

Figure 6-3-1. Observations of eddies along the Agulhas Current for May 5–22, 1980. After Szekielda and McGinnis (1985). Reproduced with permission from Prof. Dr. E. Degens, University of Hamburg.

of the subtropical convergence (Duncan, 1968; Dietrich, 1957) and have been detected with satellites (see Fig. 6-3-1) (Harris et al., 1978). Duncan (1968) observed an eddy in the subtropical convergence southwest of Africa and concluded that the subtropical convergence is not necessarily continuous, exact conditions probably depending on the strength of the seasonal variations of the Agulhas Current.

From March through August, the antarctic convergence zone and the subtropical convergence zone are well-separated and can be identified by the SST field. However, based on the location of the gradients, the convergence zones are not static but undergo significant changes in their position. In September the subtropical convergence zone is very well developed, but, as a consequence of the heating cycle, it loses its surface temperature characteristics, as seen in SST data for the following month, and only the antarctic convergence zone can be recognized by its strong temperature gradient. Figure 6-3-2 shows that the temperature gradients in some cases are as high as 9° per degree latitude. From the distribution, it is clear that the temperature gradient located at around 47°S partly covers the range for the subtropical convergence zone and the antarctic convergence zone.

A sequence of analyzed data showed that, in addition to east-west oscillations, the position of the antarctic convergence zone undergoes rather strong north-south oscillations fluctuating to the south to 50°S (Fig. 6-3-3) and oscillating back within one week.

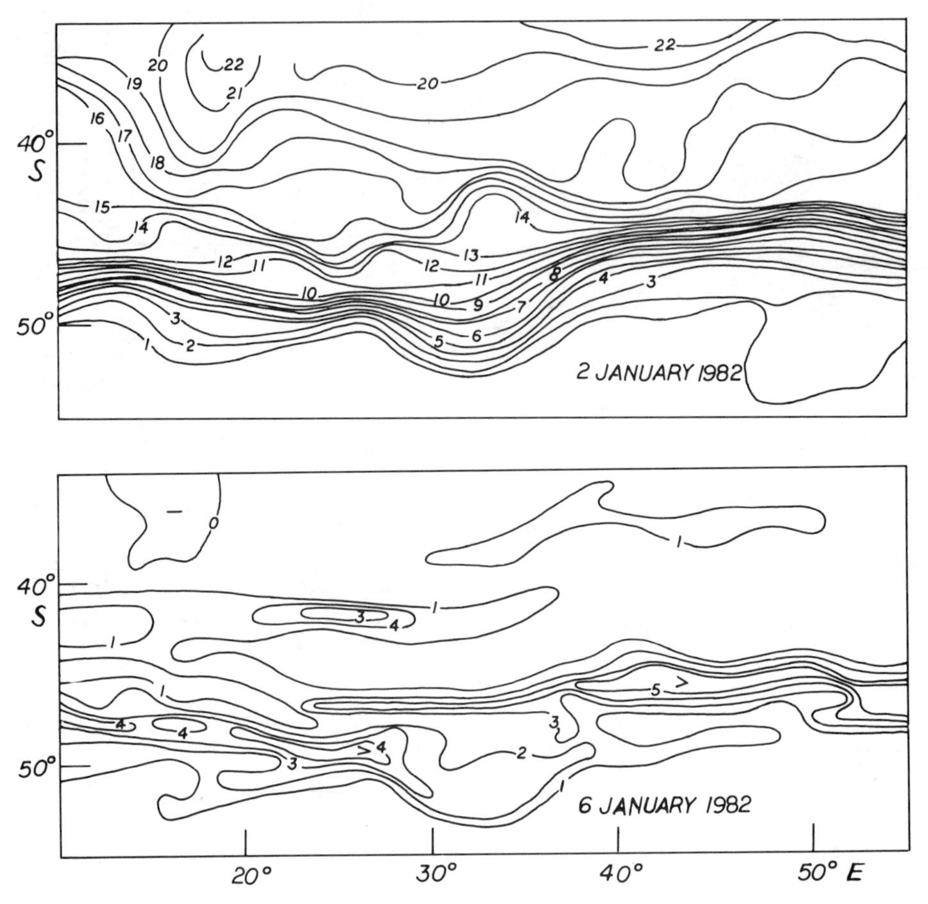

Figure 6-3-2. Temperature distribution and temperature gradients (expressed in °C per degree of latitude). After Szekielda (1983). Reproduced with permission from Deutsches Hydrographisches Institut, Hamburg.

6-4 BATHYMETRY

Bathymetry features in space imagery have been detected at an early stage of space missions. An example is Figure 6-4-1, which shows the bottom topography of the Great Bahama Banks region. Later, multispectral data allowed detailed mapping with computer techniques. It has been found that band 4 (0.5–0.6 μm) from *Landsat* MSS is the most useful for investigations and mapping of reef zones. However, water quality and the composition of the bottom alter the intensity of backscattered light, which requires specific algorithm and depth control points. Part of the same area (area in Fig. 6.4.1) has been covered by SPOT. Figure 6-4-2 shows more details of the bathymetry.

As shown by Polcyn (1976), in clear water depths to 22 m can be measured reliably, and signals from 40-m depths can be differentiated sometimes from deep water signals due to scattering only. Under certain conditions of bright bottom

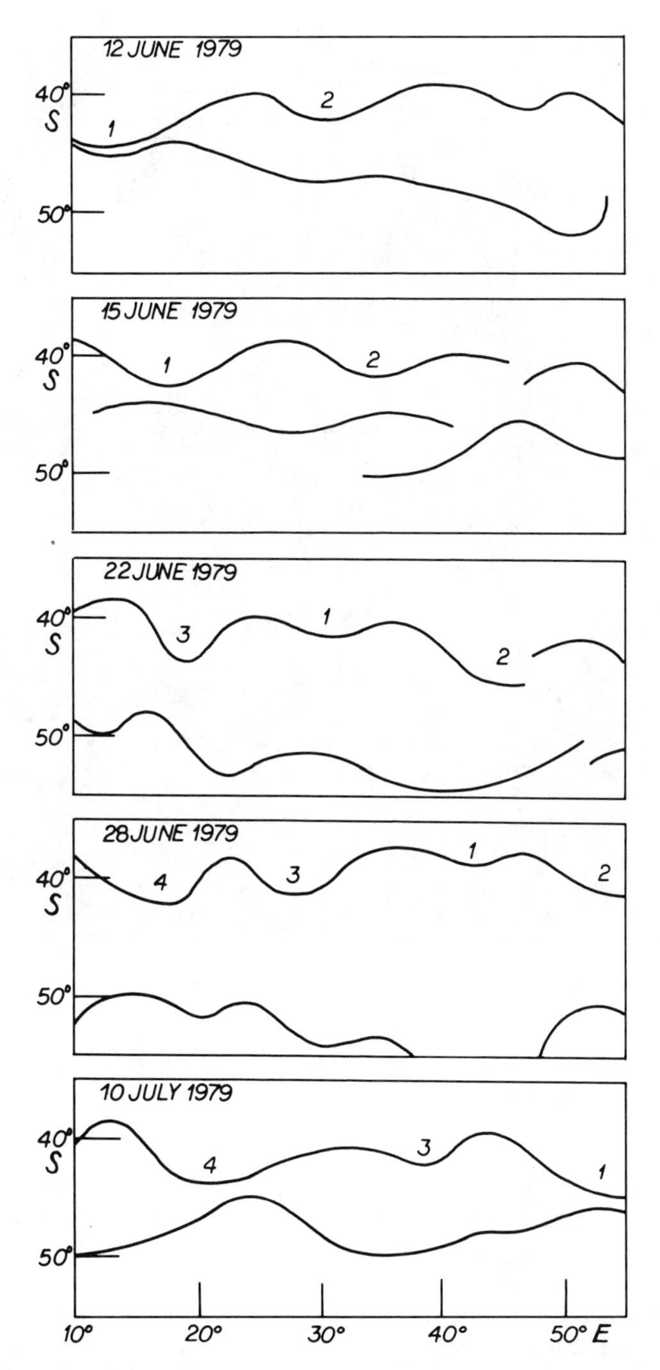

Figure 6-3-3. Location of the maximum temperature gradient in north-south direction indicating the gradients for the subtropical and antarctic convergence zones. After Szekielda (1983). Reproduced with permission from Deutsches Hydrographisches Institut, Hamburg.

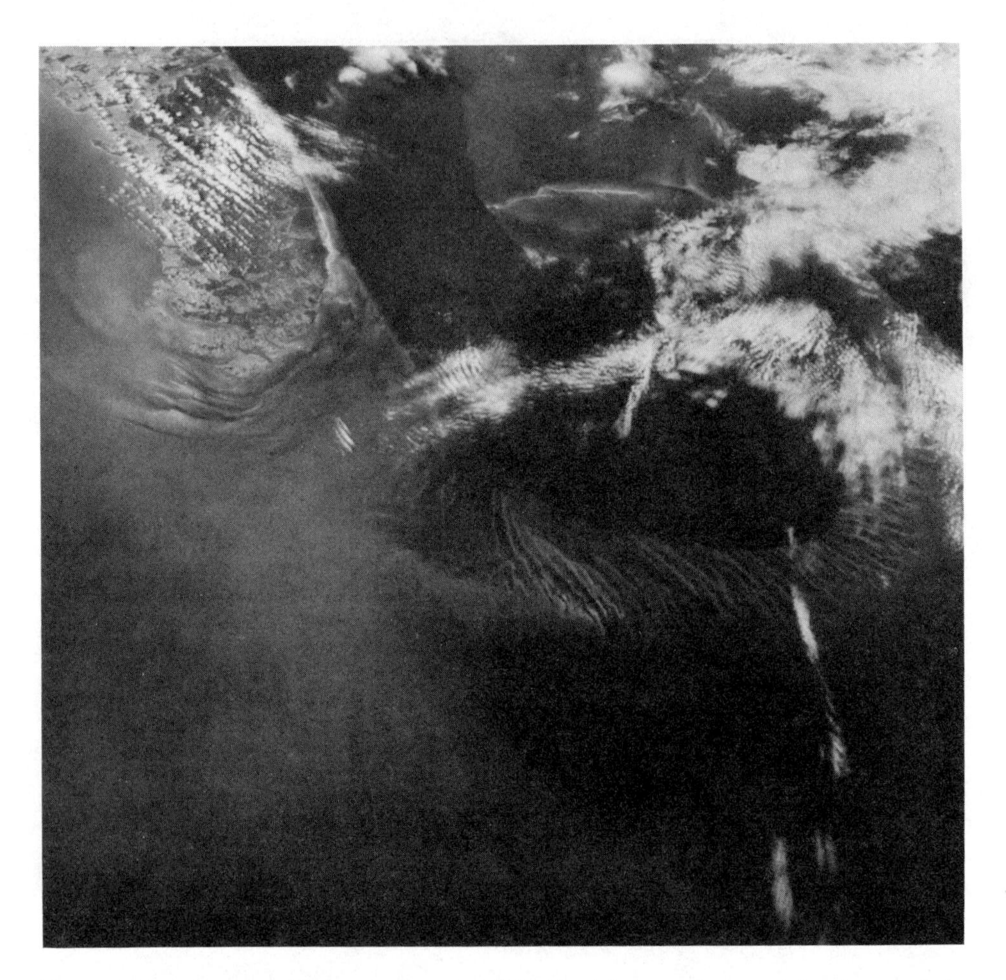

Figure 6-4-1. Example of bottom topography of the Great Bahama Banks and the Tongue of the Ocean (looking north) as revealed from space. Running diagonally across the middle from upper left to lower right is the Exuma chain of islands. The Tongue of the Ocean is over 1 mi deep; depths on the Bank vary from 6 to 30 ft. The lower left shows the bottom topography (sand gullies) at the head of the Tongue of the Ocean. Courtesy of NASA.

reflectance, 30%, or less than 1.3 m, will saturate the signal of the spectral region 0.5–0.6 μm. In this case, it is possible to use the 0.6–0.7 μm band for determining depth, because of the greater light attenuation in this spectral region.

Discrimination between water-covered features and land can be based on information provided by measurements in the reflected IR, and the information for water depth is extracted after setting a threshold level to land. If the longer-wavelength channel does not penetrate adequately, the single channel 4 (0.5–0.6 μm) of *Landsat* MSS can be used for developing a penetration algorithm. Figure 6-4-3 is an example of reefs in the near coastal area, as monitored in channel 4 in *Landsat*.

Polcyn (1976) developed an algorithm to relate the signal voltage from *Landsat* MSS and the bathymetry under the following assumption:

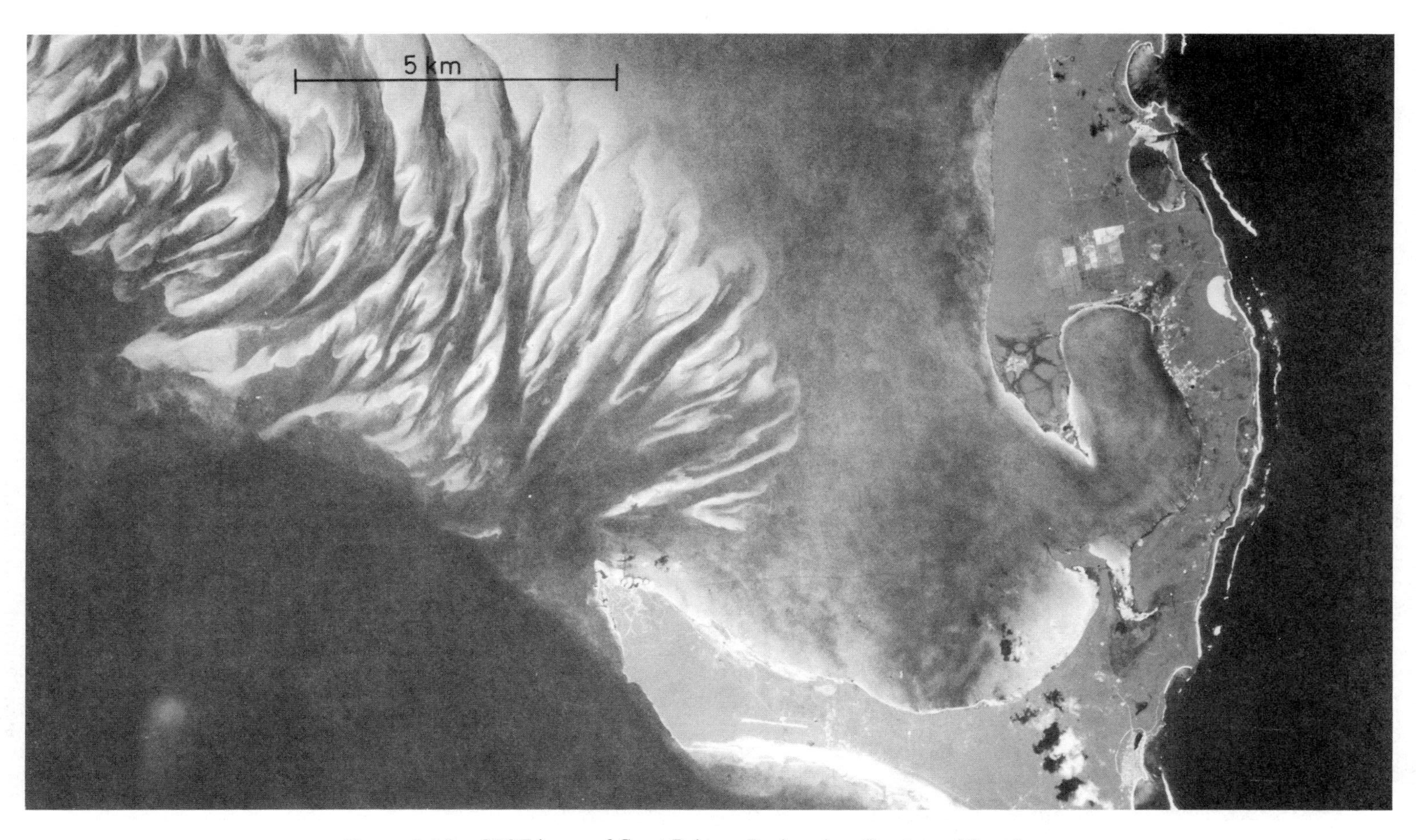

Figure 6-4-2. SPOT image of Great Bahama Bank region. Courtesy of Spot-Image.

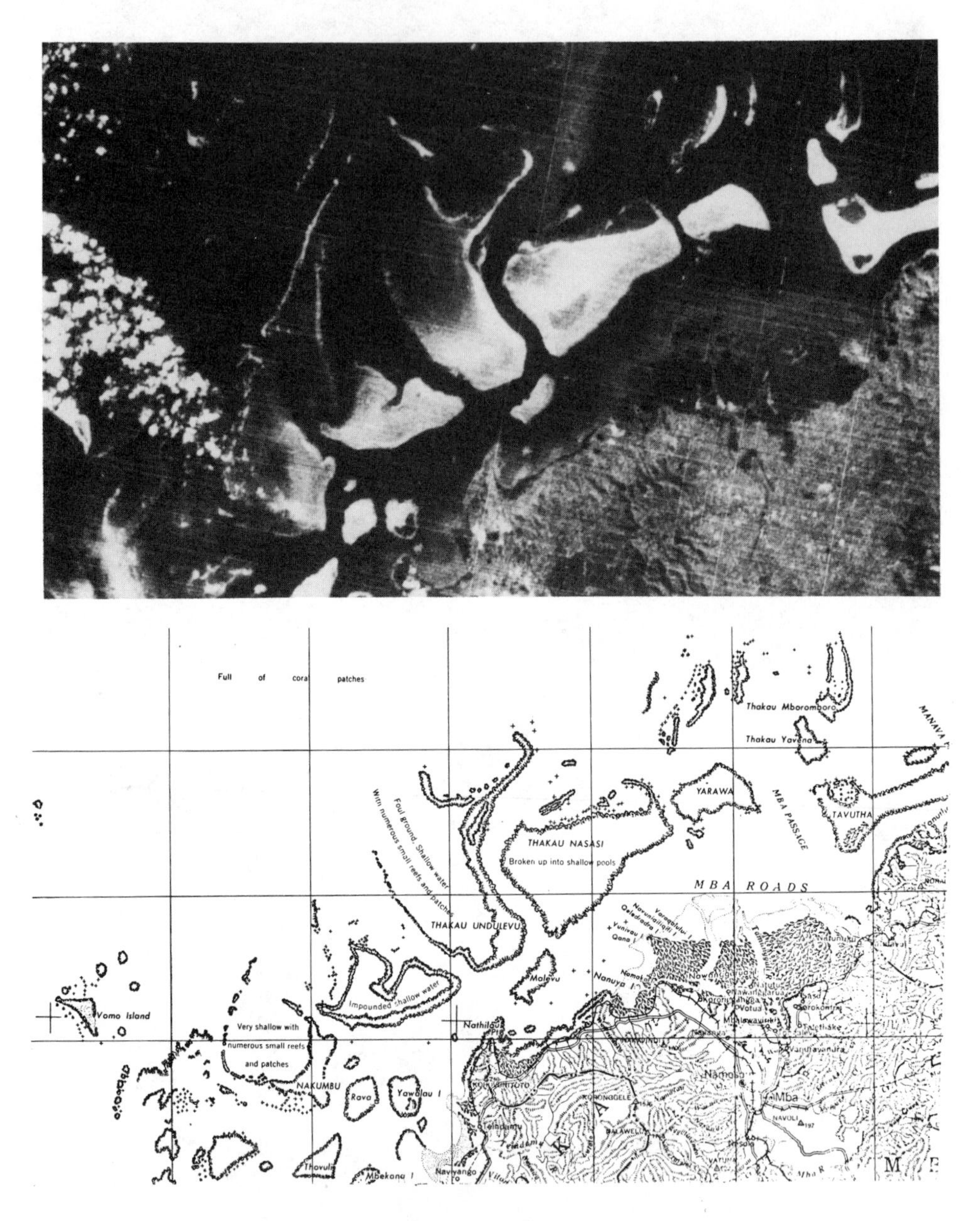

Figure 6-4-3. Reef location in the vicinity of the Fijis as monitored on channel 4 of *Landsat* on August 23, 1973.

1. Scattering in the water is neglected.
2. The signal comes only from direct solar radiation.
3. The water attenuation coefficient is independent of the radiance distribution.

Based on the model and with further assumptions taken into account, a sample of bathymetric mapping by *Landsat* is shown in Figure 6-4-4 within a nautical chart as overlay for Bahrain (Al Bahrayn).

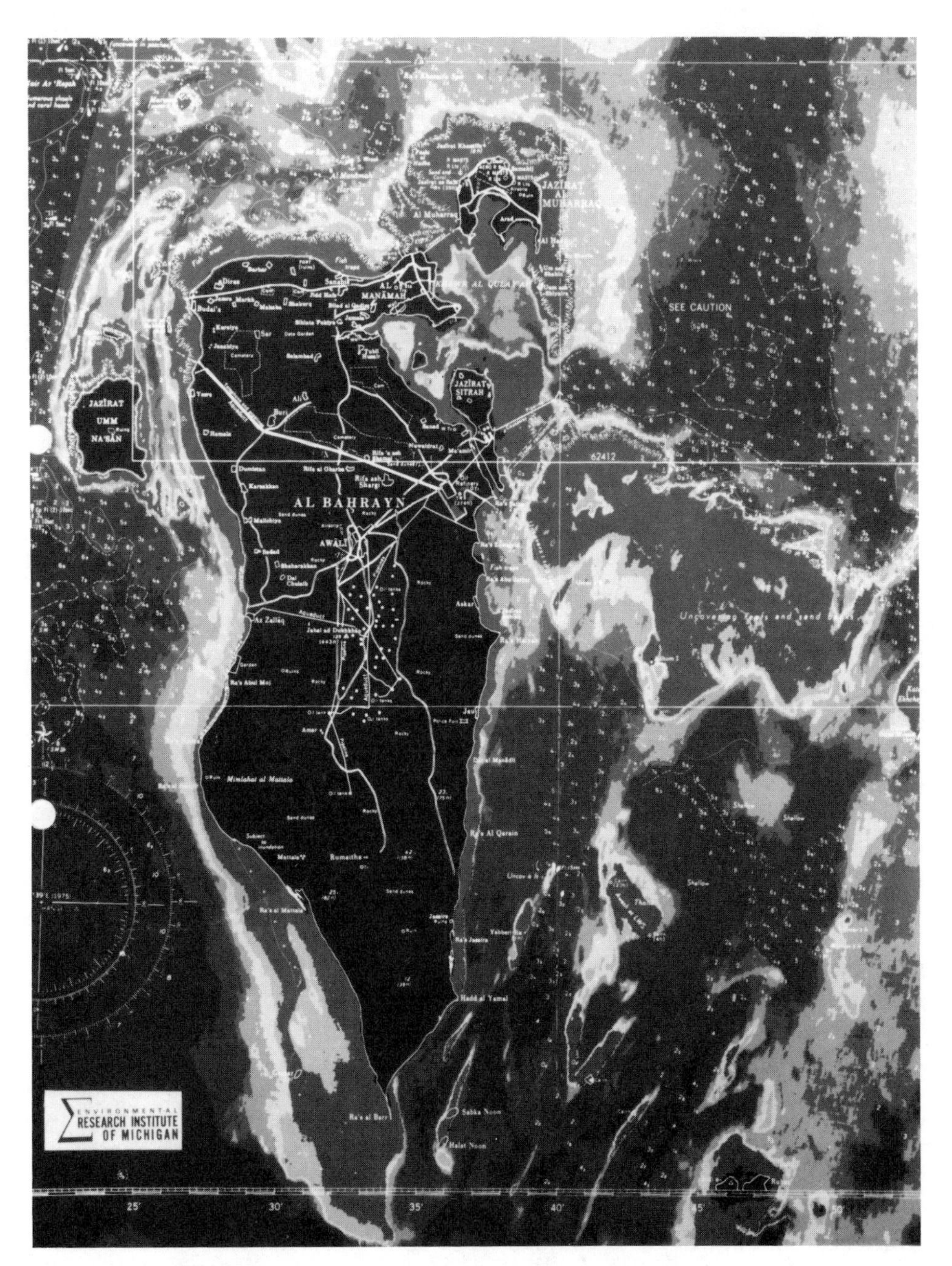

Figure 6-4-4. Bathymetric mapping with *Landsat* for Bahrain (Al Bahrayn) region. Reproduced with permission from the Environmental Research Institute of Michigan (ERIM).

6-5 SEA ICE

The lack of regular oceanographic observations in polar regions gives relatively little available data when compared with other oceanic areas. Conventional oceanographic data collection is restricted due to the near perennial ice coverage of these polar areas, but during the summer months the deterioration of the ice fields in certain areas affords limited access for purposes of data collection. As a result, the oceanographic processes, especially the vertical motions in the antarctic areas, are not documented in detail, even though the southern ocean is one of the main sources of the bottom water in all oceans.

Sea ice may be recognized in the visible, IR, and microwave regions of the EMS, but in the highest latitudes the application of data in the visible is restricted to the summer seasons. Infrared studies were therefore undertaken at an early stage in remote sensing to map arctic and antarctic sea ice, and improved techniques for ice mapping using IR data monitored by the *Nimbus 2* and *Nimbus 3* HRIR have been developed.

Sea ice undergoes large dynamic and morphologic changes within a short time and also undergoes significant seasonal changes; especially significant is the change that occurs when winter ice metamorphoses into summer ice. During most of the year, a strong temperature gradient exists in the ice cover, which separates two fluids at different temperatures—the seawater having surface temperature of −1.7°C and the surface air temperature during the winter at −30 to −40°C. During summer the ice becomes isothermal, and many melt ponds form on the upper surface, which are important for thermodynamic and morphologic reasons.

Sea ice can be identified in all the MSS spectral bands of *Landsat* because of its high reflectance, although bands in the visible range (0.5–0.6 μm and 0.6–0.7 μm) are the most useful for mapping ice edges and detecting thin ice. The near IR range (0.8–1.1 μm) is most useful for identifying features revealed by changes in the ice surface characteristics. Because high contrast exists between water and ice, linear features at least as narrow as 70 cm can be observed. *Landsat* data have been ideal for studies of lead patterns and floe size distributions. Also, the different classes of thin ice (open water, gray ice, gray-white ice) can be noted and distinguished from thicker ice. However, at present, the correlations between reflectance levels and actual thin ice thickness are qualitative. In addition, meltwater on the ice surface is indicated by its very low reflectance in the near IR, brash ice can be distinguished from other ice types, and in a few cases old ice can be separated from first-year ice.

Infrared scanners are being increasingly used in operational and research applications for sea-ice surveillance. These investigations have shown that IR band and microwave imagery provide a means of mapping gross ice boundaries during periods of polar darkness. When visible and thermal IR can be used jointly, preliminary results indicate that additional ice information, such as ice concentration, can be obtained. Thermal contrasts between ice and water are not so large as reflectance contrasts; therefore, IR imagery generally requires some special enhancement for optimal use in ice analysis. Nearly all clouds are opaque to radiation emitted at 10.5 to 12.5 μm, but many of these same clouds are partially transparent to energy in the wavelengths 0.5 to 0.7 μm; thus, delineation of ice features is more often possible by use of the visible than the IR.

The limitation in acquiring satellite data for sea ice at visible and IR wavelengths is due to cloud cover. Emission from ice surfaces at microwave wavelengths is attenuated only slightly by clouds typical of polar regions. Large brightness temperature contrasts found at ice-water boundaries are observed readily through clouds. The ice surface itself has emissivity variations, with first-year ice being somewhat more emissive than multiyear ice (Gloersen et al., 1974). The use of ESMR imagery is limited for certain types of ice studies based on the relatively coarse spatial resolution of the data.

The concentration of sea ice versus open water within a resolution element can be determined as a result of the large difference in the respective microwave emissivities (Zwally et al., 1976). The observed brightness temperature is taken to be a linear combination of the sea ice brightness temperature (ET_0) and the seawater brightness temperature ($E_w T_w$) plus atmospheric brightness temperature (A). At a 1.55-cm wavelength, $E_w T_w$ is about 120 K, A is about 15 K, and ET_0 is typically 250 K; E and T_0 are the sea ice emissivities and physical temperatures, respectively, and E_w and T_w are similar quantities for seawater. The atmospheric contribution, which is mainly due to water vapor, is assumed to decrease with ice concentration (C). It follows that

$$T_B = (ET_0)C + (E_w T_w + A)(1 - C)$$

The main uncertainties in determining C from an observed T_B at one frequency are due to the variation of E with ice type, lack of an independent measurement of T_0, and variability of the atmospheric brightness temperature. In the Antarctic Regions where most of the ice is first year, C can be determined to an estimated accuracy of $\pm 15\%$ and to better relative accuracies in localized areas where T_0 differences are minimal.

Although *Landsat* does not have the capability to directly measure the thickness of the ice cover, it can be used to distinguish between classes of ice thickness and to monitor changes within each class (Campbell, 1976b).

Figure 6-5-1*a* shows a section of the ice cover of the Arctic Ocean within the Beaufort Sea. Large floes of first-year ice or multiyear ice that vary in thickness from approximately 1.5 to 3 m can be seen separated by refrozen polynyas filled with gray ice, which measure 20 to 40 cm thick. Numerous new leads can be detected that are open or have very thin ice. In the center of the image, a large refrozen polynya that runs approximately north to south is composed of two distinct zones of gray ice that have been formed by two separate opening events of the original lead. Figure 6-5-1*b* shows approximately the same area one day later, during which time a moderate wind continued to blow from the northeast. Strong divergence of ice occurred, and the large polynya increased in width, the new lead on the eastern side of the polynya having expanded approximately 4 km. Within the polynya, four distinct zones were distinguished: (1) a zone of open water immediately adjacent to the eastern edge of the polynya, (2) a zone of recent thin gray ice formed by the freezing of the grease ice (ice slush), (3) a zone of earlier-formed gray ice 20 to 40 cm thick, and (4) a zone of older gray ice 40 to 60 cm thick.

Valuable information on ice drift within the Arctic Islands was collected by Campbell et al. (1975a and b). It was possible to obtain velocity vectors for ice movement over wide areas within the islands. Observations showed that there were very broad patterns of ice drift even within channels, with ice motion directed to the south.

La Violette and Hubertz (1975) tracked seven large floes, each approximately 25 km across, and continuous tracks of their movement were maintained. These

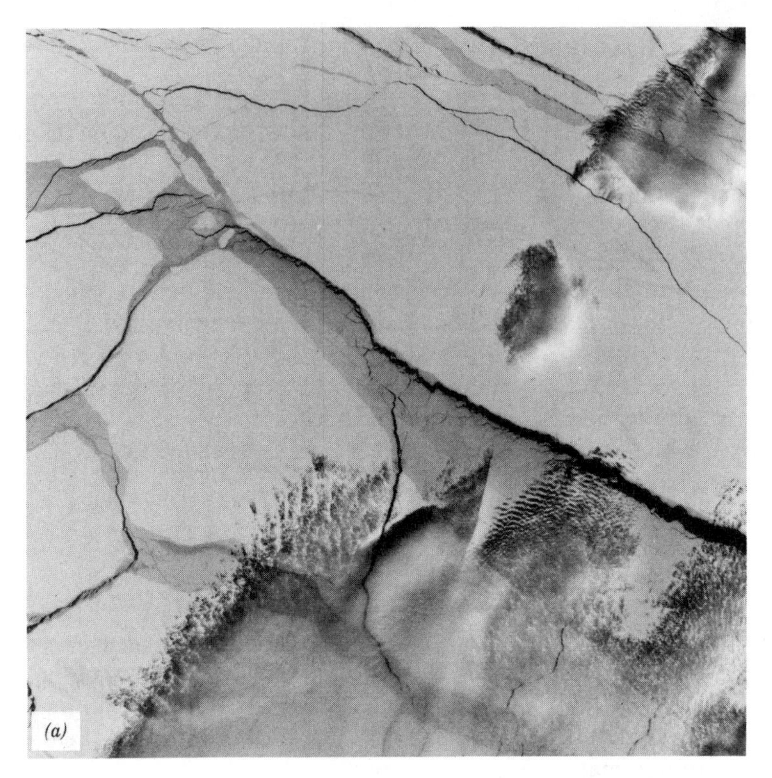

Figure 6-5-1. (*a*) *Landsat* image of the ice cover over the Arctic Ocean taken on April 5, 1973.

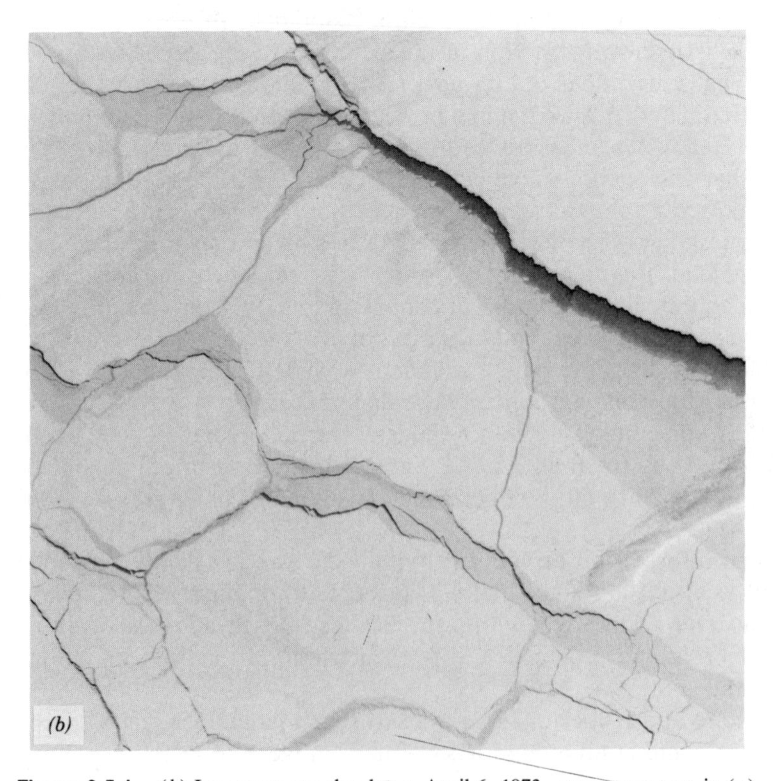

Figure 6-5-1. (*b*) Ice cover one day later, April 6, 1973, over same area in (*a*).

tracks disclosed persistent patterns in the movement of the floes and some variations in their speed. For comparison purposes, a quantitative estimate of the speed and distance covered by two of the large floes during a period of rapid movement and comparatively weak surface winds was made. This showed that floes can move at a rate of about 9 km · day^{-1} or 10.0 cm · sec^{-1}.

Campbell et al. (1975a and b) used the first passive microwave imagery from ESMR on the *Nimbus 5* satellite. The microwave brightness temperature maps of the Arctic observed by ESMR (Fig. 6-5-2) provided at that time the first synoptic maps of large-scale sea ice distribution. The ice cover of the Beaufort Sea was

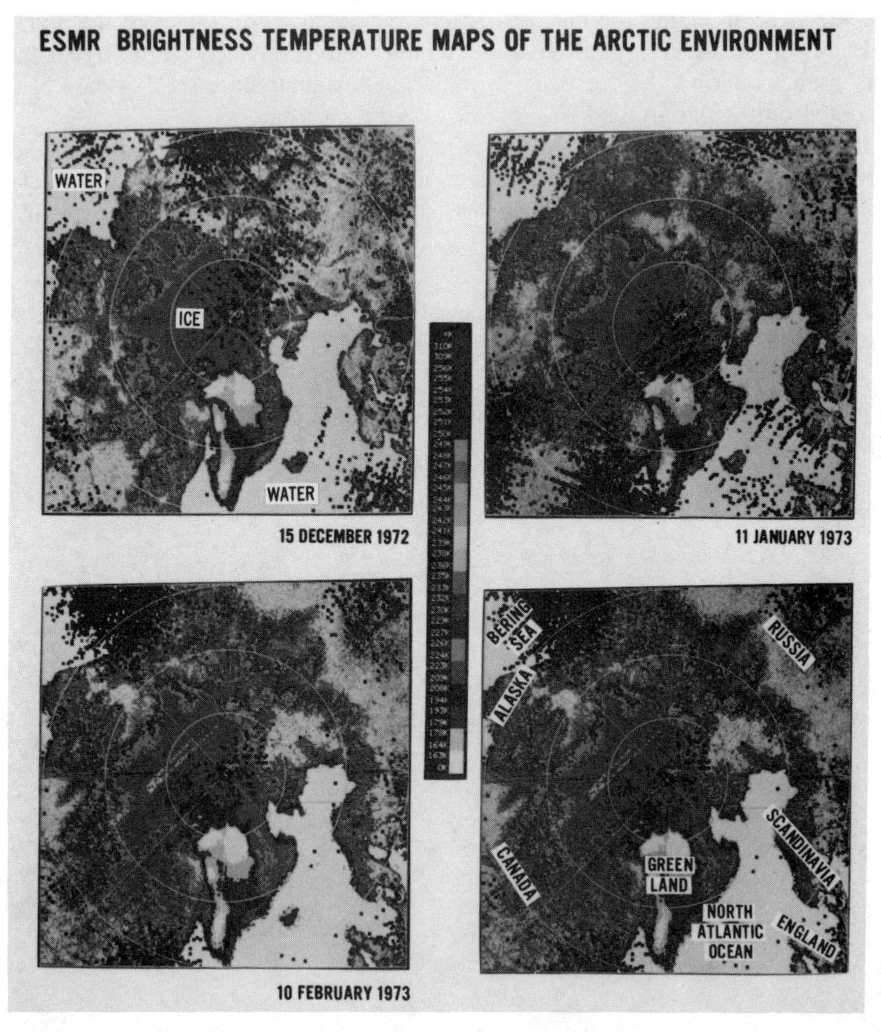

Figure 6-5-2. ESMR brightness temperature maps of the Arctic. Each image is a composite of less than 24 hr of data returned from the ESMR on *Nimbus 5*. Since the microwave sensor can penetrate through most clouds, it can record the snow and ice changes for the entire polar area. Apparent are ice changes in the Bering Strait, east of Greenland, and east of the Scandinavian Peninsula. Each image color corresponds to a brightness temperature on the temperature-color scale. Blues and greens are ice; gray is open water, except over Greenland where gray represents glacial features. The black squares are missing data points. Courtesy of NASA.

observed to be highly heterogeneous, its eastern sector being made up of large multiyear floes and its western sector being made up of a mixture of small, fragmented first-year and multiyear floes. Coverage shows that during the winter the ice canopy is composed of two general types of ice: the principally multiyear ice, as indicated by brightness temperatures between 209 and 223 K, covering the main portion of the Arctic Ocean, and either first-year or first-year/multiyear ice mixture with higher brightness temperatures covering the southern portions of the marginal seas. By the end of the melt season in September, the edge of the ice pack showed that most of the first-year/multiyear ice mixture that had covered the seas had melted. Figure 6-5-3 shows an analysis of the near-maximum ice extent. The ice areas in which the first-year/multiyear ice mixture survived the summer melt are source areas for multiyear ice, which is advected into the multiyear pack. This process allows the ice pack to maintain a steady-state size as part of it is advected out of the area by the east Greenland Current.

In a similar case for the Antarctic, Gloersen et al. (1973) studied seasonal variations of ice extent and estimated ice compactness for the summer ice pack. In November a fairly continuous ice cover surrounded Antarctica. However, in the

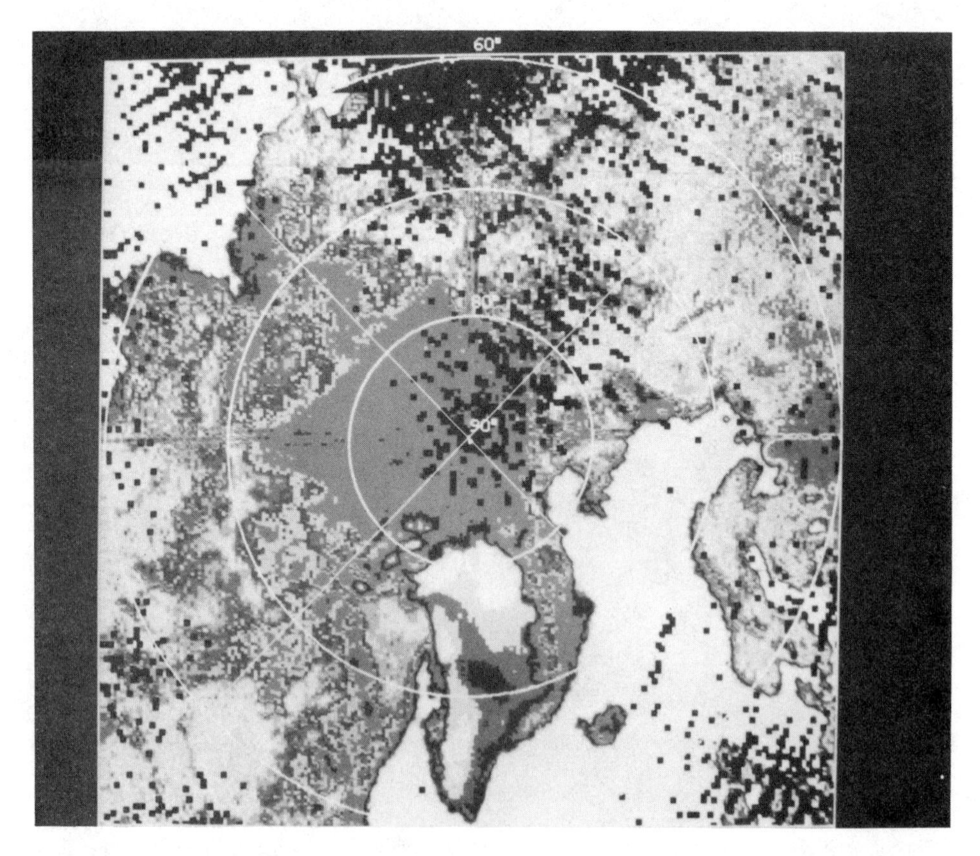

Figure 6-5-3. ESMR data of near-maximum ice extent over the Arctic. Courtesy of NASA.

Ross Sea, off the coast of Marie Byrd Land, and in the Weddell Sea, small areas of sea ice had cooler radiometric signatures than did the surrounding ice, which had small leads and polynyas that did not fill the resolution element of the radiometer.

Longtime studies on the ice caps of the Antarctic and Arctic Regions started with the ESMR on the *Nimbus* satellites. In the Antarctic it has been assumed that the overall shape of the sea-ice boundary in the winter appears to be controlled principally by the antarctic Circumpolar Current rather than directly by prevailing wind patterns. The deflections of the Current at submarine ridges are clearly mirrored in the shape of the ice boundary. Distinct corners in the boundary occur at 45°W, 10°E, 75°E, and 150°W, where the Current is deflected by the submarine ridges; straight lines between these points reasonably describe the ice boundary near its maximum extent.

Persistent areas of reduced ice concentration within the winter ice pack at two locations suggest the upwelling of warm deep water with temperatures at approximately 2°C to the surface. For the Ross Sea a band of 15 to 30% lesser concentration is indicated in Figure 6-5-4. The most dramatic evidence of upwelling is the large enlongated polynya (0.25×10^6 km^2) at 0°E, which persisted throughout the

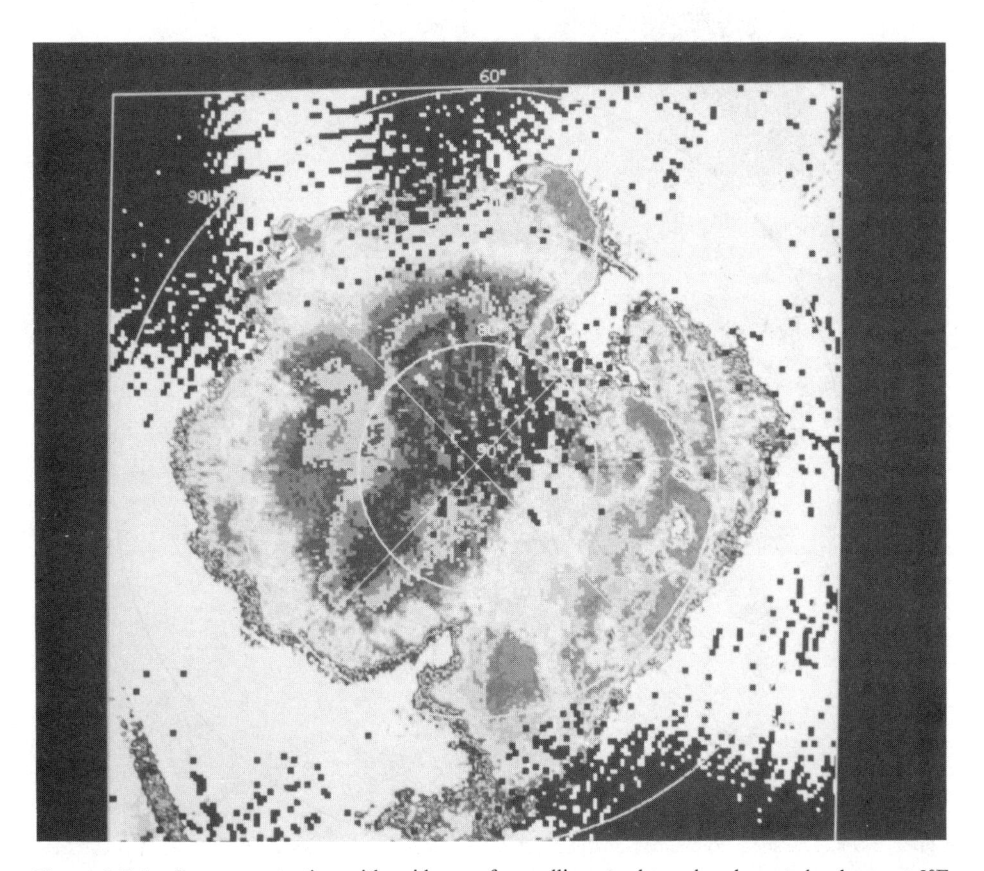

Figure 6-5-4. Ice concentration with evidence of upwelling, as shown by elongated polynya at 0°E. Courtesy of NASA.

1974 winter and throughout the 1975 winter. The ice concentration in the polynya was less than 15%. During the 1973 winter, the polynya was closed, but reduced ice concentration was observed beginning in October of that year. Both of these features appear to be manifestations in connection with the Antarctic divergence.

With the SMMR it is now possible to monitor on a quasioperational basis the ice concentration for the polar caps, which, especially with regard to the melting of the polar ice, is an important issue (see Fig. 6-5-5).

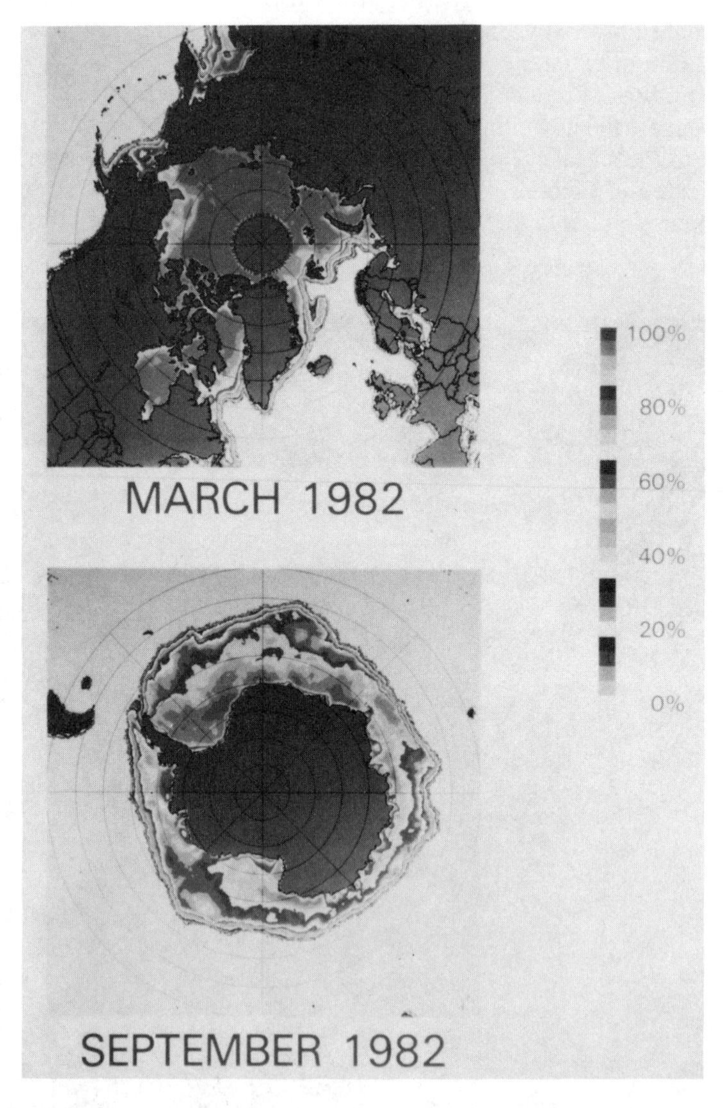

Figure 6-5-5. Sea ice concentration (percent of ocean area covered with sea ice) derived from monthly mean SMMR radiances for both the Arctic and the Antarctic for months in 1982 of maximum ice cover. Courtesy of Dr. D. J. Cavalieri, NASA.

6-6 SEA STATE, WIND SPEED, AND DIRECTION

Active and passive microwave measurements are correlated with wind speed, and backscatter has been shown to be sensitive to modifications in wind direction. The largest scattering is measured near upwind, where the measuring instrument is in the direction from which the wind is blowing. A secondary maximum occurs near downwind with minima near crosswind. Therefore, backscatter is measured for multiple azimuthal angles for each surface patch observed, and both the speed and direction of the surface winds can be estimated (Young and Moore, 1976).

Figure 6-6-1. *Seasat* radar image showing Bahía Magdalena and Bahía Almejas in dark while the offshore area shows patterns that can be assigned to different wind stress and surface waves. The dark areas generally correspond to low surface winds (less than 2 m·sec^{-1}), but current-generated interactions also produce patterns. Courtesy of NOAA.

Jones et al. (1977) described the quantity of scattering, σ^0, of the ocean. The quantity is independent of the type of radar performing the measurement, and is defined from the radar equation as

$$\sigma^0 = \frac{P_r(4\pi)^3R^4}{P_tG^2\lambda^2A_\tau}$$

where P_r = received power
P_t = transmitted power
G = antenna gain
R = slant range
λ = free-space wavelength
A_τ = effective antenna footprint on ocean surface

Ocean swell can be imaged with the *Seasat* SAR at least for certain levels of wind speed and when there is a substantial component of the swell traveling along the line of sight of the radar. An example of data from the SAR is shown in Figure 6-6-1.

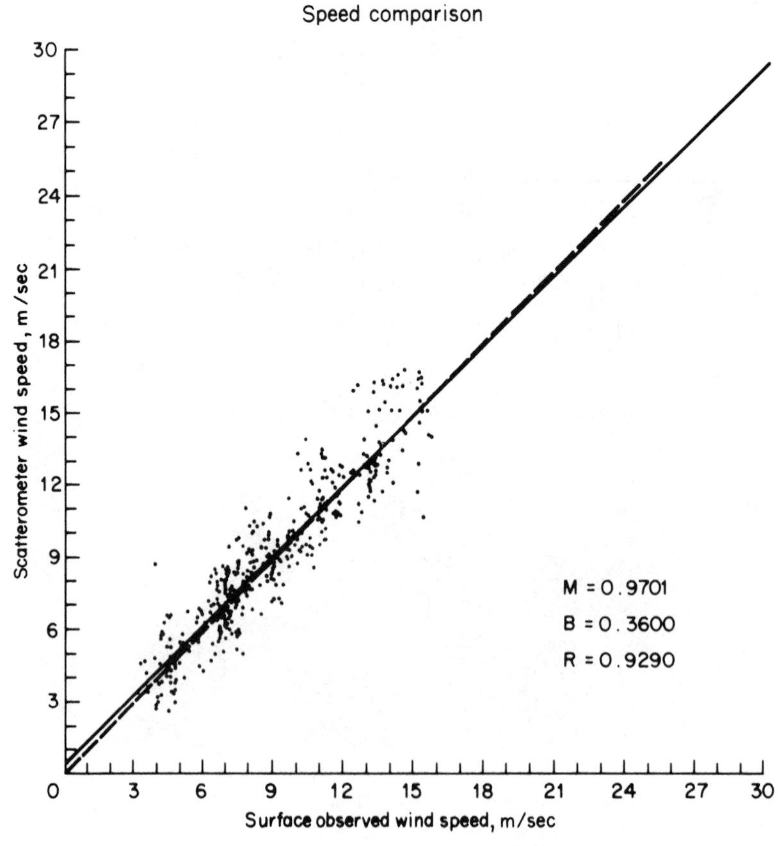

Figure 6-6-2. Comparison of *Seasat* scatterometer wind speed solutions with surface wind speed observations. After Raney (1986). Reproduced with permission from Graham & Trotman, Ltd: Szekielda, K.-H., 1986. *Satellite Remote Sensing for Resources Development.*

Comparison of surface and aircraft measurements with those from five passes of *Seasat* over the Gulf of Alaska, made by Gonzalez et al. (1979), indicated agreement to within about ±15% in wavelength and about ±25° in wave direction. These results applied to waves 100 to 250 m in length, propagating in a direction predominantly across the satellite track, in sea states with significant wave height in a range of 2 to 3.5 m.

Wind direction may be determined by measuring the apparent reflectivity in two or more directions and by solving an empirical relationship. Unfortunately, this leads to several possible wind directions.

Results from the *Seasat* scatterometer and independently observed wind speeds are given in Figure 6-6-2 (Raney, 1986). Part of the scatter in data points can be explained by noting that the two kinds of measurements are different. *Seasat* measured a spatial average of wind speed as it directly impacts the sea surface, whereas the surface observations are single-point time-averaged measures, nominally at 10-m elevation above the sea.

In 1978 the radar scatterometer (SASS) made the first global measurements of winds over the oceans, an example of which is shown in Figure 6-6-3 for the Pacific Ocean. A total of about 350,000 SASS wind measurements were obtained over three days, more than 100 times the number of wind observations available from ships and buoys during the same period. The arrows or wind vectors have a resolution of 3° latitude by 3° longitude, and the wind speed contours correspond to 1° by 1°.

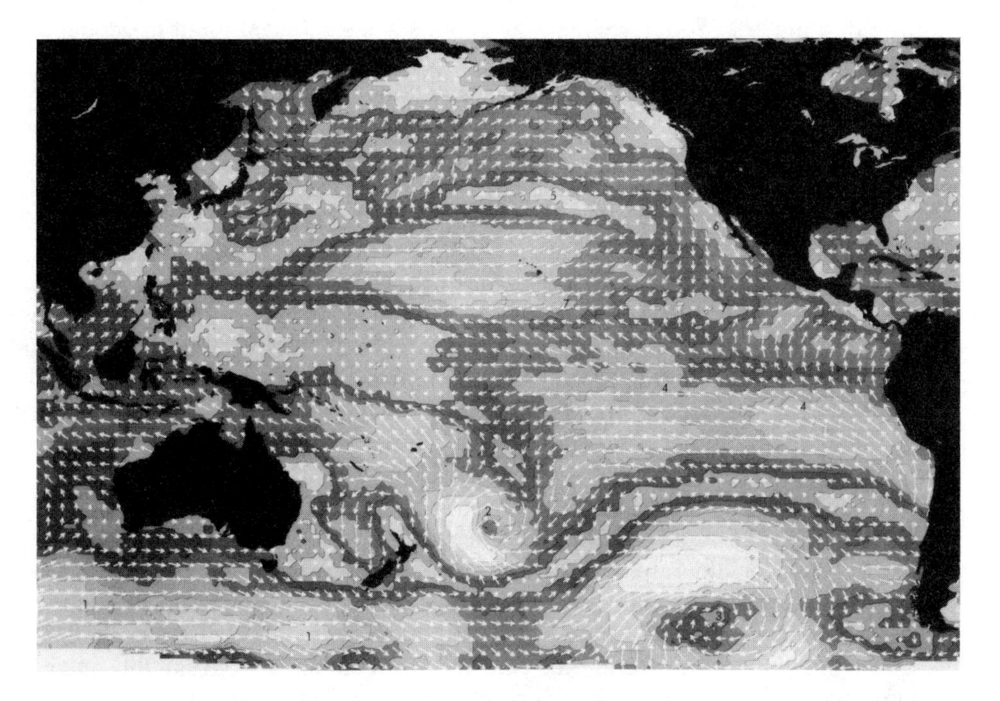

Figure 6-6-3. SASS wind measurements over the Pacific Ocean for September 6–8, 1978. Wind speed was estimated with an accuracy of ±2 $m \cdot sec^{-1}$ and wind direction to within ±20°. Produced by S. Peteherych, P. Woiceshyn, and M. Wurtele. Courtesy of NASA.

SASS wind field for the Pacific is an opportunity to see atmospheric flow patterns relatively undistorted by the influence of continents. As can be seen in Figure 6-6-3, in the southern hemisphere, strong westerlies (1) form an almost continuous band over the entire Southern Pacific Ocean, interrupted only near the tip of South America. On average, these winds are the strongest and occur in a region almost devoid of ship traffic.

Two intense cyclonic storms dominate the South Pacific: one off New Zealand (2) and the other close to South America (3). In the southern hemisphere the flow around a low-pressure or cyclonic storm system is clockwise. In the tropics, north of 20°S, the southeast trade winds (4) extend across much of the South Pacific with speeds of about 12 m $\cdot$ sec^{-1}.

Summer in the northern hemisphere is characterized by a large high-pressure or anticyclonic system, the Hawaiian high, which has a clockwise rotation about its center near 35° and 150°W (5). The eastern edge of the anticyclone provides strong flow from the north parallel to the coast of southern California and the Baja Peninsula. The northeast trade winds constitute the southern flank of the Hawaiian high. At approximately 10°N, the southeast and northeast trade winds meet in the ITC Zone (7) where rising moist tropical air leads to strong convective cloudiness and tropical squalls.

6-7 SEA SURFACE TOPOGRAPHY

The use of altimetry for determining the geoid has a significant input in geodesy, oceanography, geology, and geophysics, whereby satellite altimetry offers the most accurate method for determining the geoid, independent of gravity measurements. Recently, accuracy of several centimeters for determining the marine geoid through satellite altimetry has been achieved. This includes precise determination of spacecraft height above the ocean surface of about ± 10 cm (Tapley et al., 1979) and low noise level of the altimeter (about 5 cm).

In the oceans, sea slopes due to ocean currents cause local rise of water across the current of about 1 m and proportional to the current speed. In addition, the periodic effect of tides in the open ocean and barometric pressure could cause variations in the slope of the sea surface up to several meters. Tsunamis amplitudes in the open ocean range from a few centimeters to about 1 m.

The causes of gravity anomalies, associated with different but similar bottom and surface topographic features, are due to variations in the mass distribution in the earth's crust and mantle.

Figure 6-7-1 shows the relations between the various surfaces associated with satellite altimetry (Mourad et al., 1975). The raw altimeter range, given as TM, has to be corrected for laboratory instrumental calibration, electromagnetic effects, sea state, and periodic sea surface influences to give TS, where S represents the nonperiodic "sea level." CT and CE, the geocentric radii of the altimeter, and E, its subsatellite point on the reference ellipsoid, are computed from satellite tracking information. EG is the absolute geoidal undulation to be computed, and SG is the quasistationary departure of the mean instantaneous sea surface from the

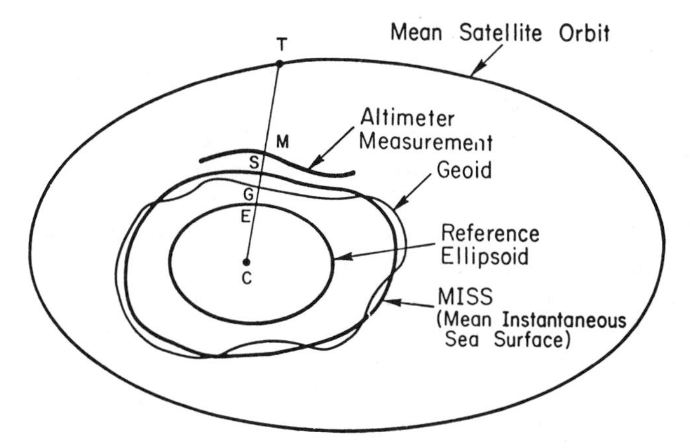

Figure 6-7-1. Surfaces associated with satellite altimetry.

geoid or the "undisturbed" mean sea level. The required geoidal undulations are given by

$$EG = ET - TM - MS - SG$$

where MS represents the sum of the calibration constants and the orbit uncertainties, if any. Since SG is considered not to vary significantly over the length of profiles corresponding to different submodes of observations, $MS + SG$ is considered as the calibration constant.

The basic requirements for determining the geoid from satellite altimetry are (1) accurate orbit determination, (2) precise altimeter instrumentation, (3) ground truth verification data, (4) methods of interpolating and extrapolating altimetry into unsurveyed areas, and (5) separation of the geoid from sea surface topographic effects.

The absolute surface topography of a given area in the ocean consists of a constant portion, which can be considered the geoid, and a time-varying part, due to tides, wind stress, and currents that introduce geostrophic height variations. Therefore, currents, storm surges, pileups, and tidal differences causing changing sea levels can be detected if accurate measurements on the order of 1 to 2 m can be carried out.

For detailed modeling and observations of the geoid, additional factors have to be considered. These factors lead to the following computation of the sea surface above the reference ellipsoid h_{ss} (Marsh et al., 1986):

$$h_{ss} = h_e(x) - h_0(x) - T_0 + C + \sigma_m$$

where $h_e(x)$ = computed height of the satellite above the ellipsoid at position x; this is obtained from the precision orbit computations based upon the *Seasat* laser and electronic tracking data and the most accurate geodetic and force models available

$h_0(x)$ = observed altimeter measurement of the height of the satellite above the instantaneous sea surface at position x
T_0 = ocean tides
σ_m = measurement noise
C = sum of all other corrections to the altimeter data

If the effect of tides, currents, and wind effects are excluded, the ocean surface follows an equipotential surface, although the geoid is irregular in shape because of the variation of the mass and density distribution of the earth. For convenience, a reference ellipsoid is adopted to mathematically describe the geoid and the deviations from the reference are defined as the geoidal undulation. Using the satellite orbit as a reference, one can derive the geoidal undulation from altimeter measurements. From the altimeter height measurements, it is obvious that any radial orbital errors will propogate directly into the geoidal undulation. Therefore, precise knowledge of the calibration constants is required, which includes knowledge of the spacecraft orbit and the sea surface topography in specified test areas. The static sea surface, which is the surface of the ocean without the time-varying effects of atmospheric pressure, surface wind, friction tides, currents, and so on, is the result of the gravitational and rotational forces.

Marsh et al. (1986) used a set of altimeter data to establish the global mean sea surface based on *Seasat* data with a 0.5° × 0.5° grid collected over a time span of 70 days. Figure 6-7-2 presents a topographic relief map of the global 0.5° × 0.5° mean sea surface, which is presented at a 1-m contour level and is referred to an ellipsoid with a_e = 6,378,137 m and $1/f$ = 298.257. The map shows the large-scale gravimetric features, such as the Indian low, the high over the western boundary of the Pacific plate, and the mid-Atlantic ridge. The deep oceanic trenches in the Pacific, which have been depicted in earlier maps, are also clearly shown. A number of short-wavelength features not resolved in earlier computations (for example, the system of fracture zones in the eastern Pacific with an amplitude of

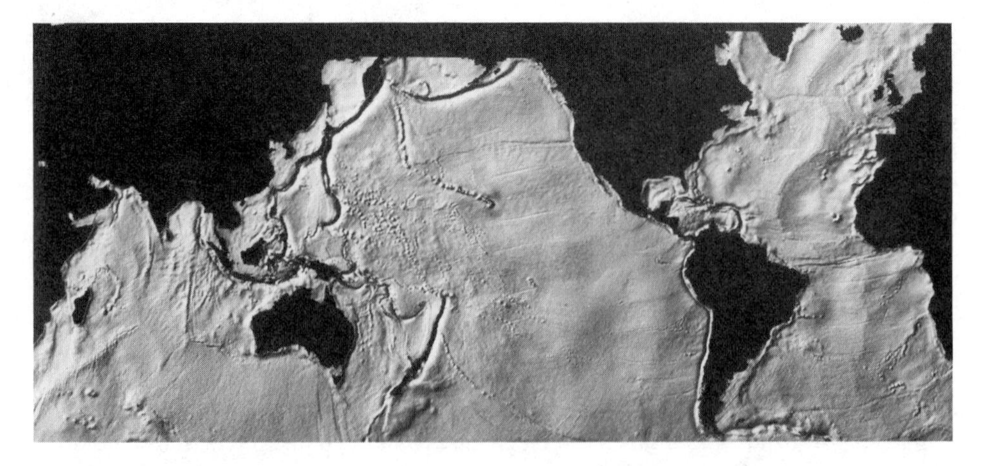

Figure 6-7-2. Topography relief map of the world based on satellite altimetry of the sea surface. Courtesy of Dr. J. G. Marsh, NASA.

about a meter) are delineated. Similar fracture-zone patterns are observed along the mid-Atlantic ridge.

If the marine geoid is accurately known, the height of the surface relative to the geoid can be determined with high accuracy. For example, Cheney and Marsh (1981) used *Seasat* altimeter data and the best available geoid model of the western Atlantic to determine the height of the sea surface, the position of the Gulf Stream and cyclonic and anticyclonic rings near the Stream, and the component of the surface geostrophic velocity perpendicular to the satellite ground tracks.

Thompson et al. (1983) worked with *Seasat* altimeter data in the Loop Current in the eastern Gulf. The Loop Current is of particular interest since it has a transport of approximately $30 \times 10^6\ m^3 \cdot sec^{-1}$ and eventually becomes a principal component of the Gulf Stream. Entering the Gulf through the Yucatan Strait, the Loop Current traces an anticyclonic path that at times nearly extends to the Mississippi Delta before turning southward and exiting through the Florida

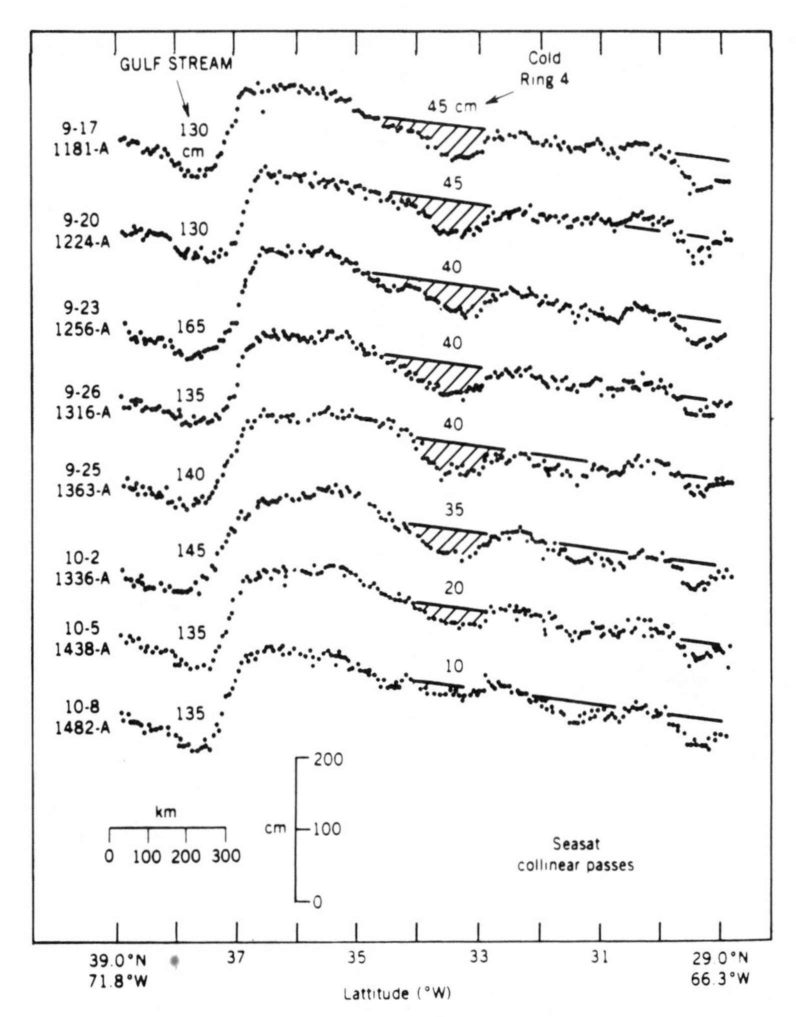

Figure 6-7-3. Successive *Seasat* altimeter profiles over the Gulf Stream system. After Elachi (1987).

Straits. Maximum geostrophic surface currents in the Loop can be greater than 150 cm · sec^{-1}, and the dynamic height across the stream sometimes exceeds 75 cm. Large anticyclonic eddies have been observed to break off from the Loop Current, following these intrusions. The major eddies typically have diameters of about 360 km and translate into the western Gulf with mean speed of 2.4 cm · sec^{-1} (Elliott, 1979).

Based on data from Thompson et al. (1983), we analyzed altimeter profiles through the Florida Straits, taking into account the large contribution of sea surface height due to the geoid. The maximum sea height change along the track is about 65 cm, with the maximum rms amplitude around 15 cm at about 84.9°W. Estimates for the geostrophic current using the maximum rms variability of the crosstrack surface geostrophic current are obtained from

$$|v| = g/f \left|\frac{\partial \zeta}{\partial n}\right|$$

where ζ is the rms variability of the sea height, f is the Coriolis parameter and n is the crosstrack direction. From September 17 to October 8 this value was about 40 cm · sec^{-1} at 84.7°W averaged over 50 km.

The tracking of a cold ring in connection with the Gulf Stream system has been accomplished with successive *Seasat* altimetric profiles, as shown in Figure 6-7-3 (Elachi, 1987). This demonstrated the capability to estimate the lifespan of cold rings with altimetry data.

7

OBSERVATIONS OVER THE CONTINENTS

7-1 GEOLOGICAL FEATURES

Imageries obtained from satellite altitude have proven to be of value for effective mapping of major geologic features, such as folds, fault offsets, intrusion outlines, volcanic flows of distinctly different rock units and many types of landforms, and in the geologic mapping of tectonics. *Landsat* data has been applicable especially for inaccessible regions and for regions for which only poor-quality maps have been produced. In fact, existing maps have been checked to correct mislocated or omitted rocks, geologic structures, and so on, and structural information, such as lineaments, have been used in search for ore deposits, oil deposits, oil accumulation, and groundwater zones. Dynamic surface processes such as interrelations between stream drainage systems and landforms are exceptionally well documented in satellite images. Deposits from glacial advances are also detectable, and moraines, drumlin fields, outwash plains, ice floe markings, and erosional features are visible in single frames or mosaics (Short and Lowman, 1973). Surface deposits in deserts and semiarid regions can be recognized from space. Regions undergoing active stream erosion can also be identified to a high degree, and sedimentation transport can be monitored more effectively than ever before. The ability to detect linear and circular features on the earth's surface is based mainly on the fact that tonally contrasted surfaces or alignments of topography are present. This includes structural elements such as metamorphic grain, fractures, joints, faults, and lineaments, which may include any of the foregoing structural elements or other tectonic features of indefinite nature.

The relation between geology and mineralogy can be seen by the fact that metallic mineral deposits are usually associated, either directly or indirectly, with igneous rocks. From a broader viewpoint, the distribution of such deposits is closely interrelated with regional geology, such as structure, lithology, and depth of exposure. For example, copper deposits, such as those of the southwestern United States, are associated with a class of young granite intrusion that is, in turn, related to the Basin and Range Province, a large zone of block-faulting. In contrast, chromium occurs in large intrusions of iron-rich igneous rock of much greater age and completely different tectonic setting. This shows that the identification of geological features from satellites may assist in the process of exploration for mineral deposits.

7-1-1 Lithology

The specific type of information extracted from space imagery for determining the distribution of rock types and soils, including alluvial material, geological structures, and landforms, is based on the chemical and physical status of the rock and soil type as well as the illumination angle and coverage by vegetation or surface deposits and topography or slope of the target.

It must be understood that changes in color, topography, and vegetation all serve to change the tone (color) and texture on space images, and it is difficult or impossible in most cases to attribute a change in tone or texture uniquely to one of the parameters alone without extensive ground data (Lee et al., 1974).

Elevation and slope direction control vegetation, which in consequence produce sharp boundaries between vegetated and nonvegetated areas and between different types of vegetation. Large tonal contrasts in the image may appear between vegetation-bedrock, vegetation-soil, vegetation-alluvium, vegetation-snow, trees-grass, and conifer-hardwoods.

Rock Types

As has been shown with spectral response data for rocks, it is rare to have specific response curves for many rocks or characteristic narrow wavelength for discriminating different rock types. The improvement in rock identification is mainly based on advanced computer processing of digital data, including contrast stretching and the use of ratios with different spectral bands. Discrimination between bedrock units is difficult and depends greatly upon the type of bedrock involved. Lithologic contrast is enhanced by the effects of differential erosion, producing long, linear outcrop bands. Tonal and textural imagery is enhanced if the spectral reflectance of adjacent units is highly contrasting, or if the units are selectively different in their support of vegetative growth.

An example of typical false-color products obtained by using three channels of the MSS of *Landsat* is given in Figure 7-1-1, which shows the Snake River Plains in South Dakota. The image in Figure 7-1-1 shows lava flows of Tertiary to Quaternary age, which appear in varying shades of dark blue, representing basalt lavas. The youngest of these flows form stable soils and appear as dark blue while other parts support vegetation such as sagebrush and grass.

Due to spectral differences, different topography, and the different types and amounts of vegetation coverage, lithologic discrimination between alluvium-covered valley floors and exposed bedrock is easily identified. Areas underlain by bedrock have moderate to high local relief and are commonly covered by coniferous forests. Alluviated areas are relatively flat and are not commonly covered by grass and small shrubs (Lee et al., 1974).

In addition to grain size, the sphericity of the grain and the moisture affect the scattering of grained components. Consequently, several factors have to be taken

Figure 7-1-1. The image of the Snake River taken with the *Landsat* MSS shows the lava flows of Tertiary to Quaternary age which appear in varying shades of dark blue, representing basalt lavas. Courtesy of NASA. See insert for color representation.

into account to properly interpret different tones in alluviated areas. The problem of discriminating lithologically dissimilar rocks from single-band images or composites of single-band images was studied by Knepper and Raines (1985). They concluded that a key factor in preparing band-ratio images of *Landsat* MSS data for discriminating rock and soil units is the determination of the proper contrast stretches to enhance spectral variations that are related to lithologic differences. In arid regions with little vegetation, histograms of band ratios show the spectral properties mostly of rocks and soils; however, as vegetation and water bodies become more abundant, the histograms no longer reflect the spectral properties of only rocks and soils, and determining appropriate contrast stretches for lithologic discrimination is more difficult.

A sample of volcanic feature superimposed on Precambrian rock and surface loose deposit is shown with the Tibesti Mountains and the surrounding Sahara in Figure 7-1-2.

Figure 7-1-2. Most of the area shown here in the Tibesti Mountains is underlain by Precambrian rocks, except for small patches of Paleozoic and younger sedimentary rocks. To the right are northwest- and northeast-trending faults appearing as straight valleys. Along the south edge lies a part of the Tibesti volcanic field; the most prominent crater is the Tarso Toon, an immense caldera. The black area at lower left is a field of andesite (plagioclase feldspar with dark ferromagnesian minerals and little or no quartz) lava surrounding the crater, Pic Touside. After Short et al. (1976).

Methods for using data from the *Landsat* MSS were identified by Blodget et al. (1978). They found that the most useful product for visual interpretation was a color-composite image constructed by using contrast-stretched ratio data. Using this imagery, they were able to visually discriminate among a wide variety of igneous and metamorphic rock classes and to identify gossans overlying known massive sulfide deposits. Color-composite enhanced imagery constructed by using principal-components and canonical-analysis algorithms permit visual separation of several major rock units, but the inherent more narrow range of color variation caused ambiguity in distinguishing among some of the rock classes that could be easily defined on the ratio image.

A test area selected by Blodget et al. (1978) was located on the southern margin of the Arabian shield with basement rocks consisting of a complex sequence of generally metamorphosed, interlayered volcanic and sedimentary Precambrian assemblages. These were locally intruded by igneous rocks ranging in composition from gabbro to syenite and in age from Precambrian to Cambrian. The volcanic rocks in the area varied in composition from andesite to thyolite and in texture from sedimentary strata that variously include sandstone, conglomerate, graywacke, shale, and limestone. The basement rocks were in part unconformably overlain by recent unconsolidated alluvial and aeolian sands and Cambro-Ordovician Wajid sandstone. The latter laps onto the shield from the east and south. This sandstone has been eroded from much of the western part of the test area and is now observed only on isolated buttes where the remnants cap the basement.

Figure 7-1-3*a* is a geologic map modified from that portion of the Asir Quadrangle and is compared with the ratio color-composite image in Figure 7-1-3*b*. The three dark blue subcircular spectral units just south of the center in Figure 7-1-3*b* correspond closely to the peralkalic granite intrusions (identified as gp in Figure 7-1-3*a*). In addition, all other granite intrusions mapped in the lower two thirds of the test area can be identified equally well. In the northern part of the scene, however, similar granites are complexly associated with basic and ultrabasic intrusive rocks, and both lithologies are in part overlain by the Wajid sandstone; thus, the margins of most of the individual rock units are not easily defined spectrally. The calc-alkalic granites are equally well defined spatially, but they cannot be visually distinguished from the peralkalic intrusions on the enhanced imagery.

The refined spectral bands selected for the TM provide the ability to distinguish among a wider variety of rock classes than was possible with the MSS data; they also provide improved definition within a specific class of rocks. This was shown by Blodget et al. (1985), who analyzed *Landsat 5* TM data in southwestern Saudi Arabia (see Fig. 7-1-4). A combination of TM band ratios 3/2, 3/7, and 5/7 displayed as red, blue, and green, respectively, provided the greatest geologic information; however, no single band, ratio, or composite devised was found to display all the geologic information in the TM data. Simultaneously, study of several enhanced images permitted good spectral separation of 10 intrusive, extrusive, and sedimentary rock classes. The Wassat Formation outcrops mainly in the eastern half of the study area. It consists primarily of metamorphosed volcanics of basaltic to dacitic composition and contains subordinate metamorphosed conglomerate, chert, and marble; facies vary significantly within and between outcrop belts. Spectral variations suggesting separability of the constituent rock units are most sharply displayed in the subtriangular Wassat outcrop designated (1) on Figure 7-1-4. On the standard color-composite image, the Wassat Formation is not separable from the Tertiary volcanics (2) or the gabbro/diorite/quartz-diorite (2) intrusions. By combining ratios 3/2, 5/7, 3/7 using red, green, and blue colors, respectively, the inselbergs forming the western margin of the Wassat outcrop display a sharp spectral difference from the main outcrop. The royal blue spectral hue associated with the inselbergs is similar to but not identical to that displayed

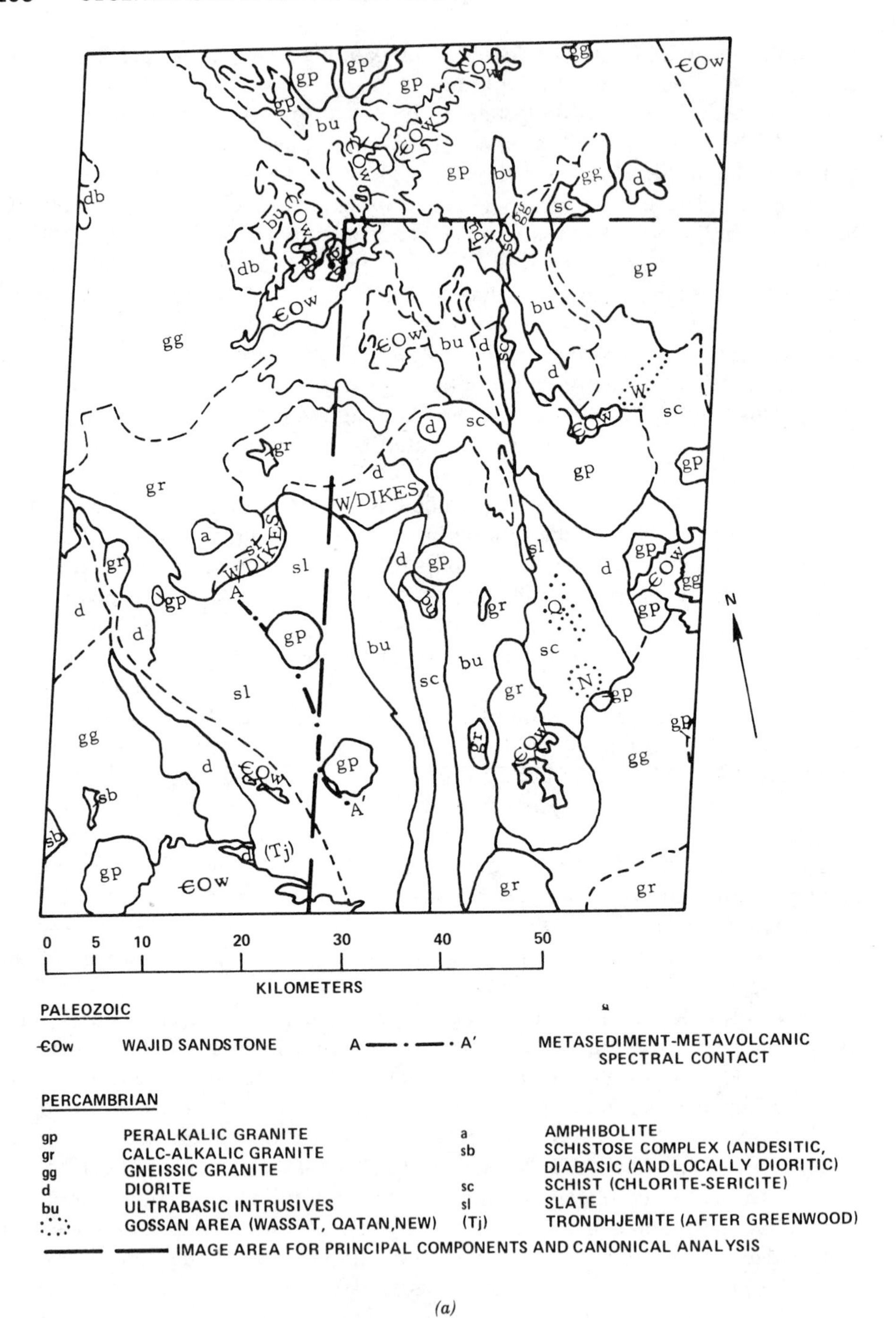

(a)

Figure 7-1-3. (*a*) Geological map of Wadi Wassat area. *Continued on next page.*

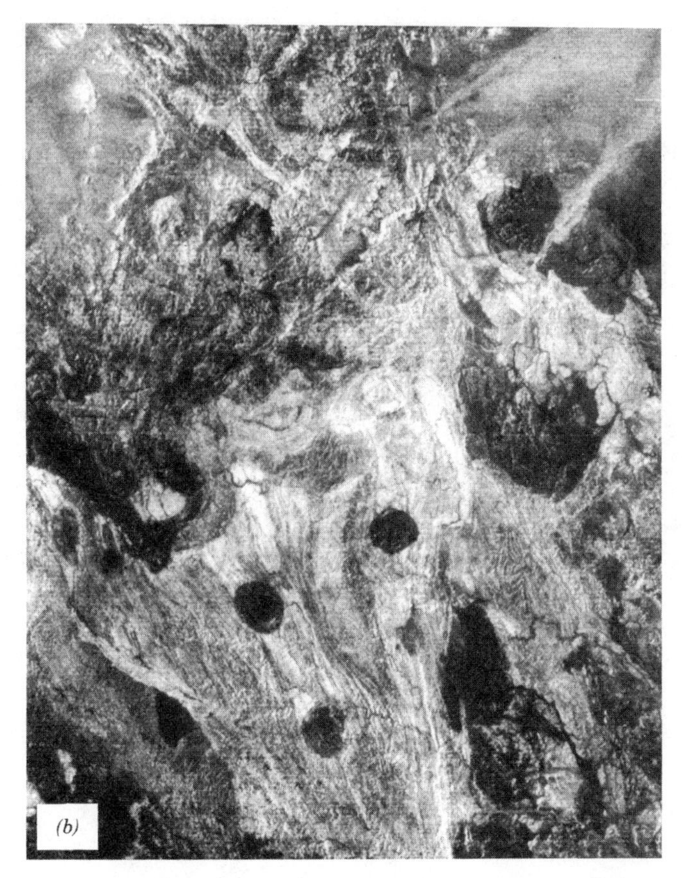

Figure 7-1-3. (*Continued*) (*b*) Contrast-enhanced ratio color-composite image of the Wadi Wassat area (digital scanner data from *Landsat* scene 1226-07011). Courtesy of Dr. H. Blodget, NASA GSFC. See insert for color representation.

by Sirat basalts, and is indistinguishable from the northern portion of the large triangular unit (3).

The SIR-A has been considered good for structural information, although detail is often difficult to observe (Koopmans, 1983). The influence of surface roughness in unvegetated areas on the radar backscatter makes terrain differentiation possible, a facility that is not attainable with *Landsat*.

The effective combination of the data sets is also reflected in approaches to merge radar images with those obtained by *Landsat* MSS. For instance, an effective method of enhancing the SIR-A imagery for geological mapping combines the radar data with data from the *Landsat* MSS. Digital registration of the two images can be facilitated inasmuch as the severe distortions encountered in imagery generated by airborne systems are not encountered in imagery generated by the SIR-A system. Color combination was found to be the best method for displaying the rock-type discrimination capabilities of computer-enhanced SIR-A imagery (Curtis et al., 1986).

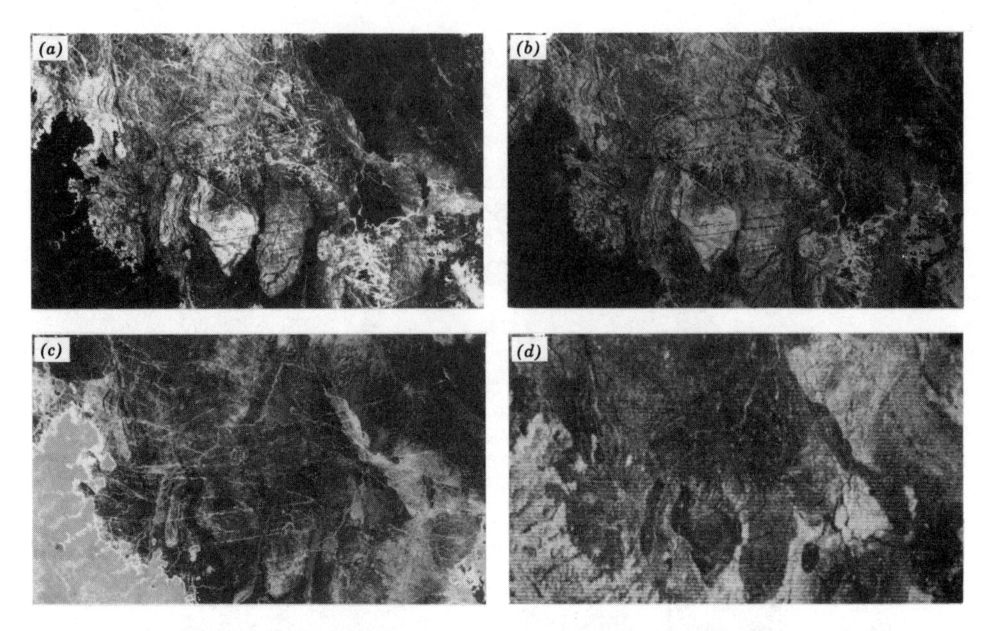

Figure 7-1-4. Enhanced TM imagery of the Sirat Mountains area, Saudi Arabia. Numbers designate areas discussed in text. (*A*) TM bands 2, 3, and 4 displayed as blue, green, and red, respectively; linear contrast stretch. (*B*) TM bands 1, 2, and 7 displayed as red, green, and blue, respectively; decorrelation stretch. (*C*) TM ratios 3/2, 3/7, and 5/7 displayed as red, blue, and green; histogram equalization stretch. (*D*) MSS ratios 4/5, 5/6 and 6/7 displayed as blue, green, and red; histogram equalization stretch. Courtesy of Dr. H. Blodget, NASA GSFC. See insert for color representation.

Visible and near IR reflectivities pertain mainly to surface chemistry, but radar at L band ($\lambda = 25$ cm) gives roughness information on a scale proportional to the radar wavelength, particle slope, and density packing (Daily et al., 1979).

Analysis of coregistered *Seasat* and *Landsat* images reveal that radar image data can contribute in rock-type discrimination over *Landsat* data alone (Blom and Daily, 1982). Incorporation of textural measures from the radar images greatly increases their value and results in discrimination ability. Other texture measures found very useful are hue saturation-intensity split spectrum processing and Fourier bandpass filtering. The added dimension of color to the radar image is a potentially powerful image enhancement tool. The bandpass filtering technique results in a radar image texture map, which separates rock units based on their characteristic spatial frequency distributions. The technique can be of considerable value in heavily vegetated terrains where only the erosional differences between various rock types can be perceived in the image.

Soil Classes

Reflectance properties of soils are a conglomerate of many factors, of which soil color and soil moisture are the most dominant ones. For instance, the medium-textured soils of the southeast United States have a reddish or yellowish hue, with reflectance minimum at around 16% moisture content. To a certain degree, vegetation cover is related to a particular soil association or soil group, which means that in many cases soil associations are delineated through the reflectance characteristics of a reasonably uniform type of vegetation possessing the same boundaries as the soil associations (Parks and Bodenheimer, 1973).

In arid areas multispectral data can be used directly for soil mapping because low percentage vegetation cover favors the detection of clay, calcareous, gypsiferous, and saline soils (Mulders and Epema, 1986).

One of the most impressive gradients produced by dry sand and moist soil covered by vegetation is shown for the Nile region in Figure 7-1-5.

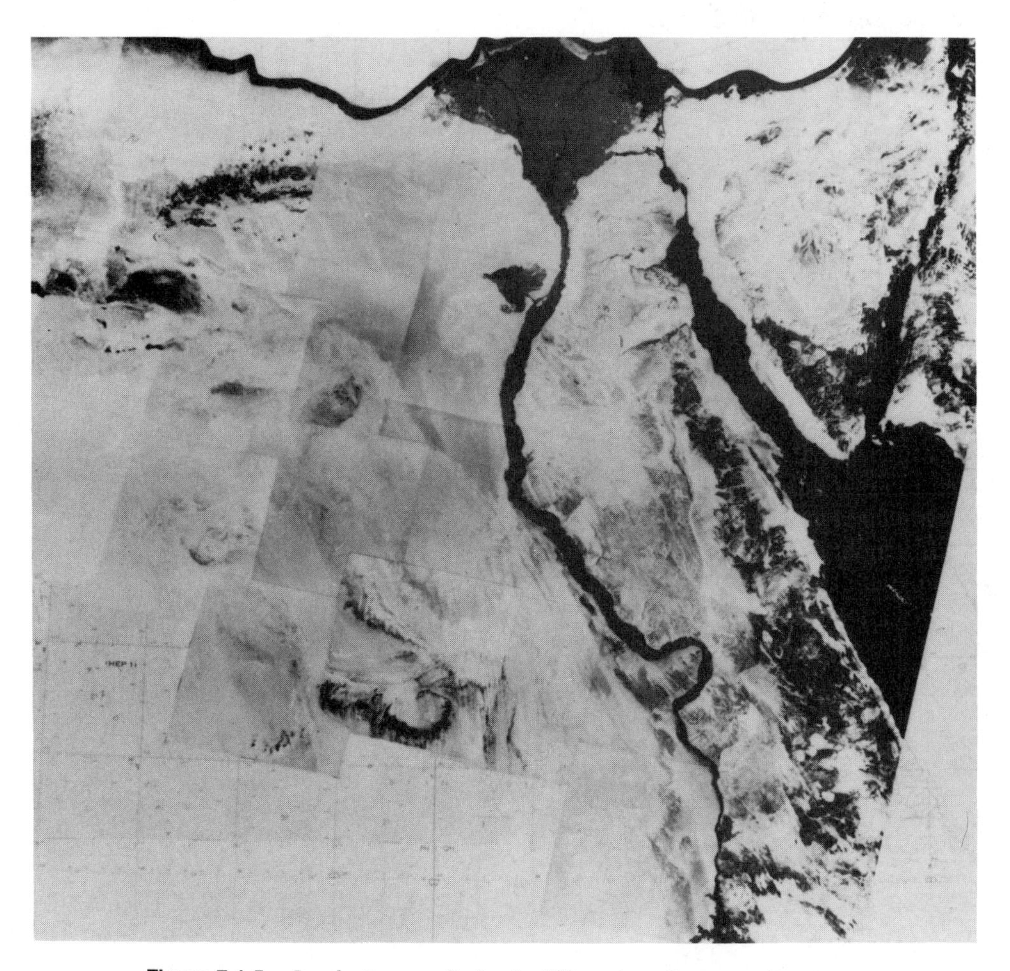

Figure 7-1-5. *Landsat* composite for the Nile region. Courtesy of NASA.

By its nature as an arid region, desert shows almost no vegetation. Therefore, the signals represent mainly information on both rocks and soils. An example of this is given in Figure 7-1-6 for the Baluchistan Desert. The Mirjawa Range (1) exposes faulted and folded sedimentary rocks of Cretaceous and Tertiary age, with some places being intruded by igneous plutons (2). Surface deposits are presented in a series of longitudinal sand dunes to the north (upper part of Fig. 7-1-6).

Another desert type of environment mainly marked by physiographic landmarks is an example of Death Valley (Fig. 7-1-7). This figure clearly shows alluvial valleys and details of rock types and geologic structures. Darker rock materials along the flanks of the uplifts are mainly gravels being washed into the basins to form sediments and alluvial fans. The central basins may contain well-developed playas consisting of fine, light-colored clays and associated salt deposits. Several major fault zones pass through the scene. The Garlock Fault marks the northern boundary of the Mojave. The Las Vegas shear zone is expressed by the sharp truncation of bedrock. Nearby, a band of light-colored rocks in Navajo sandstone overlain by older carbonate rocks at one of the best-displayed thrust faults visible in the western United States.

Figure 7-1-6. The Baluchistan Desert. Courtesy of NASA.

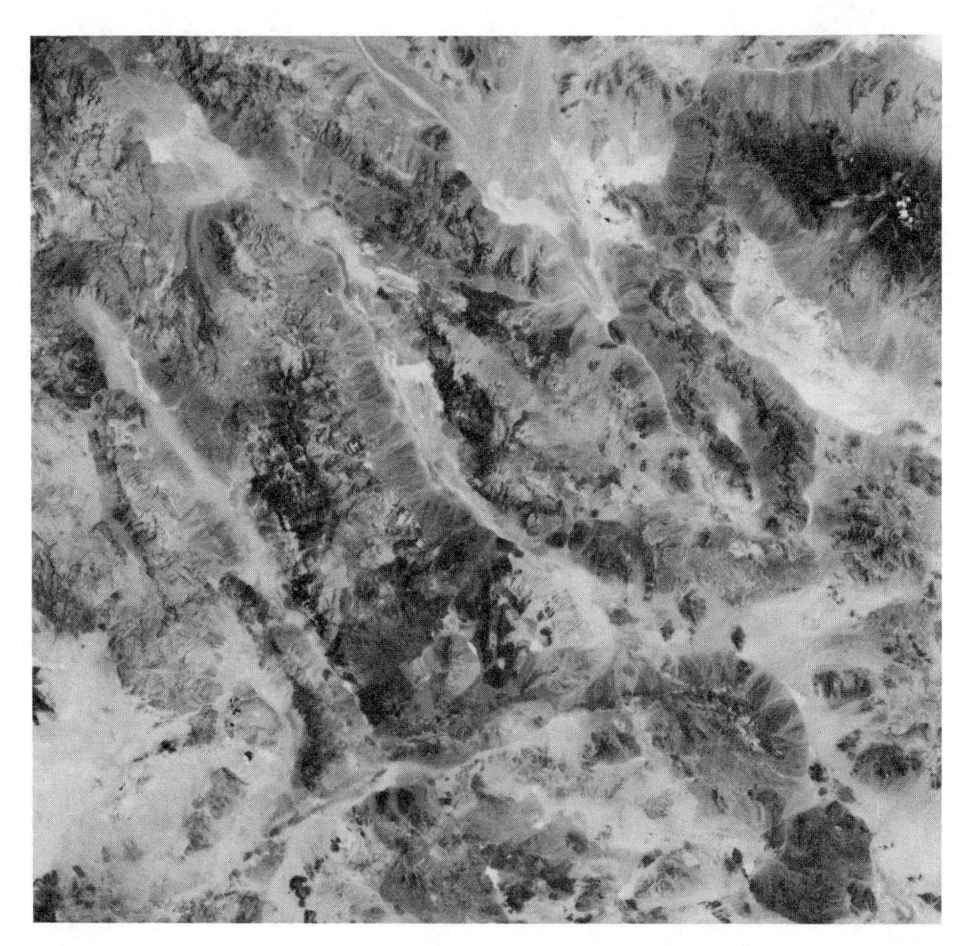

Figure 7-1-7. Death Valley region and southeastern California Amargosa Range. Courtesy of NASA.

The relationship between landforms, soil classes, and vegetation is demonstrated with a photo taken by the earth terrain camera over the Salton Sea (Fig. 7-1-8) in southern California. The Sea was created during the early part of this century when the Colorado River flooded its right bank and, aided by ditches and canals dug in an early attempt at irrigation, spilled over two dry washes and filled the lowest portion (82 m) below sea level. Subsequent evaporation produced a lake with high salinity, although groundwater input, resulting from irrigation in the valley, maintains the lake's level. The pattern on both the northwest and southwest margins of the Sea represents irrigated agriculture in the Coachella and Imperial Valleys. Mild winters permit year-round cultivation of a variety of agricultural products. Differences in seasonal planting and harvesting practices combined with variations in the intensity of land use are shown in the line of demarcation delineating the United States–Mexico boundary south of El Centro (Short et al., 1976).

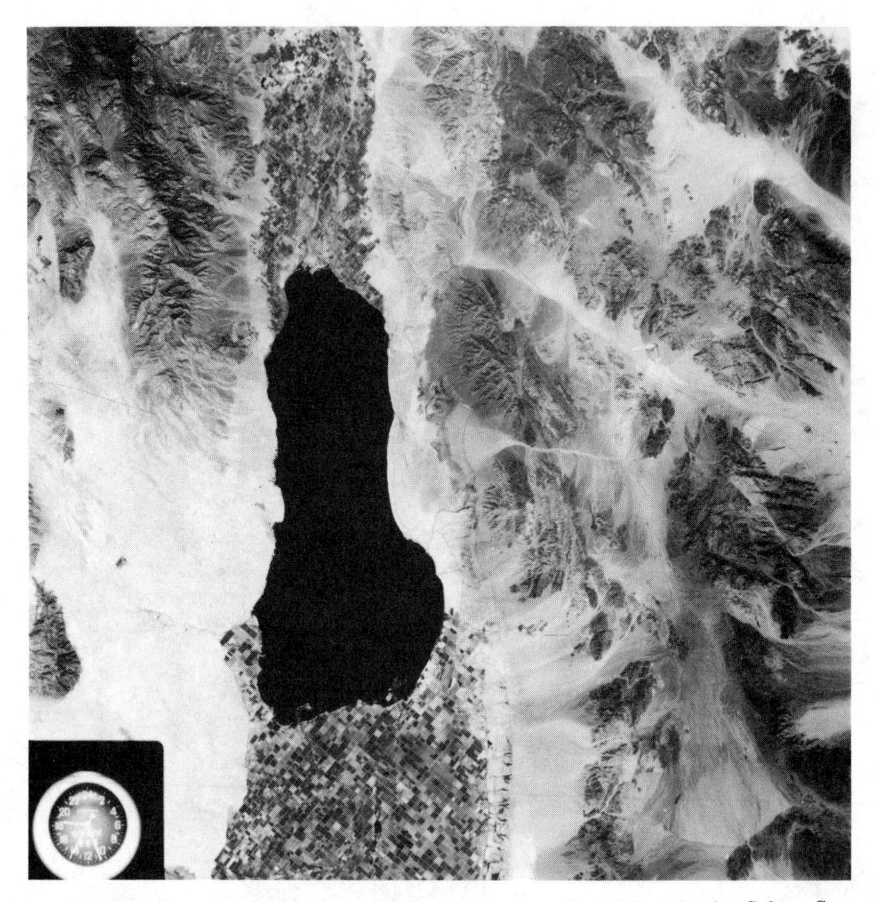

Figure 7-1-8. Earth terrain color-IR photo showing many alluvial fans in the Salton Sea area. Courtesy of NASA.

It has been observed that highly saline soils having high electrical conductivity show higher spectral reflectance than do moderately saline soils, because of the encrustation of soluble salts on the soil surface. The surface scrapping of such soils is white or light gray or pale yellow, depending on the salt content, which appears white or gray in the false-color composites. Table 7-1-1 shows the relative average reflectance levels of salt-affected soils in different *Landsat* bands in comparison with the normal soils from the same *Landsat* scene (Venkataratman, 1984).

TABLE 7-1-1. Reflectance Levels of Soils in Various *Landsat* Bands

Category	Band 4	Band 5	Band 6	Band 7
Highly saline soil	114.7	159.3	161.6	124.8
Moderate saline soil	81.2	113.5	123.0	98.0
Normal soil with crop	35.9	38.7	107.7	117.7

After Venkataratman (1984).

In some cases, based on the conditions of the crop growth, saline and alkali soils can also be differentiated, especially when multitemporal data are used. For example, wheat is not affected much by salinity, but sodicity produces sparse crop growth. When the crop is harvested, the saline and alkali land maps appear gray in the false-color imagery of *Landsat* MSS.

The concentration and type of salts in soils determine to a high degree the characteristics of soils especially in arid regions, because they affect the spectral behavior of soil throughout the EMS. For instance, the salty crust of soils may alter not only the reflective properties of the surface but also its thermal behavior. When the atmosphere is in contact with hypersaline soils, the saturation vapor pressure is lower than the value for air in equilibrium with pure water. This is of special relevance for the proper interpretation of thermal IR images of areas, such as playas, where salt minerals are a dominant soil constituent. Surface temperature patterns in these areas can be related more to the interaction between the surface of salty crusts and water vapor than to the actual bulk soil thermal properties (Menenti et al., 1986).

The effects of moisture content and thickness of surface deposit are more pronounced in the microwave range, because under certain conditions radar penetrates the surface layer on the basis of the dielectric constant of sand as a function of moisture. The power skin depth, the depth at which the power is attenuated to 37% of the value at the surface, ranges from 2 to 0.5 m, where the moisture content varies from 0.25 to 1.0% (Blom et al., 1984).

In order for subsurface imaging to occur, the following requirements must be met simultaneously:

1. The surface to be penetrated must be smooth to the incident beam (otherwise it would backscatter and be detected by the radar).
2. The subsurface to be detected must be rough so that it can backscatter the incident beam.
3. The material to be penetrated must not be too thick, or the signal will be attenuated during passage in and out of the material. The maximum depth through which an image can be formed appears to be 2–5 m.
4. The medium to be penetrated cannot have scatterers embedded within it. It must be homogeneous, fine-grained material with no cobbles or large pebbles.
5. The moisture content must be less than about 1%. A higher moisture content causes more energy to be specularly reflected away at the upper surface–air boundary and increases the signal attenuation within the penetrated medium by raising the loss tangent of the material (Blom and Dixon, 1986).

Assuming all of these conditions are met, there are two effects that increase the echo strength from the subsurface:

1. The wavelength of the incident radar beam is actually shortened, so a subsurface need not be as rough as the same surface exposed to produce the same echo strength.
2. The incoming radar beam is refracted; thus, the incidence angle is reduced at the subsurface rather than the surface.

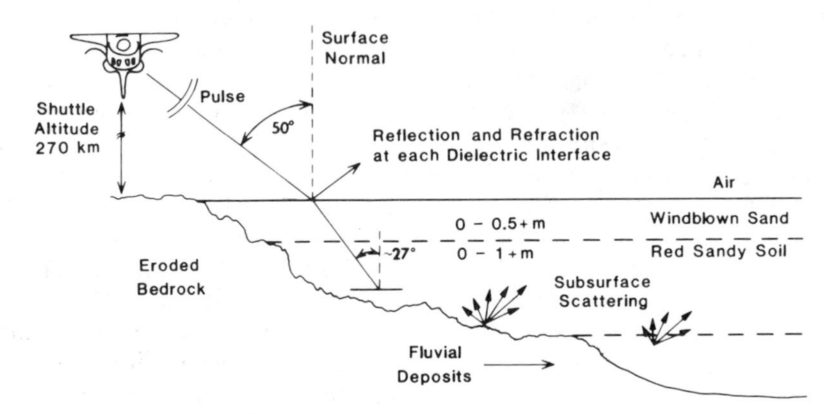

Figure 7-1-9. Diagram showing refraction and reflection of the radar wave at air-surface boundary with subsurface scattering, producing an image of the subsurface. After Blom and Dixon (1986). Reproduced with permission from Graham & Trotman, Ltd.: Szekielda, K.-H., 1986. *Satellite Remote Sensing for Resources Development.*

Figure 7-1-9 shows these effects schematically. Elachi et al. (1982) treated these phenomena and calculated that under favorable conditions a subsurface feature may have a stronger echo than if it were exposed at the surface.

Figure 7-1-10 is an SIR-A image of a hyperarid zone of northeastern Mali, the westerly end of the Adrar des Iforas region. The most conspicuous structures are the round younger granites with ring complexes, the A Idoenyan in the south, and Timedjelajen complex in the north. An extensive field of dike swarms of acid and base composition are related to these ring complexes, which cut the Lower Proterozoic granite basement. Particularly in the central valley part, many of these dikes are lightly covered by windblown sands (Koopmans, 1983).

7-1-2 Structures

In principle, features viewed on *Landsat* imagery can be grouped into the category of continuous structures, which are important to mineral exploration. Considering the ground resolution of *Landsat* MSS, only the widest fracture zones can be seen, although smaller features, such as linear structures, can indirectly be recognized due to the specific spectral signature of vegetation associated with them and other environmental conditions, which might be slightly different from those in the immediate adjacent area. Topographic features that develop preferentially along such linear fractures also reveal their presence (Hodgson, 1975).

Faults and joints in rock are important features in geologic and hydrologic studies. These rock fractures are commonly indicated by the discontinuous edge and line segments that form lineaments on remotely sensed images. Lineaments generally are mapped with a manual analysis and interpretation procedure. A manual procedure is subjective, however, and results are controversial (Moore and Waltz, 1983).

Figure 7-1-10. SIR-A imagery of Adrar des Iforas, Mali. Younger granite ring complexes are conspicuous at the top and bottom of the imagery. A field of dike swarms crosscutting in different directions is clearly visible, although they are partly covered by a thin layer of windblown sand (central part). Courtesy of USGS.

Various objective approaches to lineament enhancement are possible with digital processing. Since fractures commonly are indicated by edge and line segments that form lineaments on remotely sensed images, Moore and Waltz (1983) developed automated and objective procedures for lineament mapping. In support of this objective, a five-step digital convolution procedure was used to produce directionally enhanced images that contain few artifacts and little noise. The main limitation of this procedure, however, was that little enhancement of lineaments occurred in dissected terrain, in shadowed areas, and in flat areas with a uniform land cover. The digital convolution procedure consists of five steps:

1. Generating a low spatial frequency image with an averaging function
2. Extracting directional data with a convolution filter
3. Smoothing the directional data with an averaging or tangent function
4. Smoothing the data further by extracting directional trends in the tails of the image histograms
5. Adding the enhanced directional trends to the original image

The directional enhancement procedure can be modified to extract edge and line segments from an image. Various decision rules can then be used to connect the line segments and to produce a final lineament map. The result is an interpretive map, but one that is based on an objective extraction of lineament components by digital processing.

Lineament interpretation in the analogue mode is subjective, and results depend mostly on the scientist's experience and on the objectives for interpretation. Comparison of operator results in one study (Podwysocki et al., 1977) showed that only 0.4% of the total 785 linears were seen by all four operators, 5% by three operators, and 18% by two operators. The same study also found significant differences in the average lengths of lineaments mapped by different scientists. The patterns obtained by different operators should be similar, but wide variations occur in lineament locations and densities. Thus, lineament maps that have not been field checked can be questionable (Moore, 1982).

Another problem in lineament mapping lies in biases introduced by illumination direction and data characteristics. Lineaments with trends near the illumination direction are partly obscured, whereas lineaments with trends at angles of 60 to 120° from this direction are maximally enhanced. *Landsat* images have a southeasterly solar azimuth in the northern hemisphere, and lineament frequency diagrams generally show a northwesterly minimum and a northeasterly maximum. Another bias is caused by the nearly east-west scan lines on a *Landsat* MSS image; image interpreters tend to ignore features that have nearly the same trend as the scan lines.

The often controversial subject of linears and lineaments has received much attention. Numerous faults on geologic maps have not been identified on *Landsat* imagery. Conversely, many linears on the imagery do not correlate with mapped faults, although a number of new faults have been discovered on *Landsat* imagery, even in regions that have been thoroughly mapped. Several excellent individual investigations of *Landsat*-identified linears have been reported, but a systematic evaluation of linears expressed on different scales and types of imagery in different geologic terrains is needed.

Another example of the impact of ground resolution onto the final results of image interpretation for lineaments is shown with Figure 7-1-11, based on a comparison of MSS data with TM data shown in the image display in Figure 7-1-12.

Seasat SAR images over Jamaica allowed recognition of several new aspects of faulting in Jamaica, including a major throughgoing lineament, the Vere-Annotto lineament (see Fig. 7-1-13), and a series of curving scissors faults in the central part of the island. Both features have important implications for the Neogene structural evolution of the island. Blom and Dixon (1986) indicated that many of the lineaments visible in SAR data over Jamaica probably represent active faults, because the rapid solution of the limestone under tropical weathering conditions would quickly degrade any topographic features not actively maintained.

The advantage of satellite coverage is particularly well-demonstrated in remotely located areas. An example is given in Figure 7-1-14 which shows the Richat structure in a remote part of Mauritania. This structure is a series of concentric quartzite ridges with a diameter of 30 to 40 km averaging about 100 m in height, separated by toroidal valleys underlain by less-resistant rocks. The topography is the result of differential erosion, although the origin of the structure is not well known. The presence of igneous rock, as sills and dikes, suggests that the structure was produced by intrusion of magma. Located in the Dhar Adrar Plateau, it is essentially a flat-lying remnant of Ordovician sedimentary rocks capped by the Chinguetti sandstone. To the north is the Makteir dune field, an array of linear or seif dunes; the prevailing winds blow from upper right to lower left, as indicated by the sand streaks (Short et al., 1976).

Circular features have also been identified in *Landsat* images. For instance, Albert and Chavez (1975) mapped circular features that could be correlated with

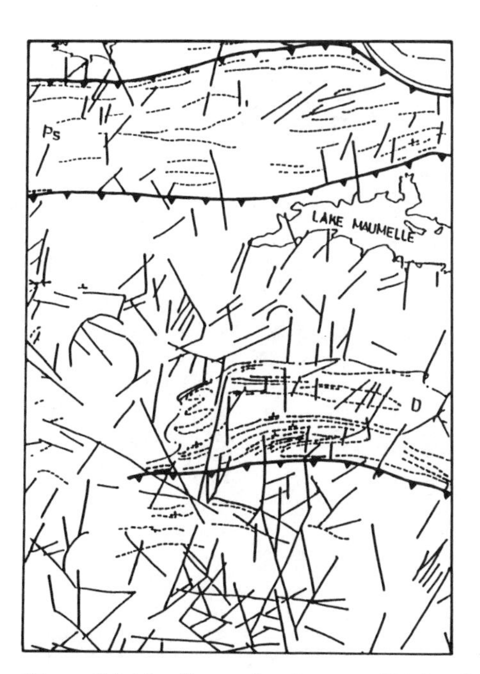

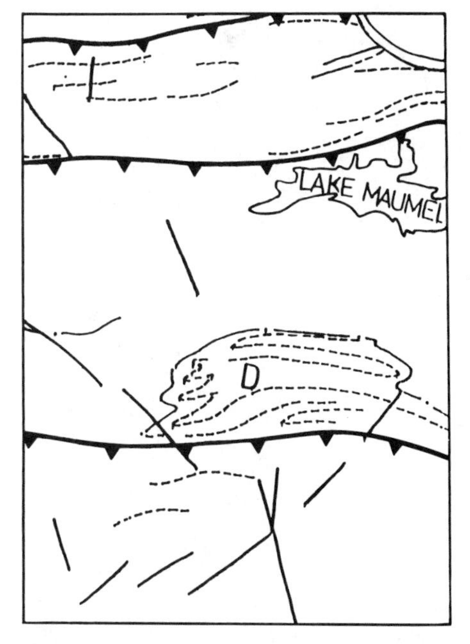

Figure 7-1-11. Comparison between TM data (left) and MSS data for lineaments interpretation of the Arkoma Basin (U.S.). Courtesy of Japex Geoscience Institute.

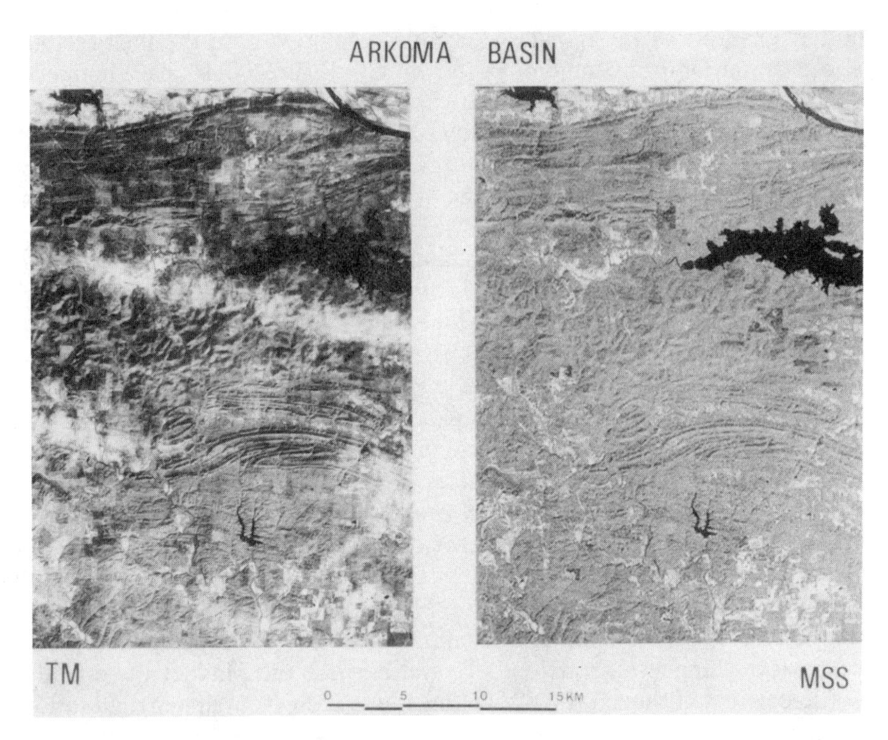

Figure 7-1-12. Image display of Arkoma Basin, comparing TM image (left) with that of MSS data.

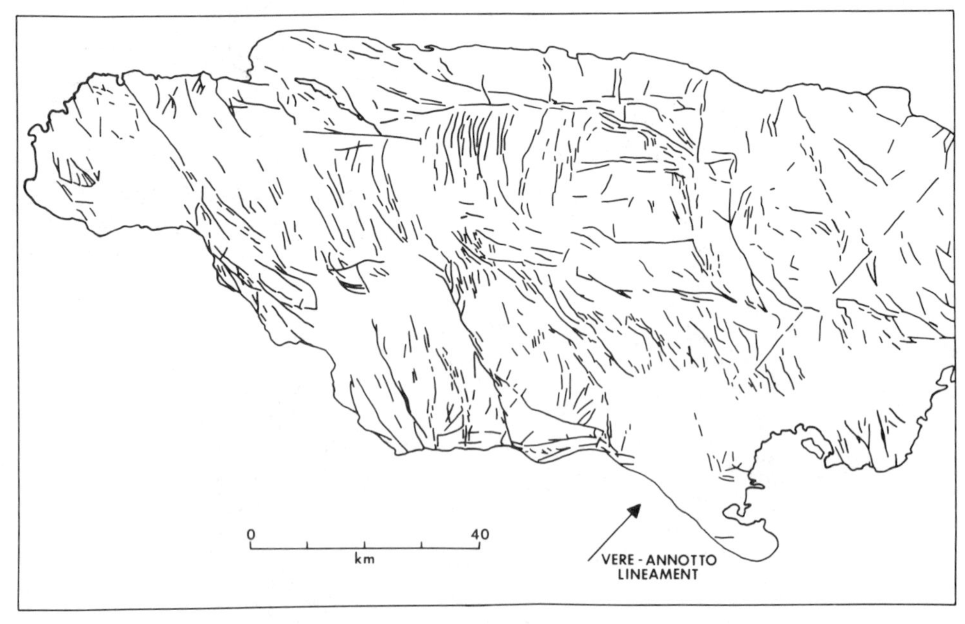

Figure 7-1-13. Lineament map of Jamaica derived from *Seasat* images. Many of lineaments are known faults; others such as the Vere-Annotto lineament, may have been an unknown structure. After Blom and Dixon (1986). Reproduced with permission from Graham & Trotman, Ltd.: Szekielda, K.-H., 1986. *Satellite Remote Sensing for Resources Development.*

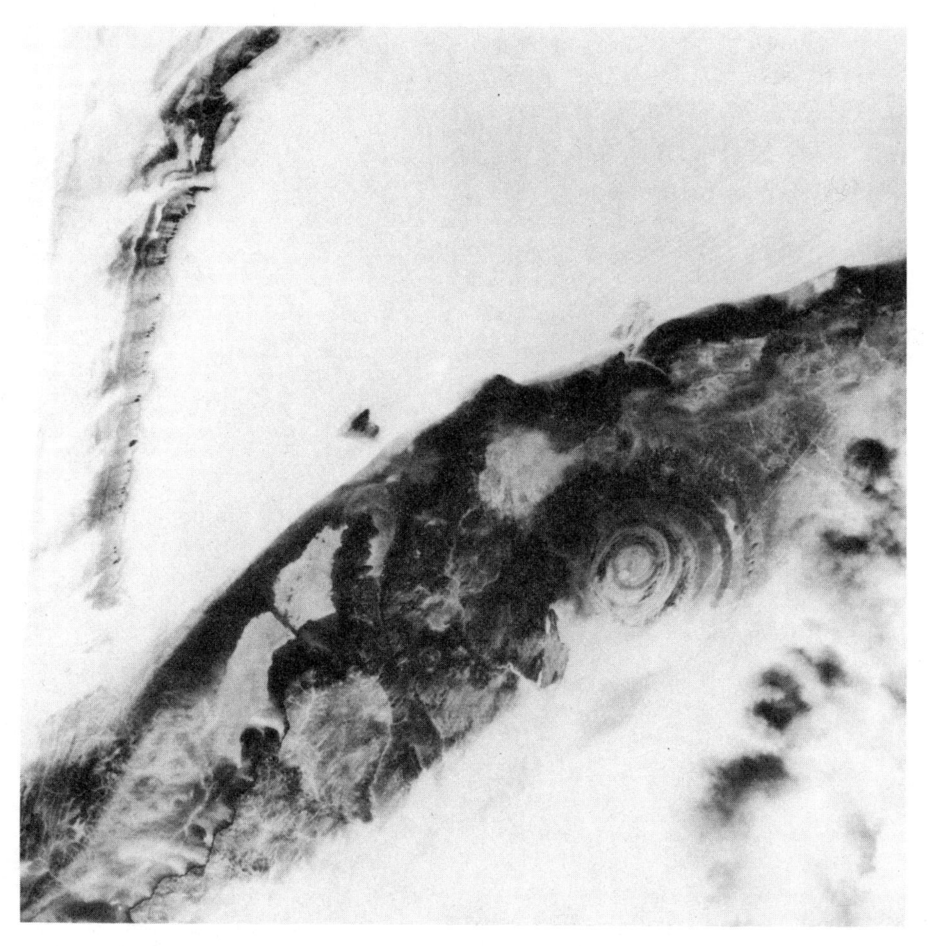

Figure 7-1-14. Richat structure, Mauritania recorded with *Landsat* MSS.

aeromagnetic and geologic data, suggesting that they may be related to concealed intrusive bodies. The circular features observed have a diameter of up to 200 km and appear to be related to volcanic activity. The relation of circular features to mineralization is, however, not clear at this point.

A circular depression about 66 km wide in east central Quebec is shown in Figure 7-1-15, where Precambrian crystalline rocks constitute most of the higher central core, but these are overlapped by Mesozoic volcanic rocks within the depressions. This geological feature is believed to be a volcano-tectonic structure, but the probability that it is a meteorite crater scar has gained wide acceptance. The circular depressions serve as a water reservoir, which is covered with ice. Adjacent hills display prominent crag-and-tail morphology resulting from Pleistocene glaciation. Such scouring also emphasizes the contrasts between the major classes of Precambrian rocks, as represented by the rugged landforms developed on crystalline, igneous and metamorphic rocks and the subdued relief of the gneisses and schists.

Folds are also expressed as lineaments, although they are not as pronounced, due to fractures. The surface expression of folds is based on topography and

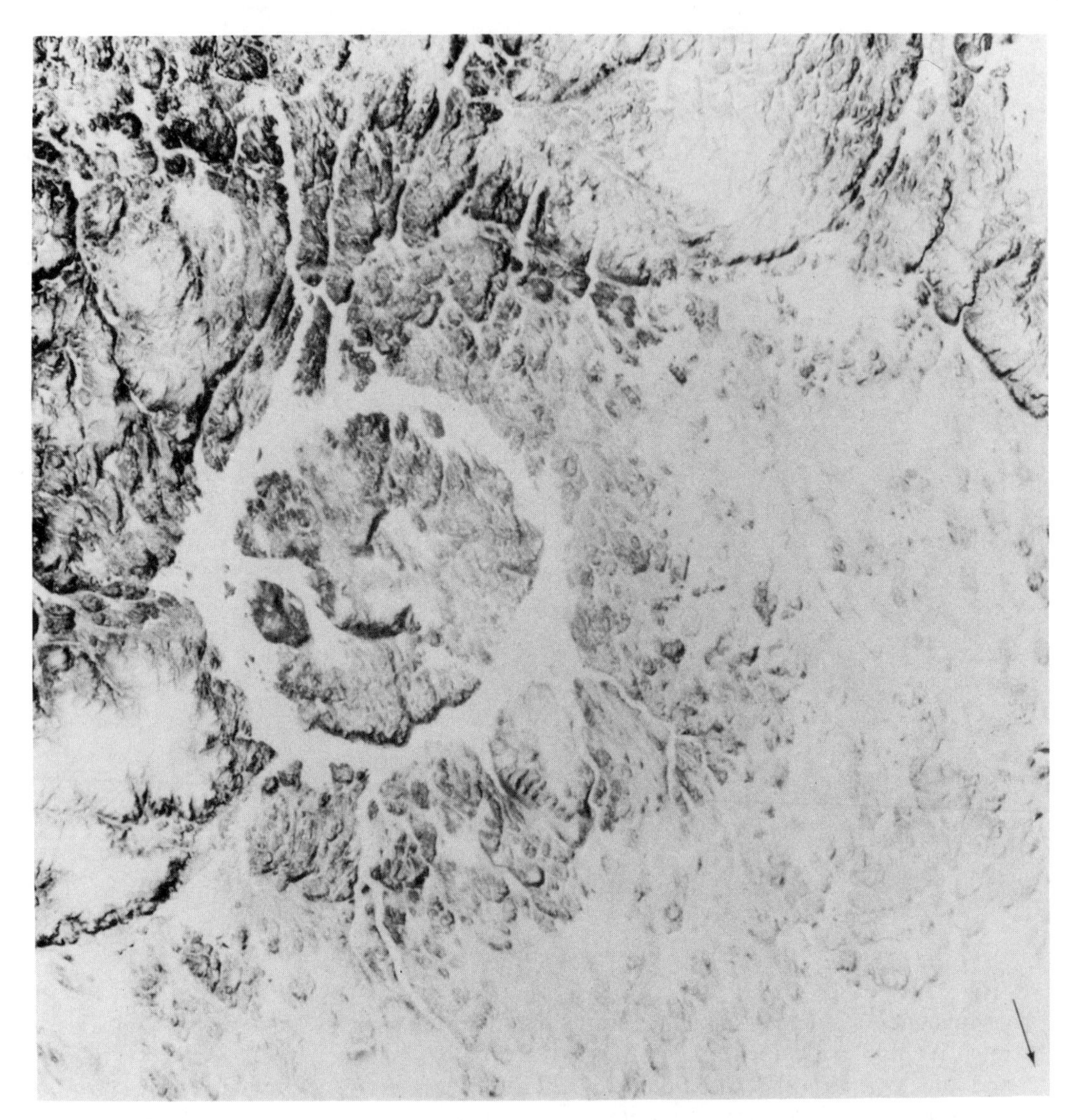

Figure 7-1-15. Manicouagan Lake, Quebec, recorded with *Landsat* MSS in March 1977. Courtesy of NASA.

related phenomena. For instance, Lee et al. (1974) pointed out that dip slopes may be directly recognized where stereo viewing is possible. Without stereo, indirect expressions helpful in slope interpretations are shadow relationships and variations of vegetation that are topographically controlled. Drainage patterns are indicators of slopes, and the drainage/strata geometry is the most consistently useful criterion for determination of dips.

An example of fold structures that have been recorded repeatedly with different sensors in different portions of the EMS is given for the Appalachians in Pennsylvania. In an IR image from the HCMM, in Figure 7-1-16, the Appalachians

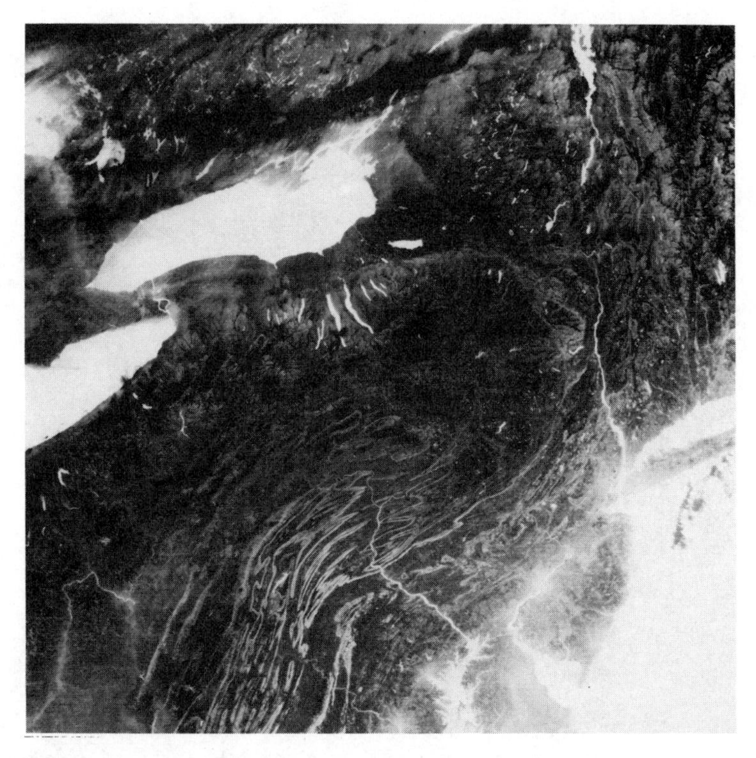

Figure 7-1-16. Northeastern United States as recorded in the IR by the HCMM during night of November 2, 1978. Courtesy of NASA.

appear as warmer ridges against the lower emission of the valleys. Topographic expressions in the hills and valleys of the Catskill Mountains, Green Mountains, and Allegheny Mountains are emphasized by the cooler valleys and slopes compared to the warmer uplands. Portions of the area can also be seen in the imagery (Fig. 7-1-17) taken with the multispectral camera aboard *Skylab,* and a portion of the area was also recorded with the *Seasat* SAR (see Fig. 7-1-18). In the *Seasat* image, the pattern of "noses" formed by the mountains and valleys represents plunging structures, such as the synclines and the anticlines. Regional drainage crosses the structural strike, as demonstrated by the Susquehanna River and its major tributary, the Juniata River. The heavily forested mountain slopes exhibit fine, smooth texture and generally low image contrast, because most of the mountains are elongated features that strike almost parallel to the direction of radar illumination. By comparison, the valleys have a wide range of image tone and texture (Ford et al., 1980).

At a higher resolution, folds are registered with more details, as shown in Figure 7-1-19 with an example over Algeria. In the image recorded by SPOT strongly folded beds appear in the northwestern part, with tones corresponding to different geological units and associated vegetation pattern. Cultivated areas and palm groves show up in bright red tones along the wadies in disseminated patches.

The continental drift, plate tectonic rifts and fault systems have been well documented with satellite data. The rift associated with Middle East plate tecton-

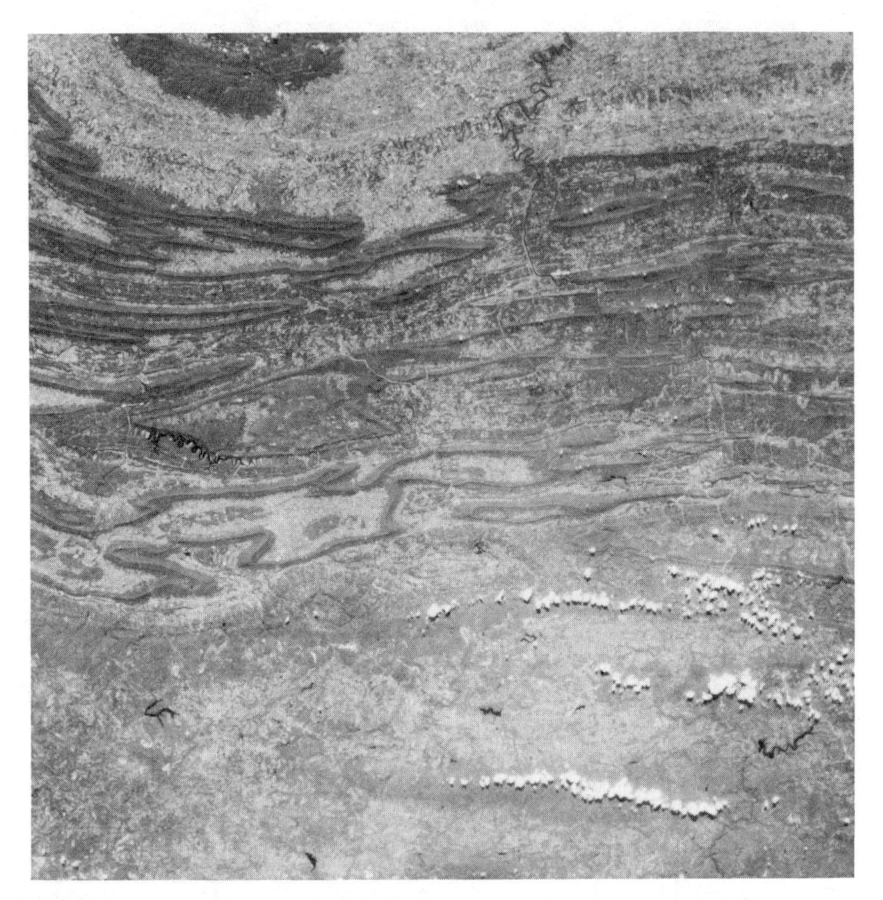

Figure 7-1-17. The Appalachians photographed with the multispectral camera aboard *Skylab*. Courtesy of NASA.

ics, for example, is shown with the multispectral camera in Figure 7-1-20*a*, and with *Landsat* in Figure 7-1-20*b*. Basically, the Sea of Galilee lies in the rift valley which is indicated by the strong gradient in reflectance. Jurassic, Cretaceous and younger sedimentary rocks are found west of the fault in Lebanon and in Syria. The Golan Heights consist of a series of late Tertiary and Quaternary basaltic flows.

The wavelength of the *Seasat* radar system is well suited for mapping geologic features that are obscure on other remote sensing images. An example is a *Seasat* image of the San Andreas Fault (Fig. 7-1-21). The San Andreas Fault extends across the image, separating the Antelope Valley and Mojave Desert to the northeast from the San Gabriel Mountains and Angeles Forest to the southwest. The fault forms a pronounced linear scarp, at the base of which occur numerous alluvial fans. Toward the top of the image in Figure 7-1-21, the San Andreas Fault is intersected by the Garlock Fault and marks the boundary between the Mojave Desert and the Tehachapi Mountains to the north. Part of the densely populated San Fernando Valley appears at the lower left of the image. The relatively smooth and level surface of the Mojave Desert produces little radar backscatter and appears dark on the image. This is notably true for the dry lake bed, Rosamond

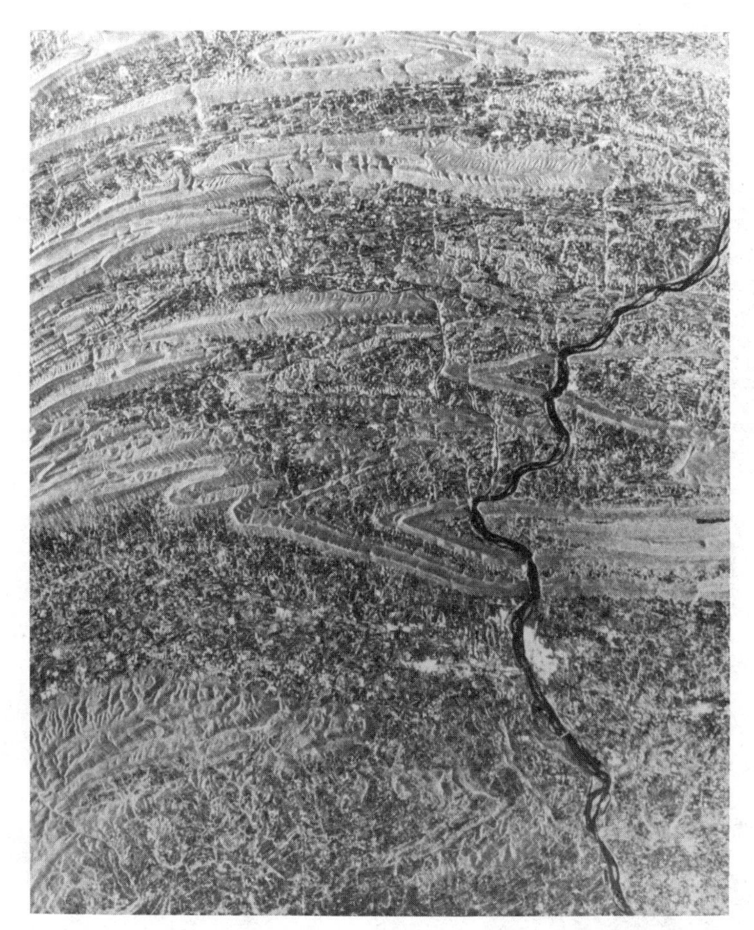

Figure 7-1-18. *Seasat* SAR image covering part of the Appalachians.

Lake. The smooth surfaces of major highways and reservoirs are likewise dark on the image. In contrast, the sloping mountain surfaces in the San Gabriel and Tehachapi Mountains that face the radar illumination direction appear very bright, due to radar layover.

Figure 7-1-22 gives a radar view over eastern California taken by *Seasat,* showing the different valley faults and other surfaces that influence reflection; the high reflection, especially at the receiving station at Goldstone, should be noted.

Volcanic lava flows are usually characterized by bright radar albedo because of their surface roughness, and recent flows may be associated with recognizable volcanic features. The SIR image (Fig. 7-1-23) of Mt. Shasta (northern California) was acquired during a mission of the Space Shuttle *Challenger*. One of the Cascade Range of volcanoes, Mt Shasta is surrounded by bright-appearing lava flows, the Shasta Valley (center), and the Klamath Mountains.

In general, a satellite image can give information on structures, rock types, and soil classes, and from this a geological interpretation can be reached, an example of which is a geological mapping of a region in the People's Republic of China, including geological subcategories (Fig. 7-1-24).

Figure 7-1-19. SPOT scene of Djebel Mour, Algeria, situated in the southern part of the Atlas Mountains. A surface of 60 × 40 km of this multispectral scene was recorded with a 20-m ground resolution. Courtesy of SPOT-Image.

7-1-3 Magnetic Anomalies

Geologic and geophysical studies utilize magnetic anomaly data for regional studies of crustal structure and composition, including possible correlation with the emplacement of natural resources and possible applications for resource exploration. A magnetic anomaly is defined to be the residual field remaining when the broader-scale earth's field is removed, which is usually computed from a spherical harmonic model. The primary interest in magnetic anomalies is to identify subsur-

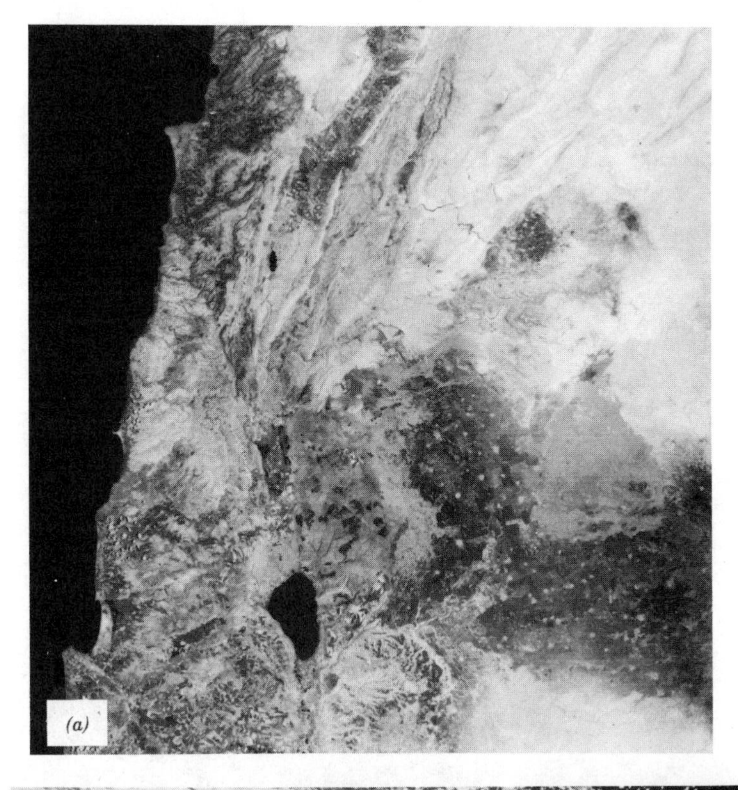

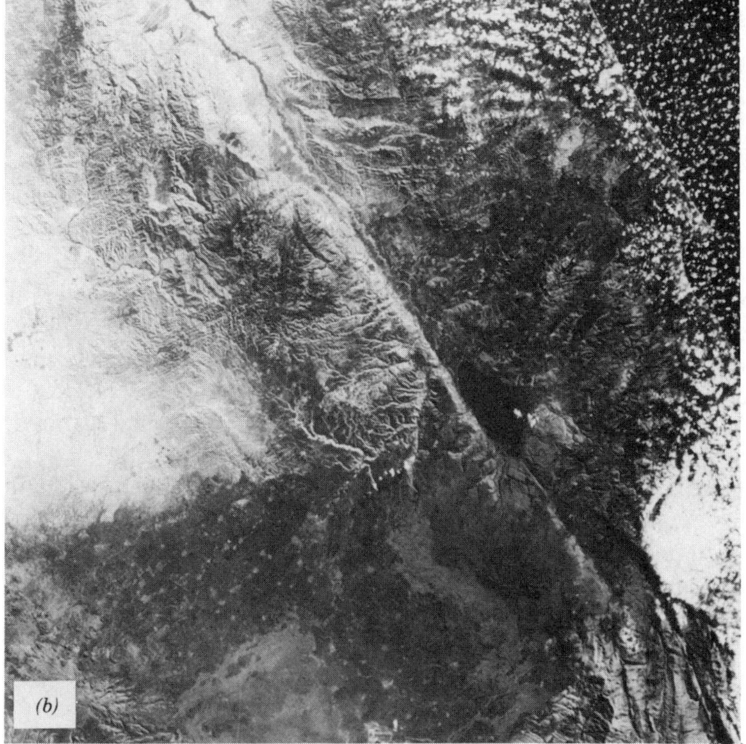

Figure 7-1-20(a). Rift associated with Middle East plate tectonics taken with the multispectral camera. **(b).** *Landsat* coverage of Levantine fault system with Sea of Galilee south of Golan Heights. Courtesy of NASA.

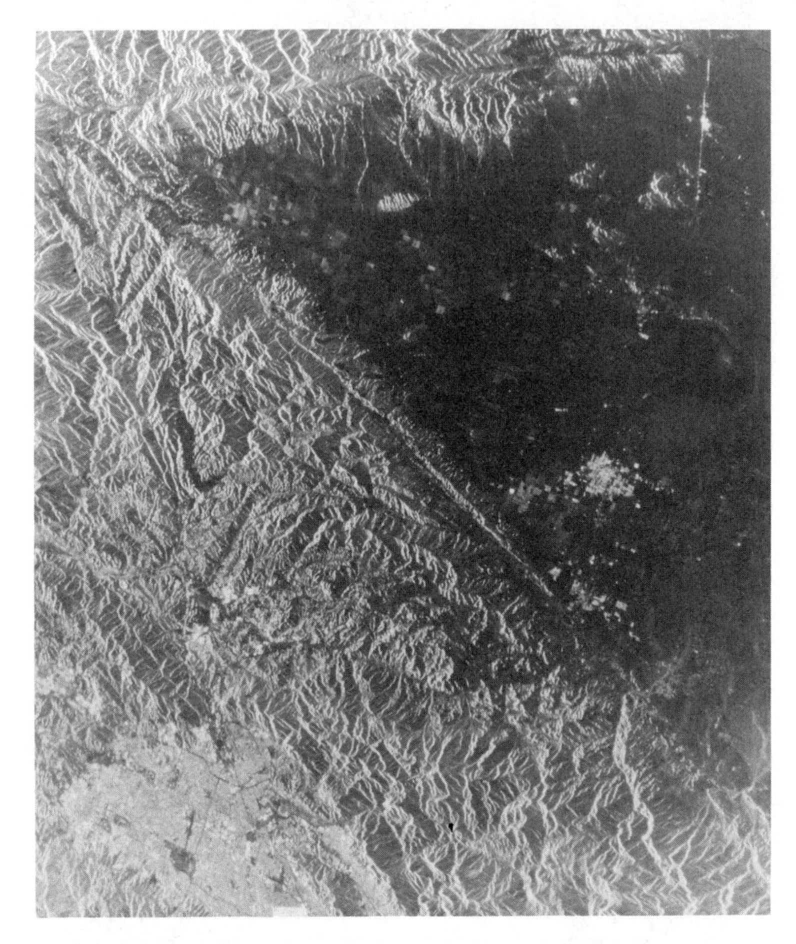

Figure 7-1-21. *Seasat* SAR image of the San Andreas Fault.

face geologic structures over distances and depths of a few kilometers for targeting large temporal changes in background fields between the epochs of adjacent local surveys and large local variations. The aeromagnetic anomaly maps cannot effectively probe broad regional geological features. However, interest is increasing, particularly in identifying and mapping these broad regional features, with the advent of the theory of plate tectonics.

Modeling and interpretation of satellite derived anomaly maps is still in its early stage. Frey (1982) pointed out that, on a global scale, there is a tendency for anomalies to be associated with large features such as shields, platforms, subduction zones (all with mainly positive anomalies), basins and abyssal plains (with mainly negative anomalies) and to be bounded by "linear" features such as sutures, rifts, folded mountains, and age province boundaries. At equatorial latitudes, these signs are reversed, unless the maps are reduced to the pole.

Magnetic field variations at the earth's surface can be caused by currents in the ionosphere or the magnetosphere. Simultaneous measurements by satellite (above the ionosphere) and the the surface (below the ionosphere) serve to determine if the source is within or beyond the ionosphere.

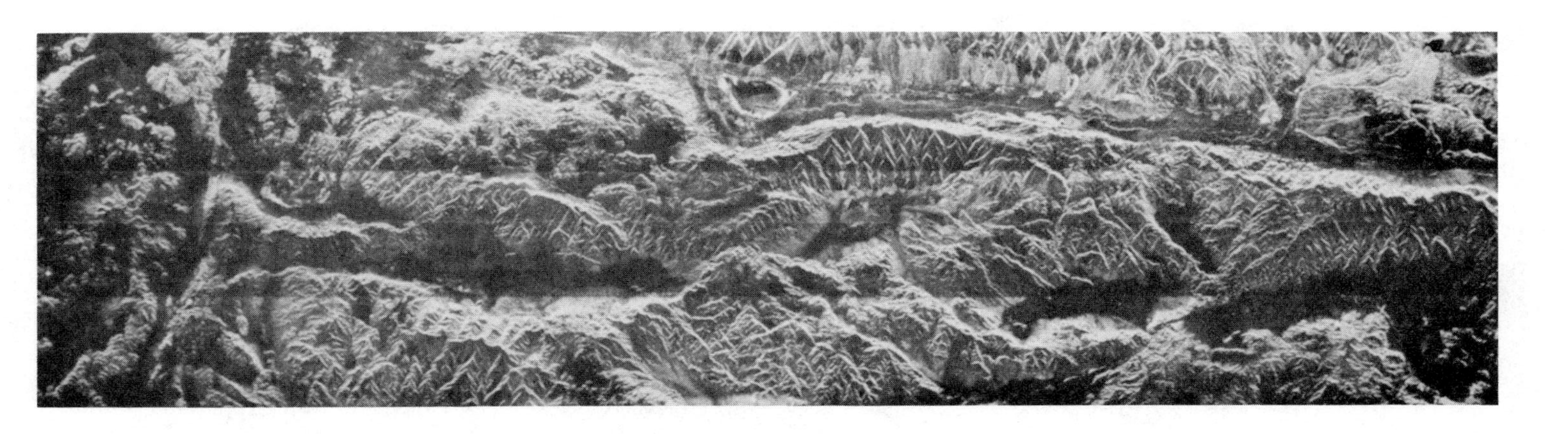

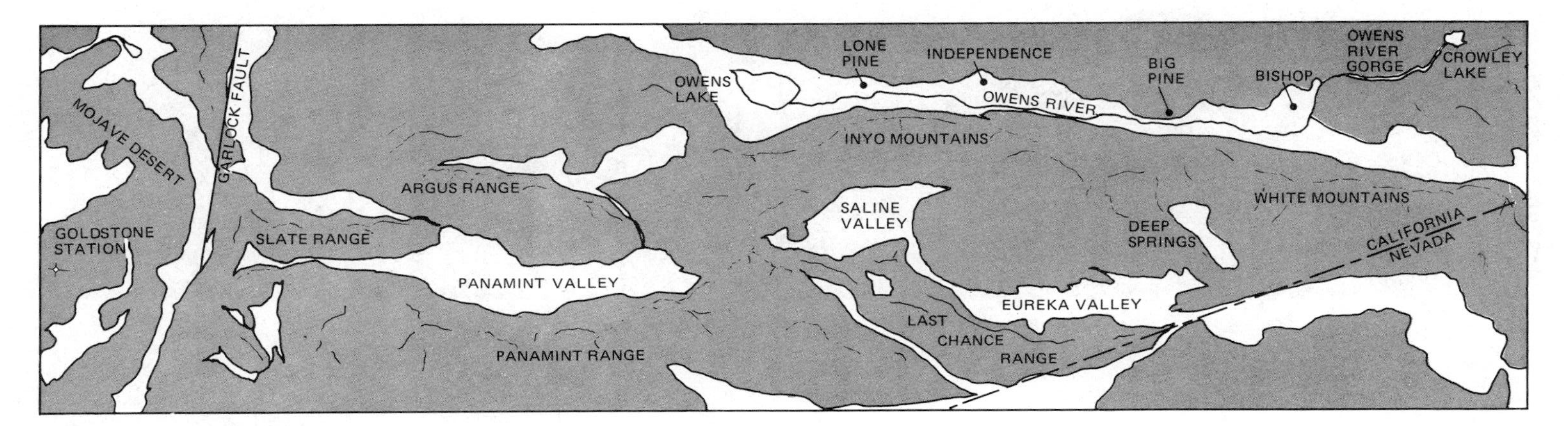

Figure 7-1-22. *Seasat* image of eastern California taken on July 7, 1978. Covering an area of 322 × 100 km, it was received in less than a minute during a sequence starting below Mexico's Baja Peninsula and ending on the British Columbia coastline. Courtesy of NASA, Jet Propulsion Laboratory.

Figure 7-1-23. SIR-B image of Mt. Shasta in northern California, recorded during a flight of the Space Shuttle *Challenger*. Area covered is 15 × 50 km acquired at a rate of about 7.5 km·sec^{-1}. Courtesy of NASA, Jet Propulsion Laboratory.

The *Magsat* mission obtained data for improved modeling of the time-varying magnetic field generated within the core of the earth, and mapped variations in the strength and vector characteristics of crustal magnetization. A major advantage of orbital magnetic field surveys is their ability to obtain a global set of magnetic field measurements of uniform precision and accuracy at a single epoch. This is particularly advantageous for deriving models of the main core field of the earth. Such a model is a necessary prerequisite for further analysis of the data (Langel, 1982). The first model from *Magsat* was based on a global vector survey of the field, and it accurately determined the fields due to the ring current and other magnetospheric currents (Langel et al., 1981).

Magsat scalar anomalies are compared on a global basis with known geologic and tectonic structures. For many anomalies, good geographic coincidences are found with some shields, cratons, and platforms, sedimentary basins, deep oceanic abyssal plains, submarine plateaus, and some subduction zones (Frey, 1982).

A preliminary map (see Fig. 7-1-25) combined high- and low-latitude equal-degree averages into a single patched-together data set between $\pm 80°$ latitude. The high latitude led to the occurrence of fictitious anomalies; the feature east of the southern tip of South America is an extreme example. In the map given in Figure 7-1-25, the zero contour was suppressed; the contour interval is 2 nT. Red and blue contours show positive and negative anomalies, respectively. The anomalies are superimposed on a map of the global tectosphere, modified by Lee-Berman and Warner, which displays different tectonic provinces by age. Correlation with crustal structure and geology is more difficult due to the greater complexity of these areas. At the scale of satellite anomalies, the crust below may be composed of small juxtaposed blocks and structures with diverse histories. The anomalies themselves may be composites of smaller features, which average to the observed values at satellite altitudes. Variations in the depth of the Curie isotherm also complicate the anomaly patterns. Comparison of anomalies from one continent to another is confused by the changing anomaly amplitude and structure with geomagnetic latitude. Some generalizations, however, can be made. Many Precambrian shields, cratons, and platforms are coincident with *Magsat* scalar anomalies; it can be seen that many of these are positive, whereas many oceanic basins and abyssal plains are overlain by negative anomalies (Frey, 1982).

7-1-4 Thermal Inertia

The term "thermal inertia" is a lumped parameter equal to $\sqrt{kC}$, where k is the thermal conductivity and C is the heat capacity per unit volume. High thermal inertia values indicate small changes in temperature, whereas the reverse is true for low thermal inertia values. Materials of low thermal inertia, such as sand, have large-amplitude radiative flux variations, whereas high thermal inertia materials, such as nonporous mafic rocks, have relatively small variations. Several factors other than thermal inertia influence the diurnal flux, including physical properties of the materials, albedo and emissivity, topographic configuration, such as slope orientation, and elevation, and atmospheric conditions.

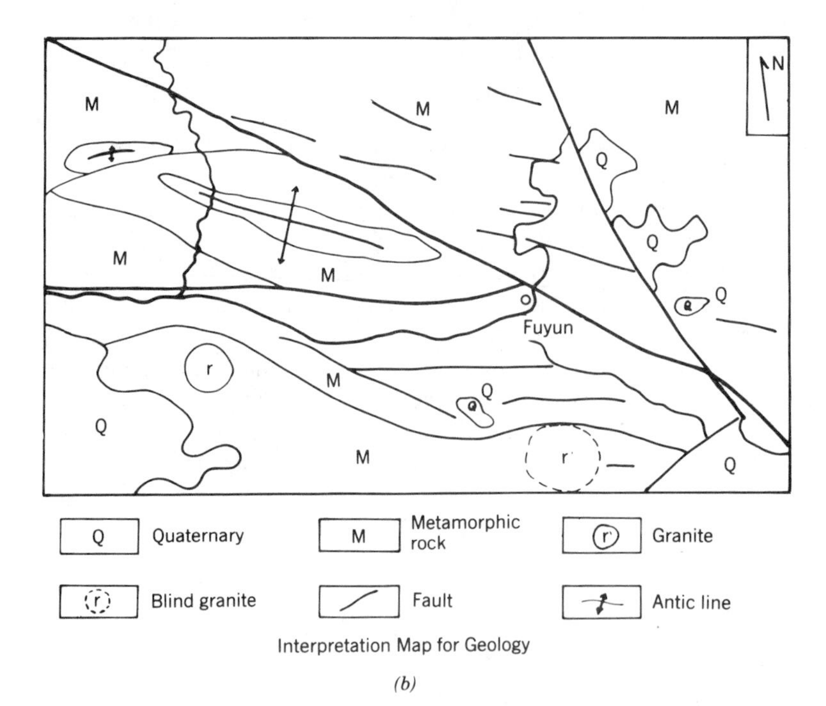

Figure 7-1-24. Fuyun area of the Xinjiang Uigar Autonomous Region in the People's Republic of China: (*a*) Digitally processed *Landsat* image; (*b*) geological interpretation of *A*. Courtesy of Prof. Yang Shiren, Academia Sinica, Beijing, People's Republic of China.

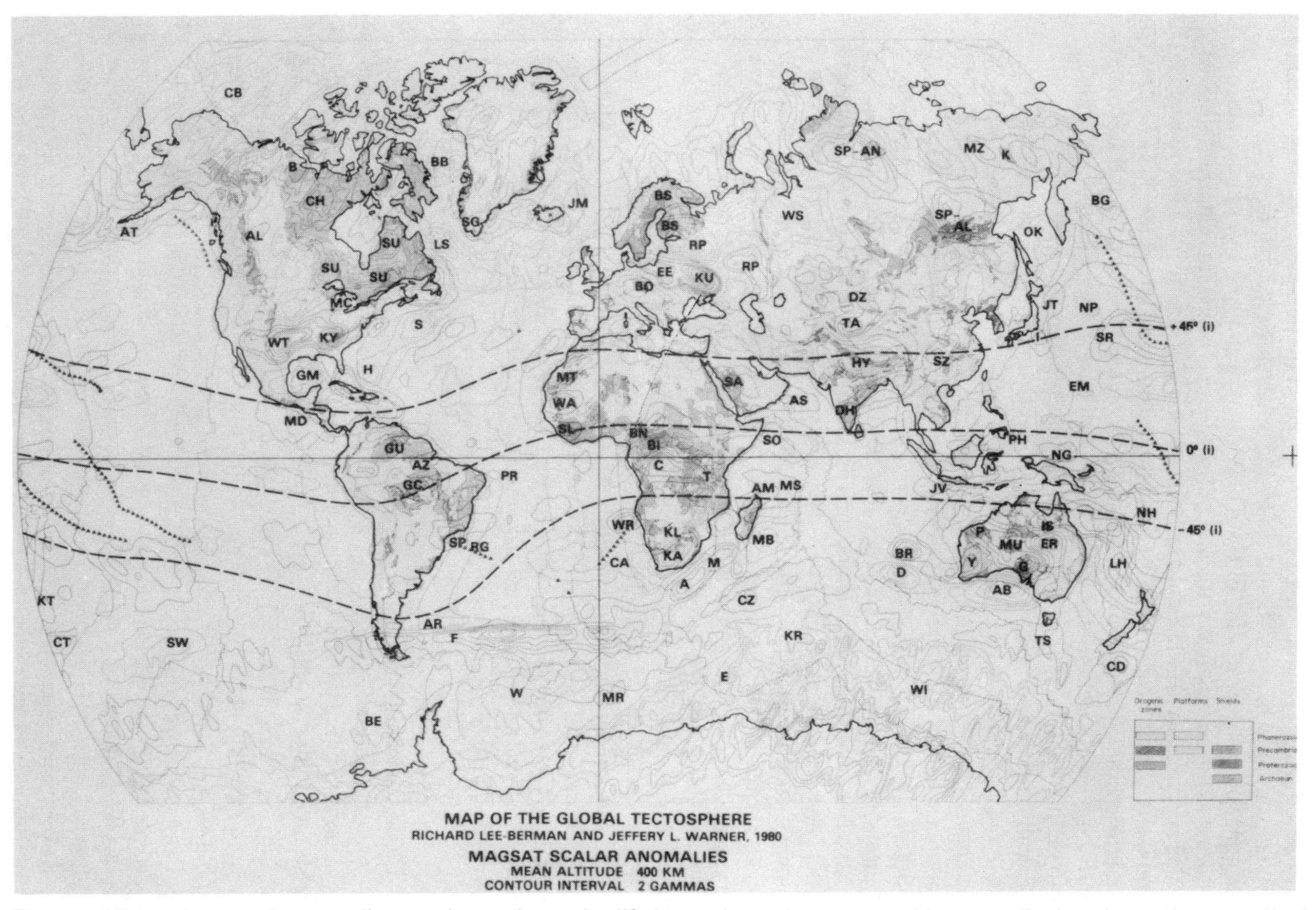

Figure 7-1-25. *Magsat* scalar anomalies superimposed on a simplified tectonic province map. Positive anomalies in red, negative anomalies in blue. Contour interval is 2 nT. Courtesy of Dr. R. A. Langel, NASA GSFC. See insert for color representation.

Mapping of soils in semiarid environments, based upon mineralogical content, is not directly possible with thermal inertia mapping techniques. However, it should be possible to detect soil changes based upon porosity and residual soil moisture content, which are functions of the mineralogical content.

As pointed out by Pratt and Ellyett (1979), mapping of relative changes in soil moisture should be possible with limited, subjective geological interpretation of soil type and porosity. If MSS data are also available, it may be possible to replace the subjective geological interpretation with an automatic machine classification procedure. With suitable ground measurements of moisture content and soil mapping, it may be possible to convert relative moisture content to absolute values and to eliminate systematic errors in the initial estimates.

Basically, the measurements of daytime and nighttime temperature differences, as an input to estimate thermal inertia, have been recorded globally with IR recordings, as shown in Figure 7-1-26, and the HCMM gave directions for future missions, especially for the need of better ground resolution in IR images (Short and Stuart, 1982).

Thermal properties of surface faces are strongly influenced by the many factors that have to be considered for image interpretation (see Table 7-1-2). It is evident that the chemical composition of the surface as well as coverage by vegetation are dominant factors in establishing image contrast. In addition, the physical shape of targets and their illumination (for instance, valleys or slopes facing the sun) have an important effect on the actual temperature sensed from the satellite. The "heating history" also contributes to the actual measured averaged temperature. This includes the cloud-influenced heating cycle during daytime and moisture distribution and varying surface roughness.

Nighttime temperature of a natural surface is a function of the solar input during daytime, and radiative and convective cooling finally cause the diurnal cycle of surface temperature of a target and determine its actual radiation. The factors of highest importance can be summarized as follows:

1. Cloud cover reflects solar energy back to space during the day, and it tends to trap radiation from the surface both night and day. Moreover, cloud cover at the time of satellite overpass obliterates the remote measurement of the surface temperature.
2. Surface winds "bleed" energy from the earth's surface by turbulent convection, resulting in a lower surface temperature than exists under quiet conditions.
3. Evaporation results in a lower temperature of a moist surface. Vegetation creates the same effect, and the vegetative canopy tends to hide the surface. In effect, dense vegetation establishes a radiating surface above ground level.
4. Absorption and reemission by atmospheric water vapor and aerosols cause a discrepancy between radiometric values observed by the HCMM and the equivalent fluxes observed at ground level. A correction (optimally based on a radiative transfer computation using meteorological sounding data) is

Figure 7-1-26. Mean daytime, nighttime surface temperature and difference of mean surface temperature on a global scale. Courtesy of Dr. K. J. Hussey, NASA, Jet Propulsion Laboratory.

TABLE 7-1-2. Characteristics of Rock Units

		HCMM Thermal Data			Landsat MSS Data	
Surface Material	Symbol	Day-IR	Night-IR	Inertia	MSS 7	Texture
SEDIMENTARY ROCKS						
Alluvial deposits	A	+	−	−−	+/++	−−
Pleistocene deposits	qC	++	O	−−	O/−	+
		O	+	O	O	−
Sandstone/quartzite sequence	Ks_2	O	O/+	O	−	−/O
Sandstone/slate sequence	hv	O	+	+/O	−−	−
Sandstone/slate sequence	Ki_4	−O	O+	O	O	O
Sandstone/slate sequence	Ki_2	O	O	O/−	O	O
Slate	si_4	−	++	+	−	++
Slate	Ks_1	−	+	O	−	−
Slate/limestone sequence	Ki_3	−	O	O/+	O	
Limestone/dolomite sequence	Ki_1	−	−/O	−/O	O	++
Limestone/marl sequence	dm/ds	O/+	O/+	O	−−	−
METAMORPHIC ROCKS						
Quartzite	si_2	O	O/+	O	O/−	+
Gneiss/migmatite	Mx_1	O	−	−	+	−
MAGMATIC ROCKS						
Gabbro	2^H	O	O/+	O/+	−−	O
Andesite, trachyte	ki	O/+	−	−−	−−	++
Rhyolite, dacite	yx_3	−	O	O/+	−	++
Diorite	Mx_2	+	O/+	O	−	−
Granodiorite	yx_2	−−	O	+	O	O
Alkali-granite	yx_1	O/−	−/−−	−/−−	O/−	O

Index:	Sign	Emissivity	Reflectivity	Texture
	−−	Very low	Very low	Very fine
	−	Low	Low	Fine
	O	Medium	Medium	Medium
	+	High	High	Rough
	++	Very high	Very high	Very rough
	/	Read "to"		

After short and Stuart (1982).

needed to permit inference of surface temperature from the satellite observations.

An example of a thermal inertia image based on recordings during the HCMM is shown in Figure 7-1-27.

7-1-5 Geodynamics

Predicted motion of the tectonic plates that make up the earth's lithosphere is based on geological and geophysical data that are averages over millions of years.

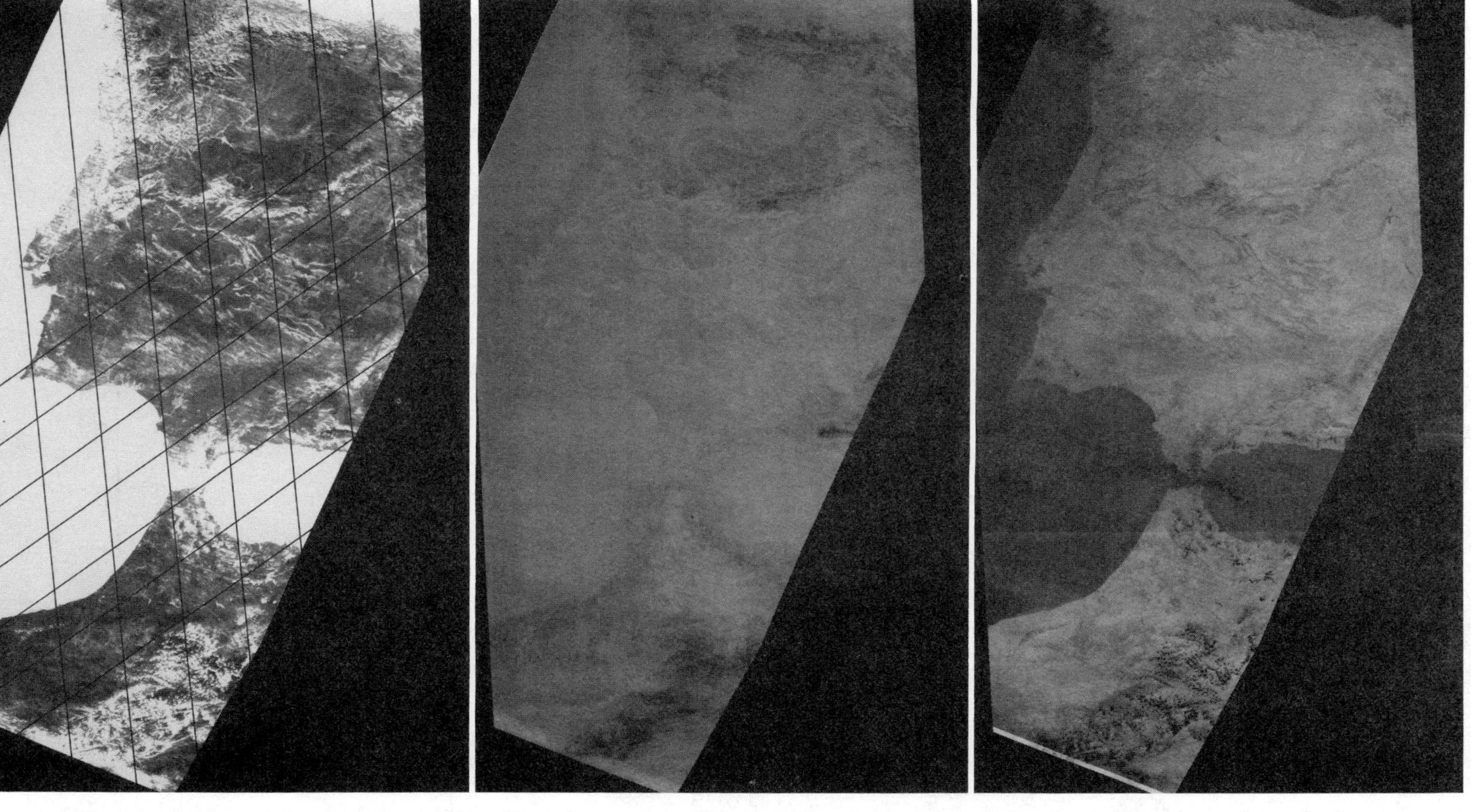

Figure 7-1-27. From left to right: thermal inertia, nighttime IR, and temperature difference. The gridding over the thermal inertia is based on an internal reference system of the satellite. Courtesy of L. Stuart, NASA GSFC.

TABLE 7-1-3. Preliminary Observations of Tectonic Plate Motion[a]

Measurement Base Line	Laser Results ($cm \cdot yr^{-1}$)	Geological Record ($cm \cdot yr^{-1}$)
North America to Pacific	4 ± 1	2
North America to South America	-1 ± 4	-1
North America to Australia	-1 ± 2	-3
Australia to Pacific	-7 ± 1	-6
Australia to South America	6 ± 3	2
South America to Pacific	5 ± 3	5

Source: NASA Geodynamic Program (1983).

[a] From 3 years of laser tracking of *LAGEOS*.

Forces that drive the plates cause complex deformation and seismic activity at the plate boundaries. Three types of plate boundaries exist: those where the plates diverge or separate (as at mid-ocean ridges), those where plates converge (at collision zones, such as the Himalayas, and subduction zones, such as the Peru-Chile Trench), and those where the plates slide past each other horizontally.

Preliminary observations of tectonic plate motion based on laser results compared with geological records is given in Table 7-1-3. Since 1972, satellite laser-ranging measurements have been made on a base line across the San Andreas Fault between Quincy and a site near San Diego, California, This 900-km base line has been decreasing about 8 $cm \cdot yr^{-1}$. The predicted decrease in distance between these sites is 5.5 $cm \cdot yr^{-1}$ (see Fig. 7-1-28).

Large-scale motions also appear in the polar regions, based on the fact that the earth is not perfectly homogeneous and symmetric, and there is a slight offset

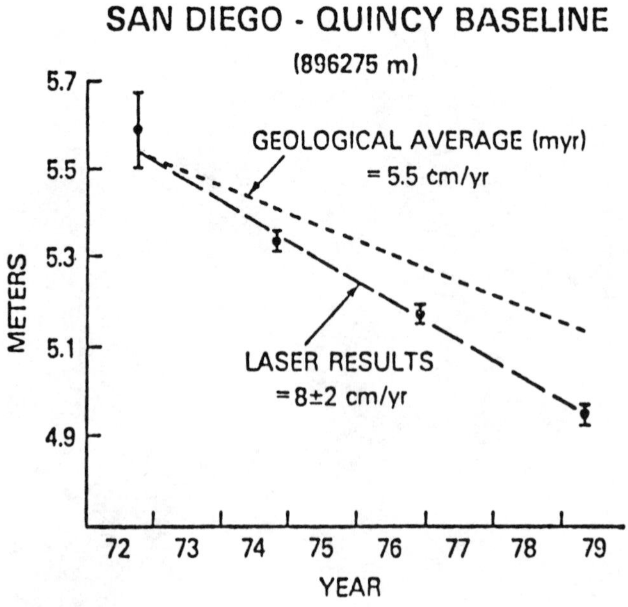

Figure 7-1-28. Motion along the San Andreas Fault. Courtesy of NASA.

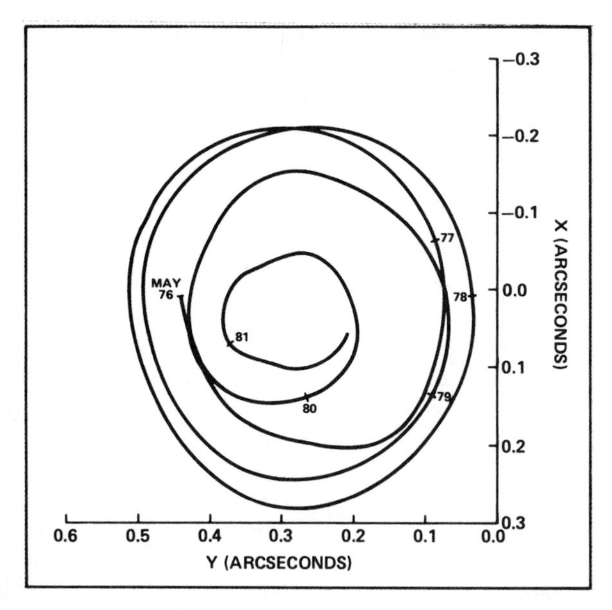

Figure 7-1-29. *LAGEOS* polar motion measurements. Source: NASA Geodynamics Program (1983).

between its axis of symmetry and axis of rotation, which leads to a movement in the orientation of the pole. Very long base-line interferometry and laser-ranging observations provide the most accurate measurement of this polar motion (see Fig. 7-1-29).

7-2 WATER ON THE CONTINENTS

The present estimate of the total volume of water on earth is about 1.4×10^9 km^3, of which 97.3% is saline and cannot be used directly for human consumption. Therefore, only 2.7% is fresh water, 77.2% of which is stored in polar ice caps and glaciers, 22.4% as groundwater and soil moisture, 0.35% in lakes and swamps, 0.04% in the atmosphere, and 0.01% is in streams. Mostly, it is surface water in rivers, streams, and lakes that basically constitutes the available water supply for man, even though groundwater has been heavily developed in many parts of the world.

The components of the hydrologic cycle provide a framework for characterizing the gross distribution of water throughout the world. Table 7-2-1 gives estimated volumes of water in storage and the average time of residence in the earth's various environments. It also shows that the amount of groundwater in storage exceeds the combined amounts in rivers, lakes, reservoirs, swamps, marshes, and the atmosphere.

7-2-1 Surface Water

Surface water can easily be detected through the low-level radiance values of water in the near IR region. For instance, Work and Gilmer (1976) mapped open

TABLE 7-2-1. Estimates of Volumes of Water in Storage and Average Time of Residence in the Earth's Environments

Environmental Parameter	Volume, km^3 (mi^3)	Average Residence Time
Atmospheric water	13,000 (3,120)	8–10 days
Oceans and open seas	$1,370 \times 10^6$ (329×10^6)	4,000+ years
Freshwater lakes and reservoirs	125,000 (30,000)	Days to years
Saline lakes and inland seas	104,000 (25,000)	
River channels	1,700 (408)	2 weeks
Swamps and marshes	3,600 (864)	Of the order of years
Biological water	700 (168)	1 week
Moisture in soil and unsaturated zone (zone of aeration)	65,000 (15,600)	2 weeks to 1 year
Groundwater	4×10^6 to 60×10^6 (96×10^4 to 144×10^5)	From days to tens of thousands of years
Frozen water	30×10^6 (720×10^4)	Tens to thousands of years

After Nace (1967).

water by analyzing *Landsat* data with a high-speed digital computer. Two different recognition techniques, a single-waveband thresholding approach and a multiple-waveband approach, termed "proportion estimation," were used. The products from the single-band technique of this processing technique were thematic maps and statistical tabulations describing open, surface water conditions.

The proportion estimation technique, which utilizes the increased information content of multispectral bands of *Landsat,* estimates the fraction of a resolution cell that may be composed of open water. This technique allowed for the recognition of small ponds and improved the apparent spatial resolution of the data by a factor of 3 compared with the single-waveband thresholding technique.

Satellite coverage is especially of interest in fluctuations of water-covered surfaces. Krinsley (1976) applied *Landsat* data to two lakes in Iran, of which the results are shown in Figure 7-2-1. Shiraz Playa and Neriz Playa occupy separate but adjacent basins within the Zagros Mountains watershed of southwest Iran. Pleistocene beaches that are similar in number and in relative vertical position and similar basin/playa ratios indicate that the two playas have had similar hydrologic responses to essentially the same climatic factors of precipitation, temperature, and evaporation.

Lake Shiraz was observed as having covered 66% of the playa area on September 20, 1972, and had an estimated average depth of 0.1 m and a volume of 16×10^6 m^3. Lake Neriz encompassed 21% of the playa area and had an estimated average depth of 0.4 m and a volume of 68×10^6 m^3. The deepest water in the lake

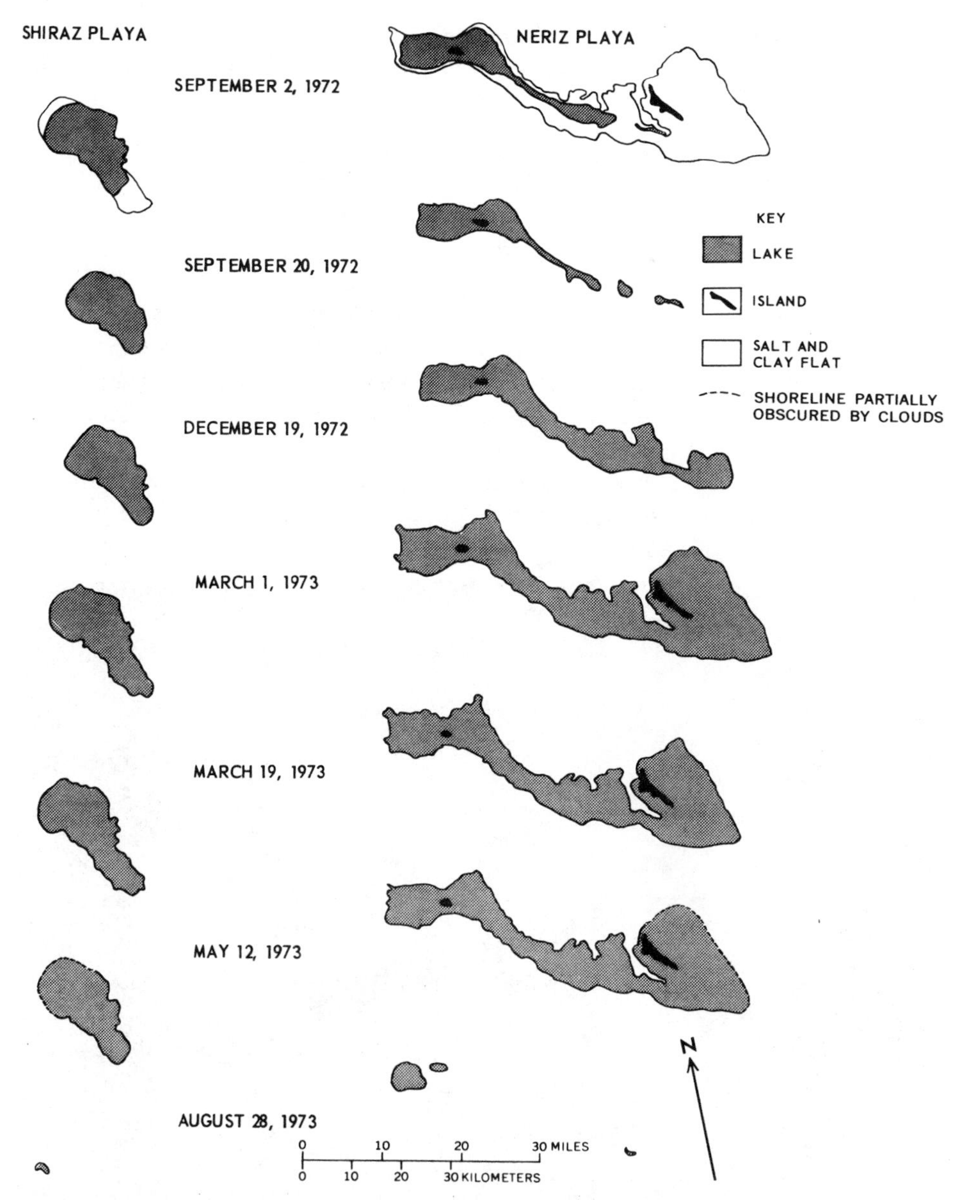

Figure 7-2-1. Diagram showing lake fluctuations at Shiraz and Neriz Playas, 1972–1973. After Krinsley (1976).

at Neriz was in the western part of the playa, whereas the long narrow central area of the lake had very shallow water. This period at the end of the long, hot summer, during which rain is generally absent and evaporation rates are highest, had the lowest groundwater levels and lake areas of the year.

Floods can be monitored, although the present state in satellite technology cannot track the progress of a flood peak with *Landsat*. Deutsch (1976) displayed temporal composites by using the IR channel of *Landsat* for preflood and flood conditions. When a nonflood image is projected as red, in register with a flood image projected as green, the composite color image is composed of the following elements:

1. Where there is surface water present in both images, the composite image receives little or no light and is therefore black, which depicts the area normally covered by the river and other surface water bodies.
2. Where the ground is not covered with water in both scenes, the composite image receives relatively equal amounts of red and green light and is therefore yellow, which depicts the area unaffected by flood waters.
3. Where there is surface water in the scene projected in green and dry soil in the scene projected in red, the composite image receives only red light and is therefore a highly saturated red color, which depicts the area of flood inundation.
4. Where there is water-saturated soil in the scene projected in green and dry soil in the scene projected in red, the composite image receives red light

Figure 7-2-2. SPOT recording of the Senegal River with its flood plain. Courtesy of SPOT-Image.

combined with a smaller amount of green light and results in a color on a continuum between yellow and red.

An example of the complicated structure of a river system with its flood plain and associated meanders and oxbows is shown in Figure 7-2-2.

Figure 7-2-3 shows the pattern of natural flooding as well as natural drainage delineated in the lower Mekong Basin. Reservoir areas affected by sedimentation, new tributary reservoirs and inundanted areas near Phnom Penh were separated by their different spectral properties.

Drainage basins between latitudes 21° and 25°S were investigated by Stoerz and Carter (1973) (see Fig. 7-2-4), based on topographic maps and *Landsat* data. Steep relief in the Andes, where most of the basins occur, reveals divides by erosion features and snow-capped ridges. An average profile of an Andean drainage basin would be an area of 200 mi^2, a basin floor altitude of about 13,500 ft, mean basin altitude of more than 14,000 ft, mean relief of more than 2000 ft, and average slopes of terrain of more than 6%.

Infrared measurements also contribute to the understanding of the dynamics of surface water. Strong (1974) analyzed NOAA IR data and compared them with airborne radiation temperature over Lake Huron and Georgian Bay. The most obvious difference in the analyses is the much smoother temperature field produced from the aircraft data. As a result of interpolation and extrapolation from the flight lines, the major discrepancy in the satellite analysis occurred along the shoreline, where temperatures were too high, probably a result of including land values from incorrectly gridding the temperature field on the base map for the satellite data. A statistical analysis performed on the satellite-aircraft data sets for data shown in Figure 7-2-5 revealed a standard deviation of slightly less than 1°C.

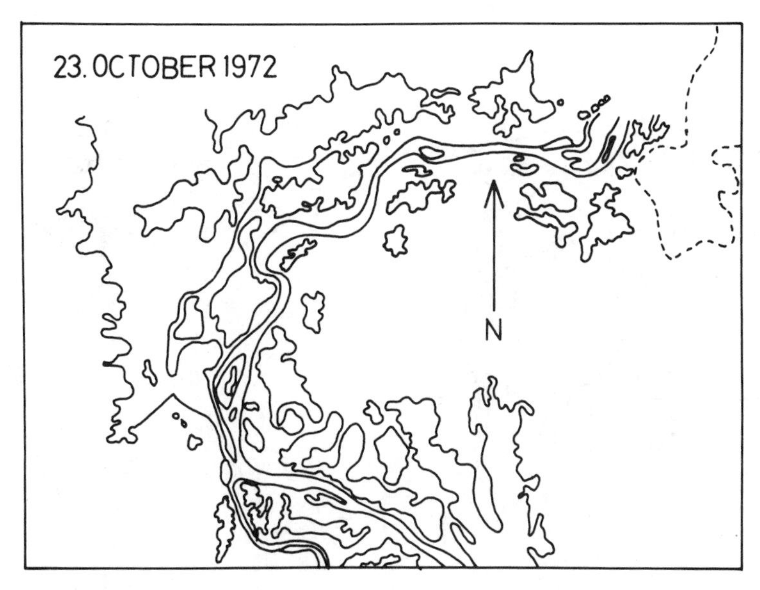

Figure 7-2-3. Flood area in Lower Mekong Basin as record on October 23, 1972 with *Landsat* MSS. Modified after Van Liere (1973). Reproduced with permission from the Environmental Research Institute of Michigan (ERIM).

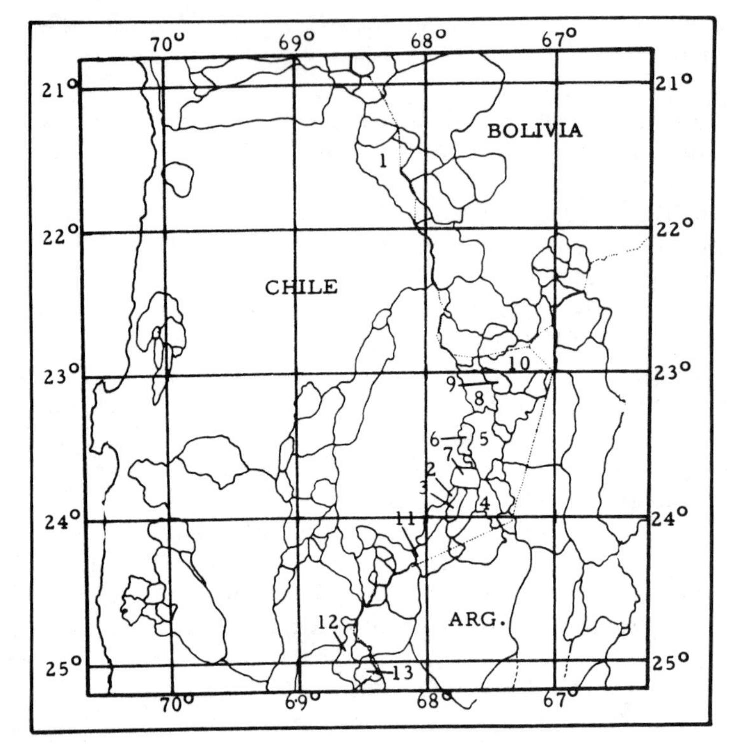

Figure 7-2-4. Closed drainage basin divides in the central Andes, 21° to 25°S. Numbers refer to the 16 drainage basins; national border shown by dotted line to distinguish from divides that coincide. Modified after Stoerz and Carter (1973). Reproduced with permission from the Environmental Research Institute of Michigan (ERIM).

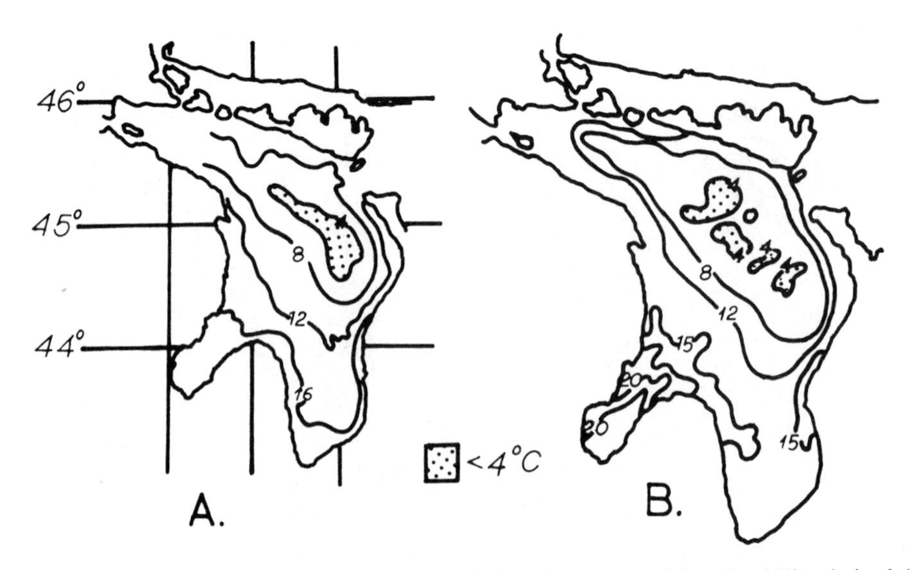

Figure 7-2-5. Comparison between airborne radiation thermometer (*a*) and satellite derived (*b*) temperatures over Lake Huron on June 14, 1973. Modified after Strong (1974).

This value was derived by first removing approximately 2°C constant bias between the two fields. Correlation coefficients (r) computed between airborne radiation thermometer and *NOAA-2* data were in excess of 0.9. Other examples of temperature distribution, as monitored from satellites, for the Great Lakes are shown in Figures 7-2-6 and 7-2-7.

Surface water shows the alterations in spectral characteristics as a function of sediment and plankton concentration. Strong (1975) observed precipitation of calcium carbonate over Lake Michigan, and similar events were noted in Lakes Erie and Ontario on imagery from *Landsat, Skylab,* and NOAA satellites. Using *Landsat* data together with NOAA thermal IR data, it was observed that whitings occur several meters below the lake surface and in relatively warm water following periods of upwelling. As the epilimnetic waters become supersaturated with Ca^{++} ions during summer, presumably biological and physical factors initiate the whiting, which could continue for several months. A combination of *Landsat* and VHRIR IR data makes it possible to make numerous interpretations of surface

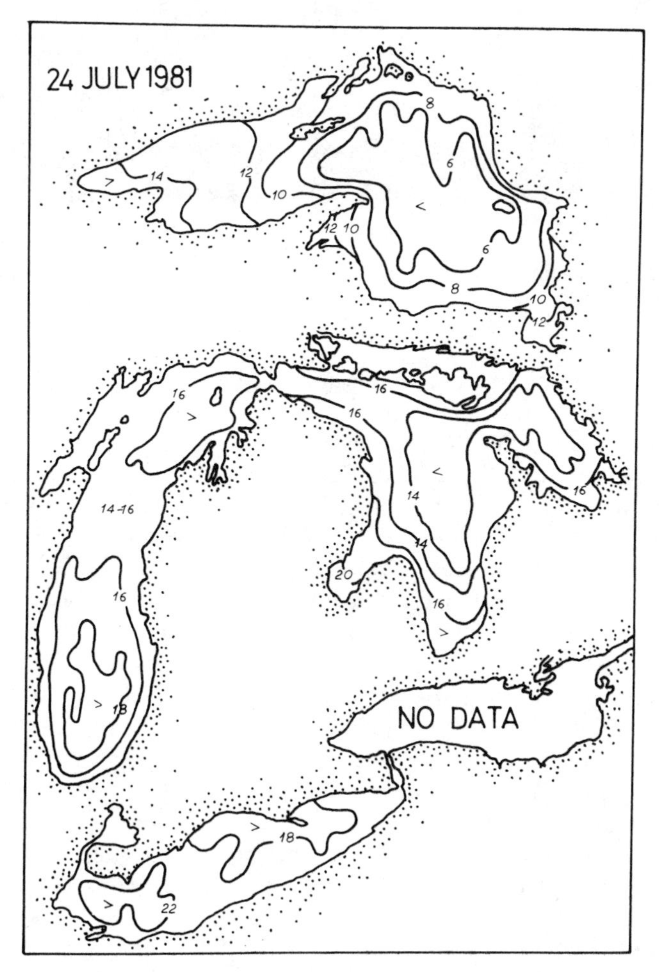

Figure 7-2-6. Temperature distribution over the Great Lakes as analyzed on July 24, 1981. Contour lines were smoothed, and some isolines close to shore were omitted. Courtesy of NOAA.

Figure 7-2-7. Examples of heterogeneity in surface temperature distribution in the Great Lakes, based on recordings from HCMM and presented in analog form. Courtesy of NASA GSFC.

currents under extreme upwelling condition in contrast to the whiting effect of lake water. Also, darker areas in the *Landsat* image are presumed to be regions of sinking or downwelling. For the most part, these vertical circulations are found on the downwind shore across the lake. Eddy circulation may be seen in the lee of points of land along the eastern shore where some of the coldest surface waters are found. An eddy in the middle of southern Lake Michigan basin is characterized by the warmest surface water observed.

Because many lakes are highly eutrophic and have high concentrations of chlorophyll (over several orders of magnitude), a good correlation is found between the signal of the *Landsat* scanner and the concentration of algae. A comparison of turbidity as a function of reflected energy in the spectral range of 0.6–0.7 μm is shown in Figure 7-2-8 (Sydor, 1976). Similar relationships can be found with other sensors working in the same spectral range. For instance, an empirical approach of relating TM data with simultaneously acquired "sea truth" data through multiple linear regression analysis was used, whereby highly significant relationships were identified between TM data and secchi disk depth, chlorophyll *a* concentrations, turbidity levels, and surface temperatures (see Lathrop and Lillesand, 1986).

The *Seasat* SAR had a capability for hydrologic mapping, even though it was primarily designed for oceanographic applications. Streams have bright signatures on the SAR imagery because the riparian vegetation produces a rough surface and, thus, high radar returns. Lakes appear relatively bright on the *Seasat* image,

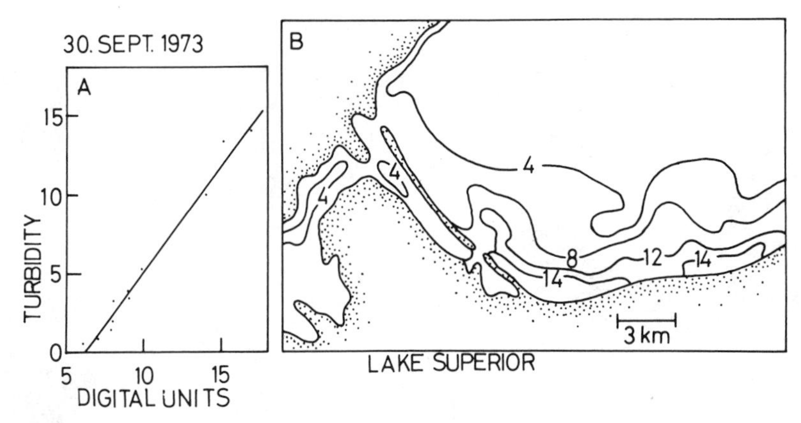

Figure 7-2-8. Turbidity measurements over Lake Superior based on *Landsat* data: (*a*) Correlation between turbidity units and digital output from band 5; (*b*) distribution of suspended solids, based on correlation in (*a*). Modified after Sydor (1976). Reproduced with permission from the Environmental Research Institute of Michigan (ERIM).

presumably in response to surface ripples and waves induced by wind action. When the wind is calm, lakes act as specular reflectors and appear smooth or dark on the imagery. On the X-band image of frozen lakes, the lakes also have a bright signature because the ice surface is rough, probably as a result of fractures, rafting, and wind action during ice formation. The SAR imagery did not reveal snow at either the 23.5 cm (L-band) or 2.8-cm (X-band) wavelengths (Foster et al., 1980). Although the drainage detail extracted from the radar imagery is similar to that which can be extracted with the visible wavelengths, much more information regarding presence of snow can be acquired from the visible wavelengths than from L-band and X-band SAR.

7-2-2 Snow and Ice

A constant problem is the presence of clouds in images, since their reflectance commonly matches that of snow, although cloud patterns change from image to image. Therefore, computer programs have been used to produce five-day composite minimum brightness charts, which in effect "removed the clouds" by computer selection of minimum brightness.

Snow depth estimates correlate well with snow brightness. The parabolic relationship in Figure 7-2-9 indicates that a rapid increase in brightness corresponds to

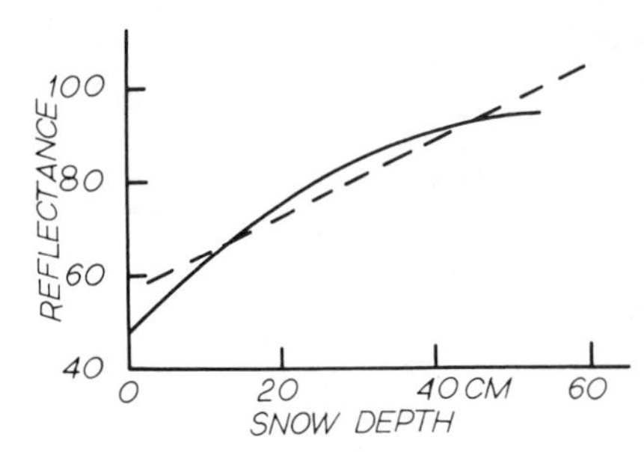

Figure 7-2-9. Comparison between snow depth and brightness. After McGinnis et al. (1975).

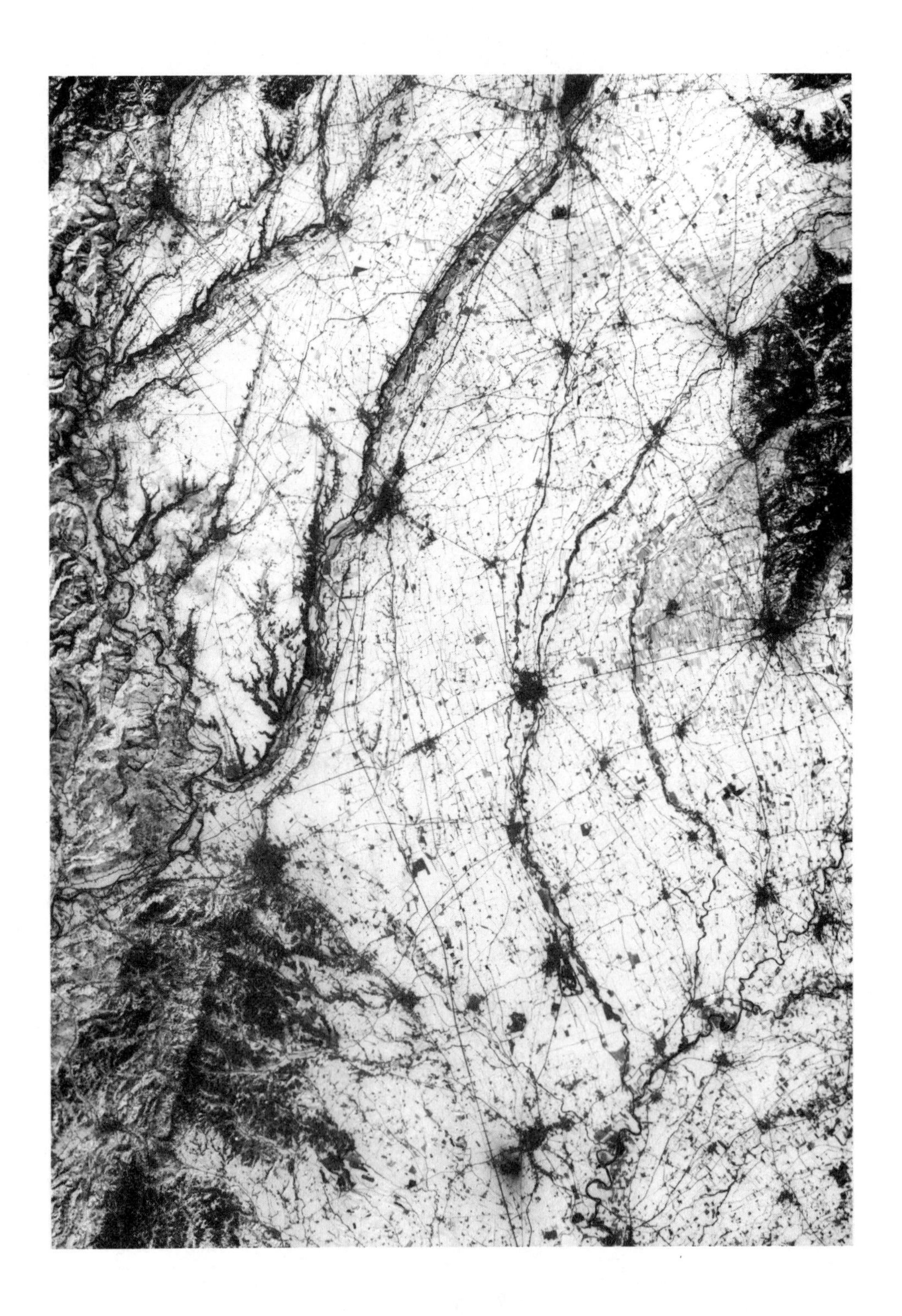

increases in snow depths when depths are less than 25 cm; for depths exceeding 25 cm, much smaller brightness increases are encountered as depths increase.

McGinnis et al. (1975), using 1-km resolution, fitted a grid of 201 VHRR squares, and the greatest snow depth for each square was matched with the greatest brightness value. Linear and parabolic best fits were calculated for the data pairs with brightness as the dependent variable. Correlation coefficients were 0.80 for the straight line and 0.84 for the parabolic curve. The parabolic fit of the data appears more consistent with the apparent relationship of brightness to snow depth. Small increases in depth, when snow is only several centimeters deep, produce large increases in brightness. This may increase contrast between snow-covered and noncovered surfaces, as can be seen with a SPOT image in Figure 7-2-10. Once the snow accumulates to about 25 cm, most small plants are covered; only the larger shrubs and plants remain visible, except in some mountainous areas where extremely heavy snows occur.

Some radiation penetrating into a snow pack, if not absorbed by underlying vegetation and soil, eventually returns to the surface and contributes to the albedo. Absorption from an underlying black surface greatly reduces the albedo of aged snow until snow depths exceeded 20 cm. Thus, small increases in satellite observed brightness may be observed even when tops of vegetation are covered with 10 to 20 cm of snow.

Snow-line altitudes can be obtained by superimposing a topographic map on *Landsat* or weather satellite imagery. An example of the use of *Landsat* is shown with an image of Anchorage, Alaska (Fig. 7-2-11), where the snow-line altitude ranges from less than 400 m near Prince William Sound to highs of over 1200 m west of Kenai Lake and in the higher parts of the Chugach Mountains. The freezing level is therefore lowest along the coast and highest further inland. Over most of the area shown in Figure 7-2-11, the altitude of the snow line could be determined confidently to the nearest contour on the map (contour interval 200 ft = 61 m). In some local areas, the snow line could not be mapped without considerable subjective decision, and large variations were found between operators and between spectral bands. However, the snow-line altitudes could be determined to within 200 m.

The impact of snow on three lakes in central Saskatchewan (Canada) near 55°N, 110°W, were analyzed by Schneider et al. (1981). On April 12, when the regional snow line was located to the north of the lakes, all three lakes appeared bright in both the visible and near IR images. By April 27, when almost all of the Saskatchewan snow cover had disappeared, the lakes still appeared snow covered or ice covered in the visible, but barely discernible on the near IR. Visible and near IR profiles for the three lakes are shown for April 12 and April 27 in Figure 7-2-12. On April 12, the visible and near IR albedos are 0.64–0.70 and 0.51–0.58, respectively, compared with 0.15–0.17 and 0.20–0.24 for the adjacent land. By April 27, albedos over the lakes had dropped to 0.24–0.40 for the visible and 0.19–

Figure 7-2-10. SPOT panchromatic scene with 10-m resolution of Pô River Plain (Italy) (60 × 40 km^2 area) taken on February 23, 1986. Upper part of the Plain is covered by snow, bounded by the Alps to the West and the Apennin Mountains to the east. Numerous scattered farms (small black dots) and cities linked by a heavy network of roads can be seen. Despite snow cover, the patchwork of small agricultural plots show up clearly, and tracks are well identified. Courtesy of SPOT-Image.

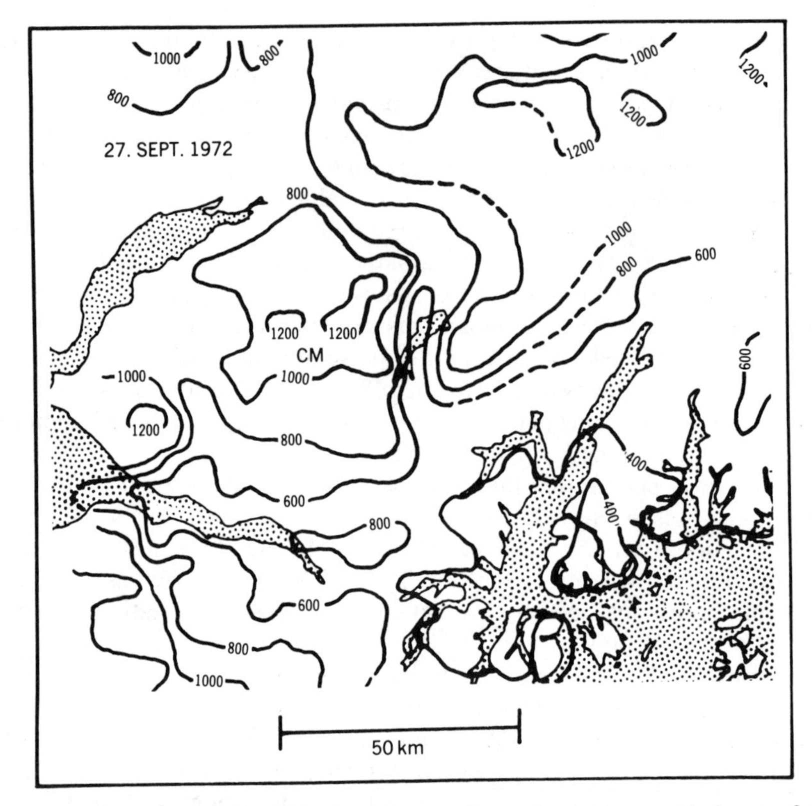

Figure 7-2-11. Map showing the altitude of the snowline in the Anchorage Alaska area from data derived from ERTS (*Landsat*) on September 27, 1972. Modified after Meier (1973a). Reproduced with permission from the Environmental Research Institute of Michigan (ERIM).

0.29 for the near IR. On this date, the westernmost lake, Egg Lake, was less reflective in the near IR than was the adjoining land, which was most probably caused by liquid water on the snow and ice cover.

Monitoring of snow and ice coverage is in an operational mode with weekly charts of snow and ice coverage, based on analyses of visible imagery for the northern hemisphere. From daily photographic data, changes are noted in extent or brightness of the snow and ice fields when cloud free. Ice charts from the NOAA/U.S. Navy Joint Center and surface weather reports help confirm the interpretation of satellite data. After seven days of data are compiled, a polar stereographic projection map with a scale of 1 : 50 m is prepared. At the end of each month, a monthly mean of snow and ice boundaries is drawn. This product is used in the areas of snow and ice limit tracking, albedo studies, long-range weather forecasting, and various research projects. An example of an ice coverage map is shown in Figure 7-2-13.

Associated with snow cover is the prediction of runoff as a consequence of snowmelt. Large changes in snowcapped areas are often visible in successive satellite imagery over relatively short periods of time. Measurements of the satellite derived snow cover areas have been related to seasonal streamflow, and

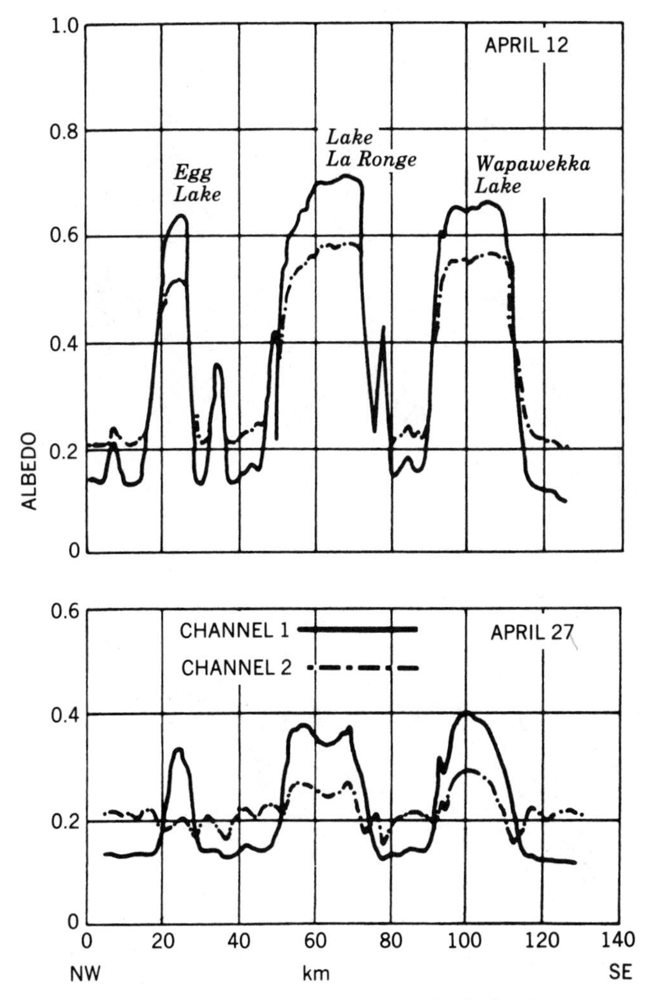

Figure 7-2-12. Visible and near IR profiles for three lakes in Saskatchewan area. Modified after Schneider et al. (1981).

results indicate that snow cover is a potentially important index parameter for reducing error in runoff forecasts. As a result from comparing snow cover areas from satellite images and seasonal runoff predictions, a situation can be simulated as shown in Figure 7-2-14.

Although currently operating satellite sensors cannot provide a direct measurement of the thickness or density of the snowpack, several studies have shown a significant correlation between the areal extent of snow cover and the corresponding runoff. A study was carried out over the western Himalayas in northern Pakistan and neighboring countries, where *Landsat* MSS band 5 data were extensively used for estimation of the areal extent of snow cover employing both visual interpretation methods and computer-aided processing (study by Pakistan's Space and Upper Atmospheric Research Commission reported in *Remote Sensing*

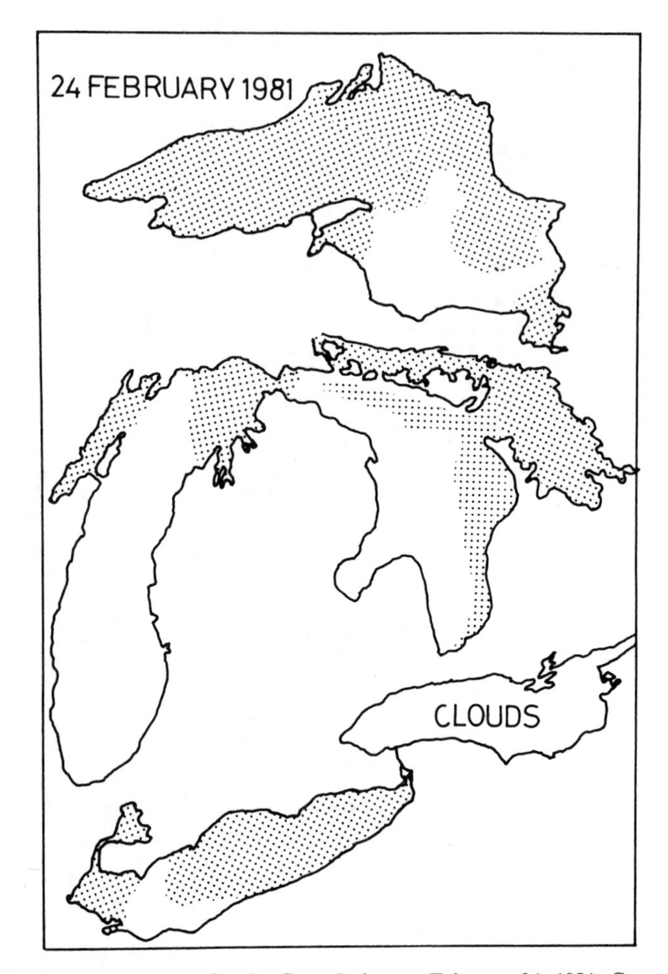

Figure 7-2-13. Ice coverage map for the Great Lakes on February 24, 1981. Courtesy of NOAA.

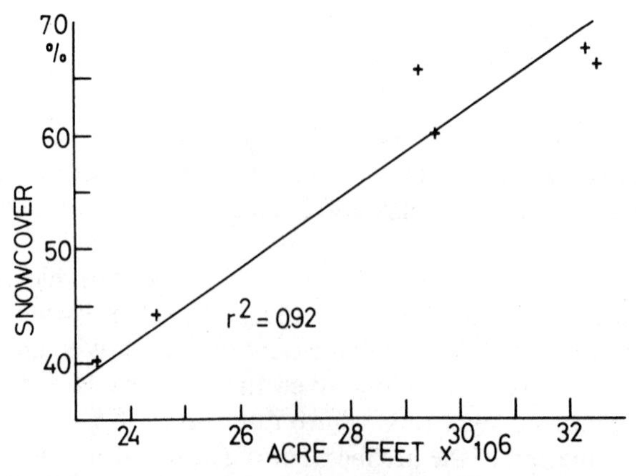

Figure 7-2-14. Runoff prediction based on satellite data. Satellite derived average snow cover (S) estimates versus measured runoff (R) for the Indus River, 1967–1972 for April to June yield. $R = (0.3S + 11.52) \times 10^6$.

Newsletter, 1986). Examples of snow cover data are given in Figures 7-2-15 and 7-2-16. Correlation between snow cover extent at the beginning of the melting season and subsequent river runoff, for four different years, was attempted, leading to a set of linear regression equations.

Foster et al. (1980) demonstrated that statistically significant regression relationships exist between snow depth and the data obtained by the ESMR. The strength of the relationship and the form of the regression line result to a large degree from the homogeneity of the area and the structure and condition of the snowpack. The microwave brightness temperature (T_B) is a measure of the emissivity and the physical temperature of the snowpack. Snow particles act as scattering centers for microwave radiation. Computational results indicate that scattering from individual snow particles within a snowpack is the dominant source of

Figure 7-2-15. Pakistan recorded with VHRR on May 17, 1982. Courtesy of NOAA.

Figure 7-2-16. Pakistan recorded with VHRR on July 17, 1982. Courtesy of NOAA.

upwelling emission in the case of dry snow. Because a deep snowpack has more crystals or grains to scatter microwave radiation than a shallow snowpack does, the T_B is lower. Snow particle size and shape, liquid water in the snowpack, snow density, and the condition of the underlying surface also affect the amount of scattering and, consequently, the emissivity of the snowpack. The physical temperature of the snowpack, especially the upper layers of snow, is primarily a result of the surface air temperature (T_{air}).

Results in Table 7-2-2 show that R^2 values obtained for air temperature versus snow depth and the ratio of microwave brightness temperature and air temperature versus snow depth were not as high as the R^2 values obtained by simply plotting microwave brightness temperature versus snow depth. Multiple regression analysis provided only marginal improvement over simple linear regression.

TABLE 7-2-2. R^2 Values for Correlations between Snow Depth, Brightness Temperature and Air Temperature

	Russia	Montana–North Dakota	Canada
R^2 Values Obtained for Air Temperature Versus Snow Depth and the Ratio of Microwave Brightness Temperature and Air Temperature Versus Snow Depth			
1.55-cm T_B vs. snow depth	0.52	0.81	0.76
0.81-cm T_B vs. snow depth	0.60	0.88	0.86
Air temp (T_{ave}) vs. snow depth	0.44	0.72	0.73
1.55-cm T_B/T_{ave} vs. snow depth	0.37	0.74	0.63
0.81-cm T_B/T_{ave} vs. snow depth	0.56	0.81	0.83
R^2 Values Obtained by Plotting Microwave Brightness Temperature Versus Snow Depth			
1.55-cm $T_B + T_{ave}$ vs. snow depth	0.52	0.84	0.81
0.81-cm $T_B + T_{ave}$ vs. snow depth	0.60	0.91	0.86
1.55-cm T_B + 0.81-cm $T_B + T_{ave}$ vs. snow depth	0.60	0.91	0.86

After Foster et al. (1980).

Foster et al. (1980) applied these relationships to areas of the Canadian high plains, the Montana and North Dakota high plains, and the steppes of central Russia in an effort to determine the utility of spaceborne microwave radiometers for monitoring snow depths in different geographic areas. In each of the study areas, *Nimbus 6* (0.81-cm) ESMR data produced higher correlations than *Nimbus 5* (1.55-cm) ESMR data in relating microwave brightness temperature to snow depth. It was also found that it is difficult to extrapolate relationships between microwave brightness temperature and snow depth from one area to another because different geographic areas are likely to have different snowpack conditions.

Short-term variations in the microwave response to snow water equivalent can be related to variations in the surface structure and, to some degree, the temperature of the snow cover. Especially for small snow depths, the microwave response is also affected by the temperature of the underlying ground. Annual variations were observed by Hallikainen and Jolma (1986) to correlate in addition to snow parameters, the water content, and state of the underlying ground. The microwave response to snow water equivalent was found to depend substantially on the land cover category. The responses are clearly higher for open areas than for forested areas.

Glacier surges can be monitored using *Landsat* imagery. The causes of glacier surges are of scientific interest, because this type of periodic sudden slippage is common to many other phenomena in nature. Surging glaciers may advance over large areas and cause devastating floods by blocking and suddenly releasing large quantities of meltwater. *Landsat* data have also shown sequential changes in large tidal glaciers.

A quantitative approach was undertaken by Post et al. (1976) to measure the motion of the Lowell and Tweeds surging glaciers of British Columbia, of which the results in displacement vectors and changes in medial moraines are shown in Figure 7-2-17.

As has been demonstrated in the use of remote sensing for the polar regions, microwave remote sensing has a wide application for ice studies. The SAR on

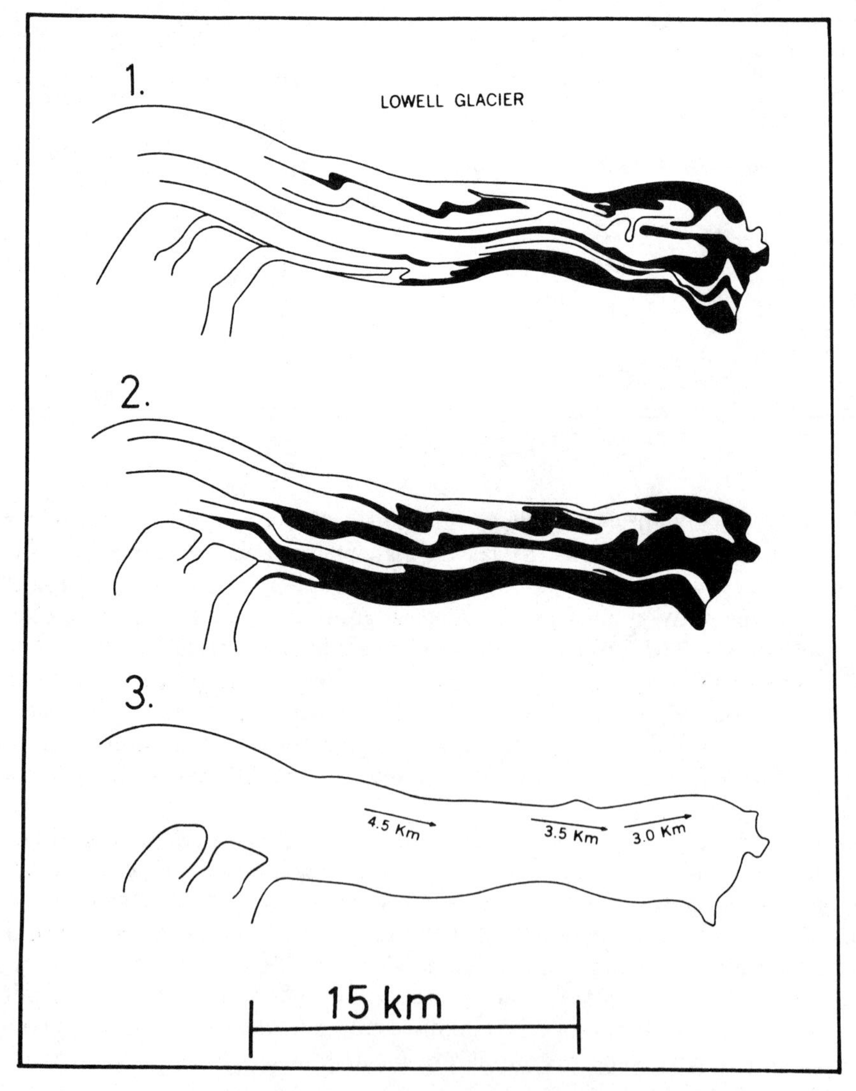

Figure 7-2-17. Changes in medial moraines of the Lowell Glacier (British Columbia) 1954–1973 from late summer 1954 aerial photography (1), late summer 1973 *Landsat* image (2), and map showing displacement vectors of the motion of the glacier. After Post et al. (1976).

board *Seasat* demonstrated that high-resolution radar is able to monitor detailed structures on glaciers and snow-covered areas (see Fig. 7-2-18).

When considering the brightness temperature images of firn-covered glacier ice, one must bear in mind that the history of the physical temperature in the firn affects its physical structure, and thus its emissivity, and that the brightness temperature is approximately proportional to the physical temperature at the time of the emission ($T_B = \varepsilon < T$). Because spatial variations in emissivity over the ice sheets tend to be larger than the spatial variations in physical temperature, the different T_B values observed in these images are usually closely related to variations in emissivity and therefore to variations in accumulation rates, mean annual temperature, or melting effects.

Snow and ice cover charts have been prepared for the northern hemisphere since November 1966 by NOAA/NESDIS (Matson and Wiesnet, 1981). These charts show the areal extent and brightness of continental snow cover, but do not indicate snow depth. The analysis is based on satellite imagery, and the snowline represents the latest cloud-free image of that area. From the digitized data, monthly anomaly and climatological snow cover maps can be created.

Snow transition zones (STZs) and snow cover frequency charts for February

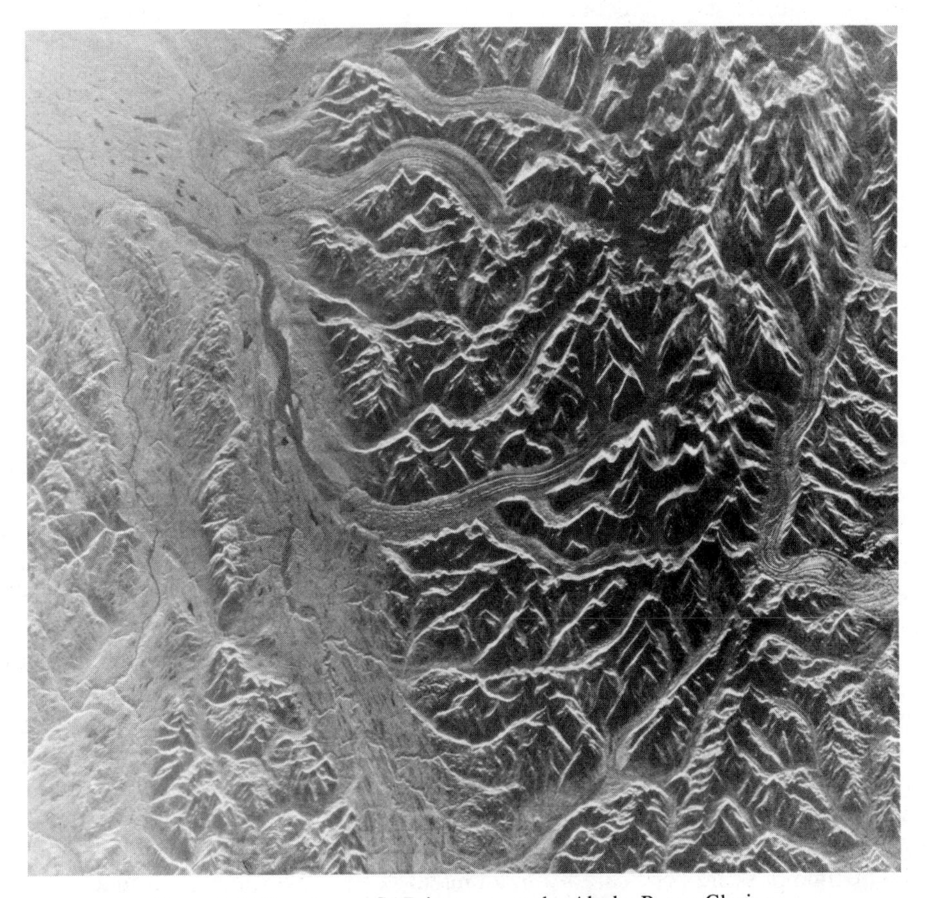

Figure 7-2-18. *Seasat* SAR image over the Alaska Range Glacier.

and October are shown in Figure 7-2-19. The symbols "+" and "−" on the charts fill the regions "always" or "never" snow covered during the periods of record. The zones between these symbols are thus the STZs, which are annotated by the numbers 0 through 9 and the letter *T*, denoting increasing frequencies of snow cover. Specifically, "0" denotes cover greater than 0 but less than or equal to 5% of the number of archived weeks during the period of record; "1" denotes greater than 5 but less than or equal to 25%, and so forth, up to and including "9." Finally, "T" denotes greater than 95% but less than 100%. The boundaries of the various STZs are the outlines of record periods of snow cover that occurred during the time span of the archive. The northern boundaries are the imprints of extreme sparse periods, the southern boundaries of record abundant periods of snow cover (Morse, 1983).

7-2-3 Groundwater and Soil Moisture

The presence of groundwater can indirectly be located by interpreting in satellite images geological structures, landforms along with soil associations and land cover. Within the hydrological cycle, most of the groundwater originates at the surface through precipitation or snow melting. *Landsat* data assist to identify the most probable areas of recharge or discharge, along with good aquifers. Additional information has to be furnished from topographic maps and prospecting in the field, which means that with the help of satellite images, favorable areas for potential ground are distinguished, thus reducing time and loss for prospecting.

In principle, interpretation of remote sensing data for ground water exploration is based on four fundamental steps:

1. Fractures in hard rocks with bedding or cleavage planes are conduits for water percolation, and have to be marked on the image or transferred on a base map.
2. Landforms reflect dynamic processes such as rainfall and transportation of eroded material and deposit of weathered material, which means that the type of soil deposited also determines the water percolation and its capability to recharge.
3. Geological-lithological mapping for hard rocks, intrusives, sedimentary, metamorphic, unconsolidated alluvium, and so on, are distinguished.
4. Groundwater conditions are highly related to land cover, whether naturally existing or cultivated in irrigated areas; dense vegetation, for instance, indicates promotion for groundwater recharge.

Features on *Landsat* images that are indicators of groundwater occurrence are landforms, drainage characteristics, snowmelt patterns, vegetation types and associations, outcrop patterns, soil tones, lake patterns, and land cover types. Some features imply the occurrence of shallow sands and gravels; other features indicate rock types or the presence of folds and fractures. A color-composite image is the most useful for groundwater interpretations, but lineaments and some drainage characteristics are seen best on black-and-white IR images (Moore, 1977).

The time of year is critical for obtaining maximum hydrologic information from *Landsat* images because some features are enhanced by a low sun elevation angle.

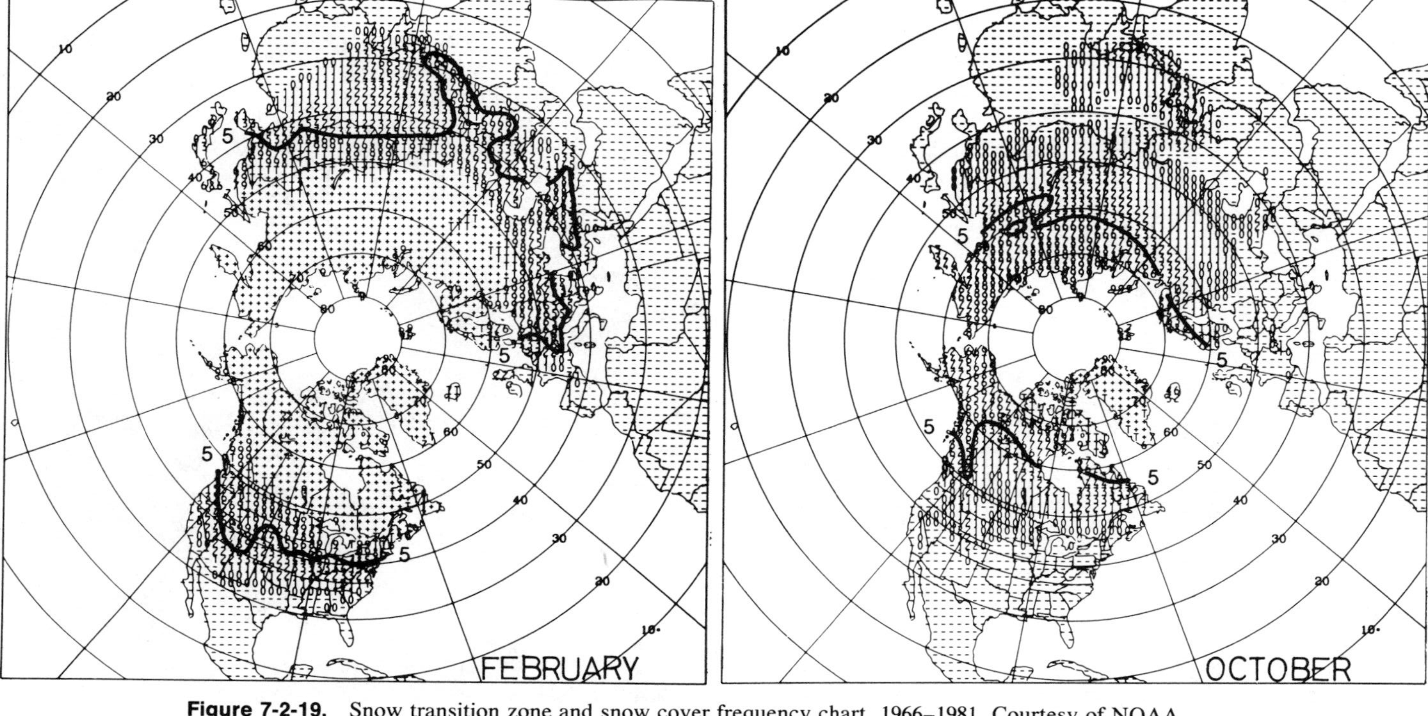

Figure 7-2-19. Snow transition zone and snow cover frequency chart, 1966–1981. Courtesy of NOAA.

Better times of year depend on local cycles of wet to dry, warm to cool, and bare soil to mature agricultural crop.

Shallow sand and gravel aquifers are based on fluvial deposits, such as sorted sands and gravels, which appear in the form of stream channel deposits and valley fills or alluvial fans. Moore (1977) tabulated a key for detection of shallow sand and gravel aquifers on *Landsat* images (see Table 7-2-3), and the table can also be applied to imagery from other satellite systems.

In arid regions, the presence of springs and seeps can be indicated by anomalous snowmelt, soil moisture, or vegetation patterns. If the surface conditions are homogenous, snow first melts in areas where groundwater is at or near the surface; hence, vegetation may also begin earlier growth in these areas.

A case study on targeting groundwater exploration in southeastern Arizona using *Landsat* imagery was described by Taranik et al. (1976); see Figures 7-2-20*a* and *b*. The main stream channels of the Tucson basin are normally dry, and flow occurs only in response to intense showers. In the foothills, a few small springs flow almost perennially, but the water seeps into alluvium a short distance downstream. Near the mountains, groundwater normally is found at shallow depths, and the water table occasionally rises to intercept the stream during prolonged periods of surface flow. Infiltration to the groundwater reservoir is greatest in late winter and early spring when runoff from melting snowpack combines with rainfall. The only important groundwater aquifer is the valley alluvium, although small

Figure 7-2-20. (*a*) *Landsat* image over Southeastern Arizona. *Continued on next page.*

amounts of groundwater can also be obtained from fractured rocks in mountain areas. The coarse-grained facies are hydraulically connected and form a single aquifer in the Tucson basin; however, the occurrence of groundwater may be affected by faults that enhance or impede groundwater movement.

Landsat has been used to a certain degree to study soil moisture conditions. Darch (1979) combined the use of additive viewing and digital processing of *Landsat 2* imagery for the Pantanal of Brazil, which allowed detailed maps of the drainage network. Distribution of wet and dry areas, including differentiations of clear water, water containing suspended sediments, and categories of land with differing moisture conditions, were mapped (see Fig. 7-2-21).

Many experiments on soil moisture content have been conducted with the use of remote sensing at different wavelengths for passive and active sensors. In the microwave region, experiments have provided basic understanding of the effects of soil texture, surface roughness, and vegetation cover on the responses of microwave radiometer and radar systems. Theoretical models, both rigorous and

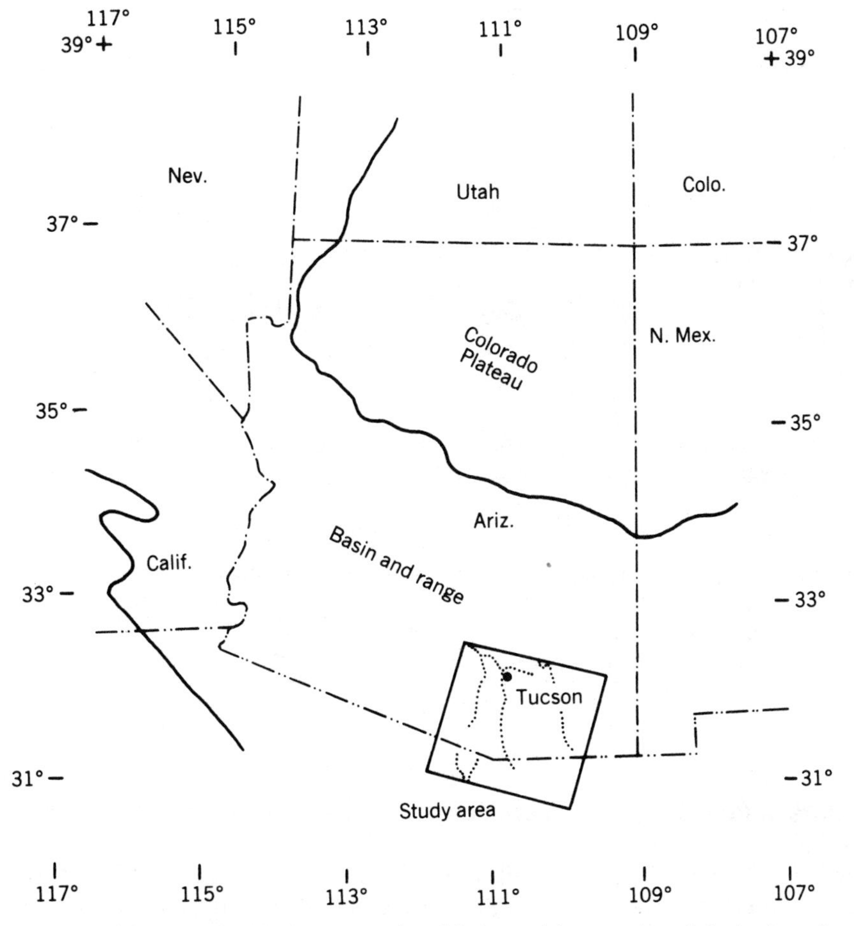

Figure 7-2-20. (*Continued*) (*b*) The area analyzed is located in a portion of the basin and range physiographic province in southeastern Arizona. After Taranik et al. (1976).

TABLE 7-2-3. Features that Indicate the Occurrence of Shallow Sand and Gravel Aquifers on *Landsat* MSS Images

Alluvial Features

1. Coarse drainage texture (arid region) or low drainage density (humid region).
2. Alluvial cone or fan; coalescing fans; bajada. Splay pattern representing partly formed or dissected fan. Dendritic, parallel, or fan drainage. Disappearance of drainage lines on image.
3. Broad and relatively flat-bottomed valley. Valley is wider than the meander belt.
4. Meander loop showing location of point bar. Meander scar or braided drainage scar in floodplain; oxbow lake.
5. Natural levee; levee complex.
6. Underfit or abandoned valley. Valley may contain ponded drainage or have a stream meander wavelength smaller than that of the floodplain and terraces.
7. Elongate lake or aligned lakes and ponds, representing remnants of a former drainage network.

Coastal Features

1. Arc delta (coarsest material); composite and estuarine deltas.
2. Cheniers or beach ridges.
3. Fan-shaped area of different soil, vegetation type, or land use—indicating a landlocked arc delta.
4. Aligned oblong areas of different soil, vegetation, or land use—indicating landlocked bar, spit, or beach. Drainage deviation or centripetal pattern at upstream boundary.

Glacial Features

1. Esker; valley train.
2. Outwash plain.
3. Splay soil tone or vegetation pattern, representing partly formed alluvial fan.
4. Other alluvial features (above).

Eolian Features

1. Longitudinal, transverse, parabolic, and star dunes (wind streaks, shown as parallel soil tone or vegetation pattern in image, generally indicate thin or fine-grained soil).

Residual Features

1. Smooth topographic surface that lacks bedrock outcrops. Low relief.
2. Coarse drainage texture. Dendritic pattern of small tributaries.
3. Light-toned soil.
4. Wetlands at approximately the same elevation as nearby major streams.

After Moore (1977).

phenomenological, have also been developed to account for these effects based on experimental results.

Studies of the *Nimbus 5* satellite ESMR data at 1.55-cm wavelength showed that it has the capability for soil moisture detection, although limitation is caused by a vegetative canopy over the soil. For situations where there is a significant amount of bare ground, the ESMR brightness temperatures have shown signifi-

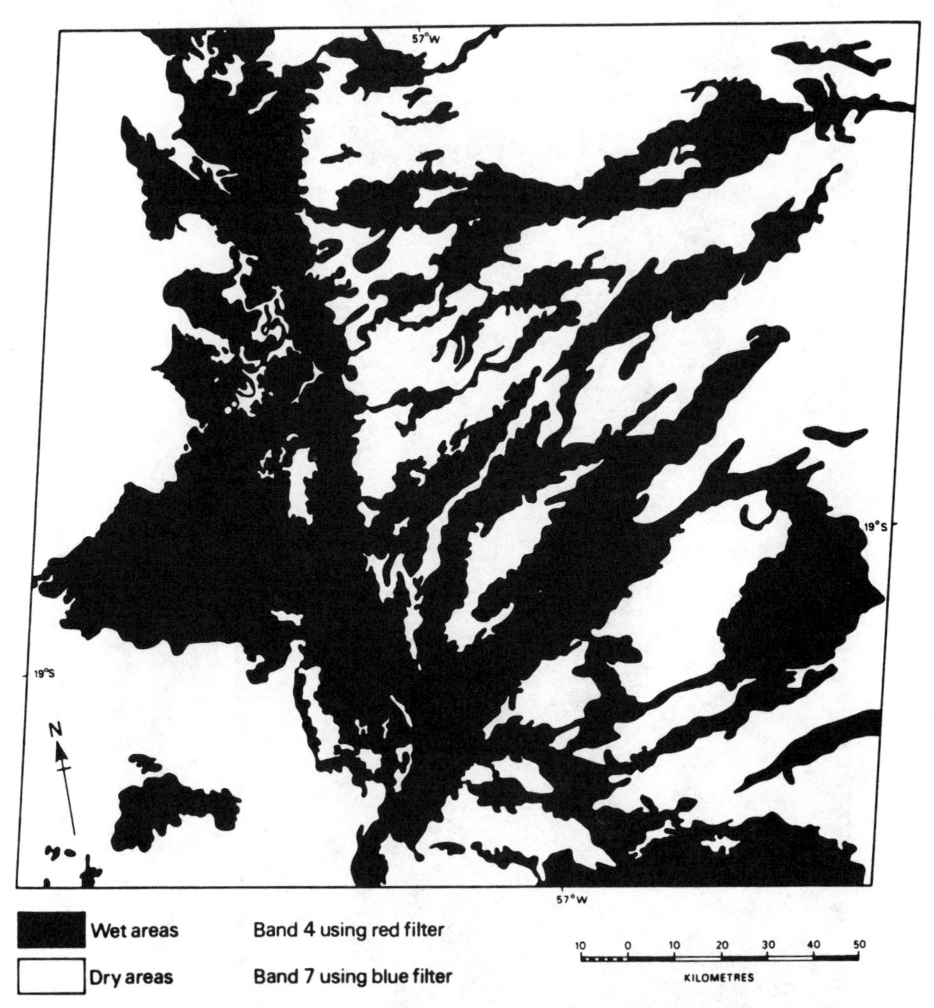

Figure 7-2-21. Wet and dry area differentiation over the Pantanal of Brazil. After Darch (1979). Reproduced with permission of Elsevier Publishing Co., Inc.

cant correlations with soil moisture (McFarland and Blanchard, 1977; Schmugge et al., 1977). When vegetation is present, the sensitivity to soil moisture changes is significantly decreased due to the screening effects of vegetation. In this context, the global coverage of *Nimbus 5* ESMR data shown in Figure 7-2-22 have to be evaluated accordingly.

The effect of rainfall on soil moisture conditions and its distribution pattern can be demonstrated with a *Seasat* pass over central Iowa, when a large frontal storm moved into the vicinity from the west and the northwest. Typical mid-latitude frontal convective storms occurred for several hours approximately 10 to 12 hr prior to the acquisition of the image (Fig. 7-2-23). Prior to the passage of the satellite, the storm broke up and several isolated cells moved to the east and northeast, distributing rain along the respective ground tracks. The result is seen in Figure 7-2-23, where the light areas to the west reflect the general location of the major frontal precipitation activity.

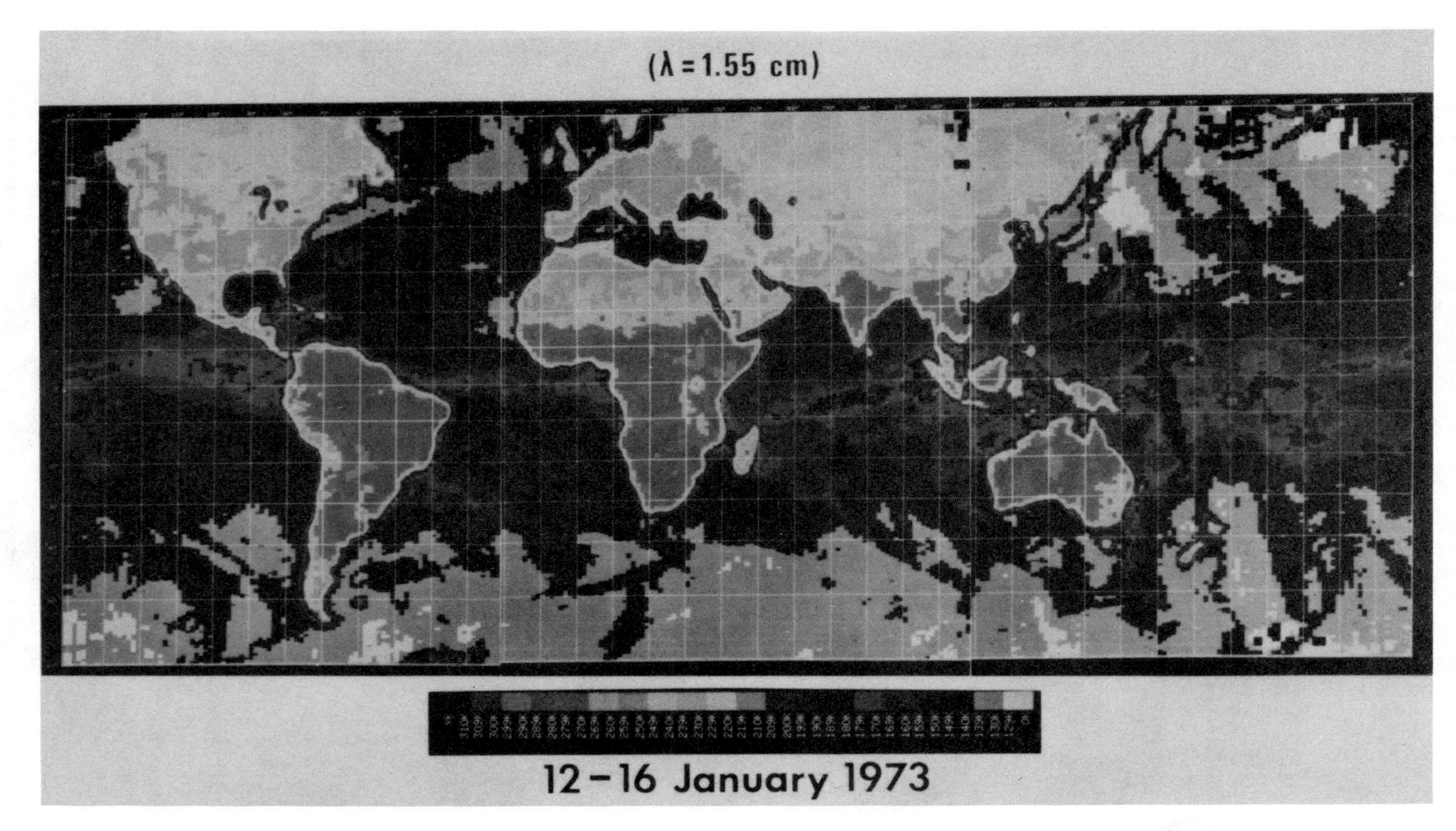

Figure 7-2-22. Global coverage of radio brightness recorded by *Nimbus 5* ESMR. Courtesy of NASA, Nimbus Project.

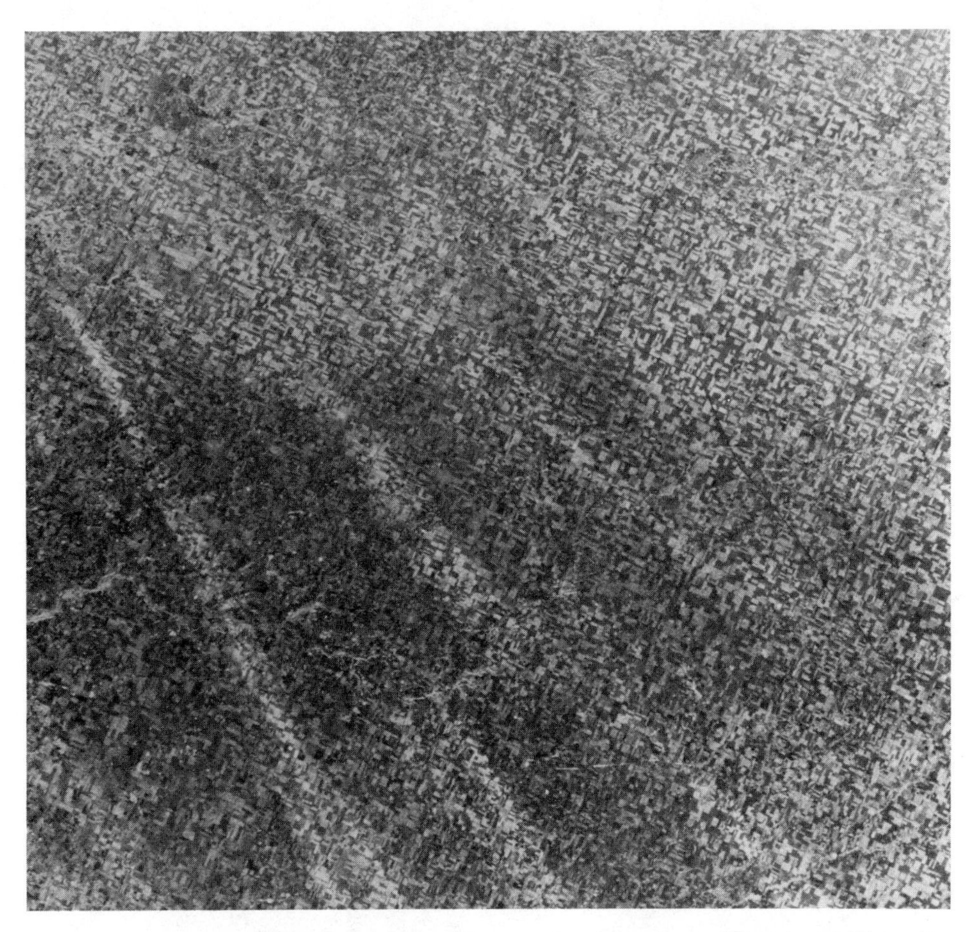

Figure 7-2-23. *Seasat* radar image showing different soil moisture due to different rainfall intensity. Courtesy of NASA, Jet Propulsion Laboratory.

7-2-4 River Discharge

River runoff with the ocean forms a low salinity in the upper layer of an estuary and tends to move seaward, and the seawater entrained into the surface water is replaced by deeper seawater with a movement toward the river. Lateral and vertical mixing causes seawater to be added to the effluent, which is carried in the offshore direction. In moving seaward through the estuary, the upper layer continuously increases in volume and salt content. Deep water rises to replace that which is carried seaward in the surface layers, whether entrained into the overlying surface layers or frictionally driven by the stress exerted by the overlying seaward-flowing surface layers. This process describes river-induced upwelling. In addition, wind-induced upwelling may drastically influence the hydrographic conditions in an estuary or the characteristics of the effluents.

Photographic material obtained under the manned space program revealed the complex distribution pattern of suspended material, especially in the near coastal areas where river discharge or upwelling are present. *Landsat,* with its separated

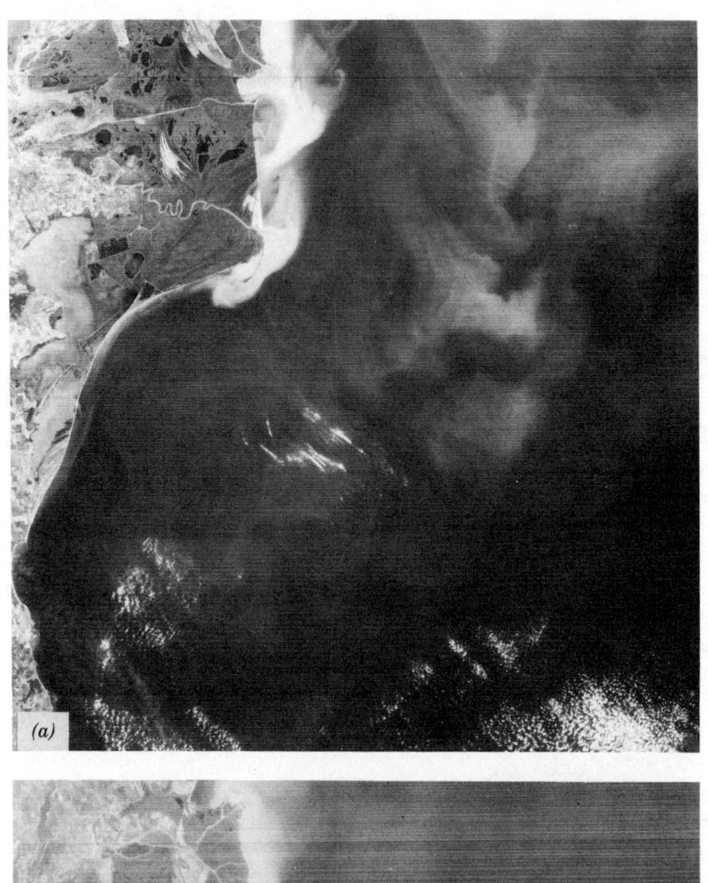

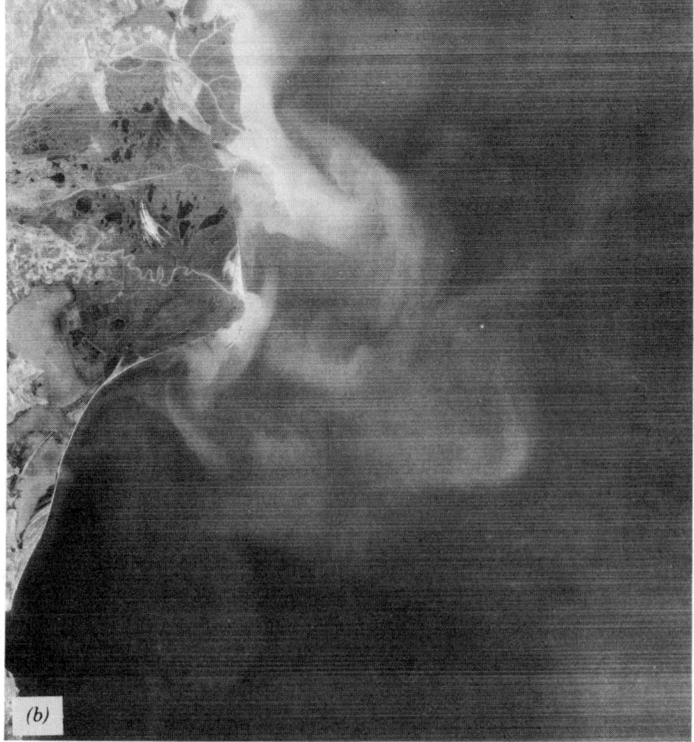

Figure 7-2-24. Effluent from the Danube River recorded at the spectral interval 0.5–0.6 μm on June 18, 1975 (*a*) and on July 24, 1975 (*b*).

spectral recordings, shows sediments mainly in the spectral intervals of 0.5–0.6 μm and 0.6–0.7 μm. Samples of recordings in the 0.5–0.6 μm channel over the Danube estuary are given in Figures 7-2-24*a* and *b*. It is obvious that the actual river discharge is limited to the very near coastal areas, as characterized by the high reflectance values. Further offshore, isolated patches in suspended material can be recognized.

The Amazon, the world's largest river, has a maximum flow rate in May and June of 2.25×10^5 m$^3 \cdot$ sec^{-1} and a minimum discharge of 0.9×10^5 m$^3 \cdot$ sec^{-1}. Accordingly, the river plume may extend as much as 200 km offshore and 700 km alongshore northwest as far as Surinam and Guyana, with a strong horizontal gradient in salinity (see Fig. 7-2-25). Therefore, the interaction between the Amazon and ocean water, which was investigated by Szekielda et al. (1983), undergoes different seasonal cycles with respect to the biochemical regimes. At the start of mixing between the river and ocean water, suspended matter from the Amazon settles rapidly and the resulting water is sufficiently transparent for increasing photosynthesis to commerce at around 7‰ salinity. As a result, high diatom productivity has been observed at a salinity between 7–5‰, which depletes dissolved nutrients. The Amazon effluent has a very thin layer of brackish water, which overlays ocean water with $S > 35$‰. This stratification can easily be broken into separate lenses with freshwater from the surface or high saline water from below. In a *Landsat* image (see Fig. 7-2-26) two processes can be recognized. The first is the injection of ocean water from below the surface; the second process is the separation of plumes from the major river effluent into the ocean. Both processes may have the same or similar impact on the biogeochemistry of the ocean/

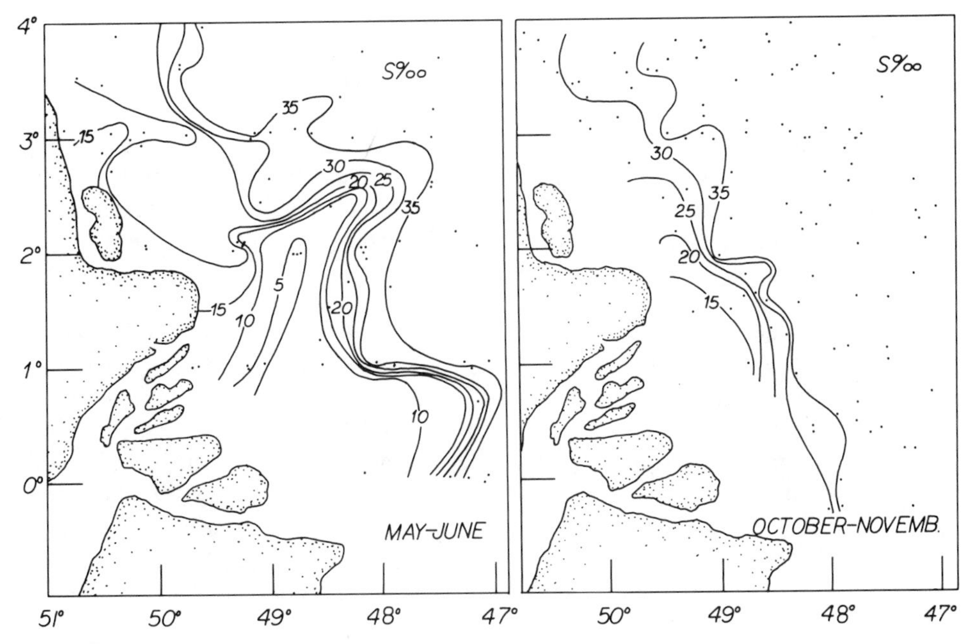

Figure 7-2-25. Salinity distribution in the effluent of the Amazon. After Szekielda et al. (1983). Reproduced with permission from Prof. Dr. E. Degens, University of Hamburg.

Figure 7-2-26. *Landsat* data over the northern edge of the Amazon effluent. Darker area in the river effluent show lower concentration of particles. Data was recorded at wavelength 0.6–0.7 μm on October 4, 1972.

river system although the intrusion of high-saline water to the surface may be regarded as an upwelling process.

Evidence for eutrophication in the river effluent is demonstrated with satellite data in the reflected IR. Keeping in mind the decreasing penetration depth of light as a function of increasing wavelength, one can qualitatively give differences in the suspended matter at the surface. Figure 7-2-27 shows that the surface water in the coastal region within the effluent of the Amazon has higher reflectance levels than the river itself.

The Changjiang is another river that has been intensively studied with satellite data (Szekielda and McGinnis, 1985). The maximum runoff of the Changjiang over the past 30 years has been recorded to be 92,600 $m \cdot sec^{-1}$ at peak discharge, with an annual average discharge of 9.25×10^{11} m^3 and an annual average sediment discharge of 486×10^8 tons per year. The sediment discharge during the flood season amounts to 78% of the annual total sediment discharge, with highest discharge in July of about 22% and the lowest monthly total representing 0.6% of the annual total in February.

Shen et al. (1983) showed that more than 50% of the sediments from the Changjiang River Basin are deposited at the mouth, forming a wide subaqueous delta area outside the south passage and the north branch. The region between 122°30′–123°E is an important boundary region, because little suspended sediment dis-

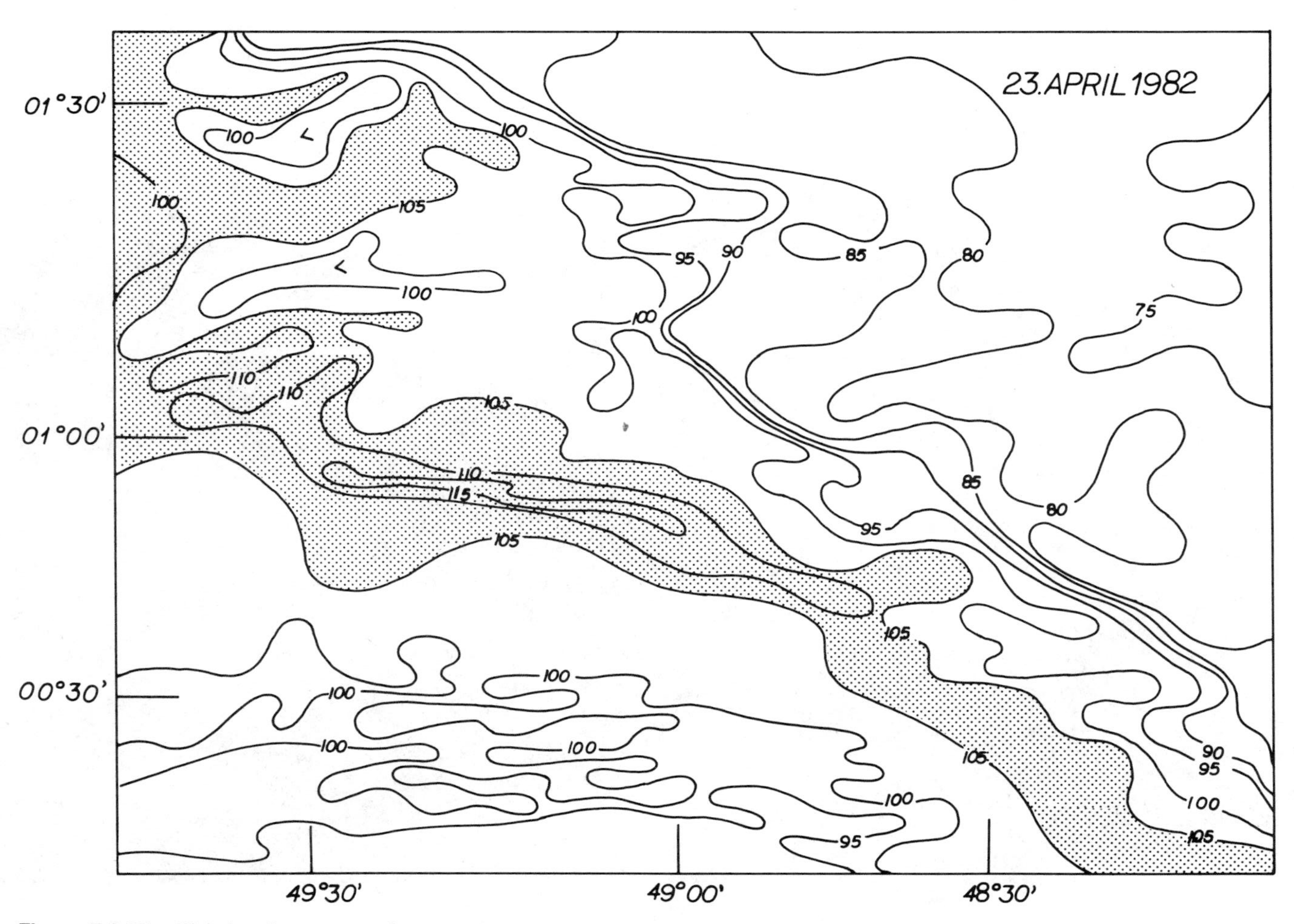

Figure 7-2-27. Digital radiance map of apparently cloud-free regions over the Amazon effluent. Shaded areas are those with highest eutrophication, while south of the eutrophication lower radiance can be interpreted in terms of sedimentation and consequently of lower reflectance.

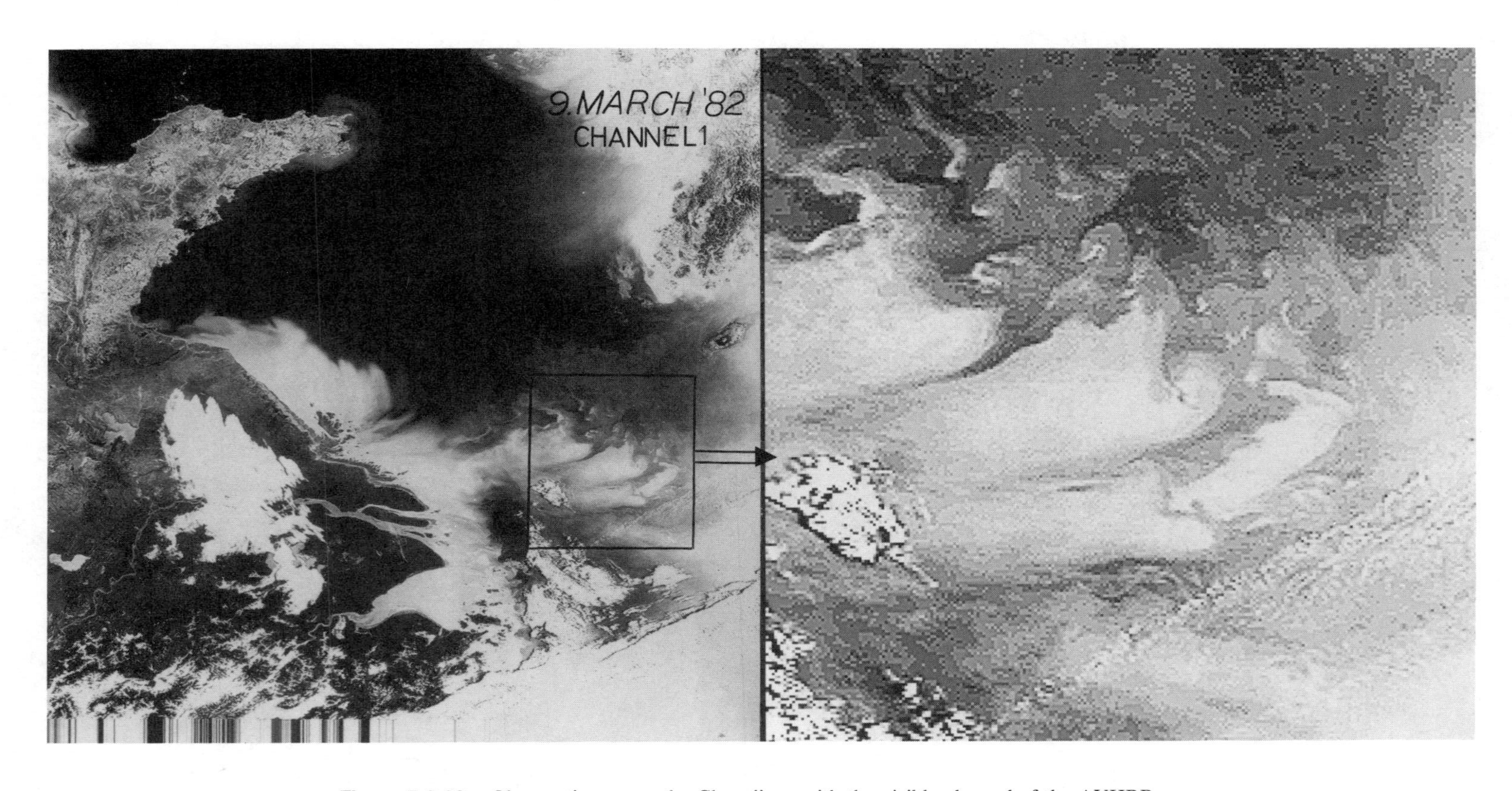

Figure 7-2-28. Observations over the Changjiang with the visible channel of the AVHRR.

perses beyond it. This line is nearly identical with the front edge of the subaqueous delta. Sediment exchange occurs between the Changjiang estuary and the northern Jiangsu coast, with some sediments from the northern Jiangsu coast transported to the north branch during the dry season. In summertime, the freshwater discharge near the mouth of the Changjiang exhibits a bimodal distribution. The freshest water $S < 18‰$, extends in a band to the south along the coast at depths less than 12 m, in a relatively shallow, low-salinity ($18‰ < S < 26‰$) plumelike structure, extending offshore on average toward the northeast, as can be seen in Figure 7-2-28. The very fresh water found to the south is clearly of Changjiang River origin. The origin of the low-salinity plume found to the north is

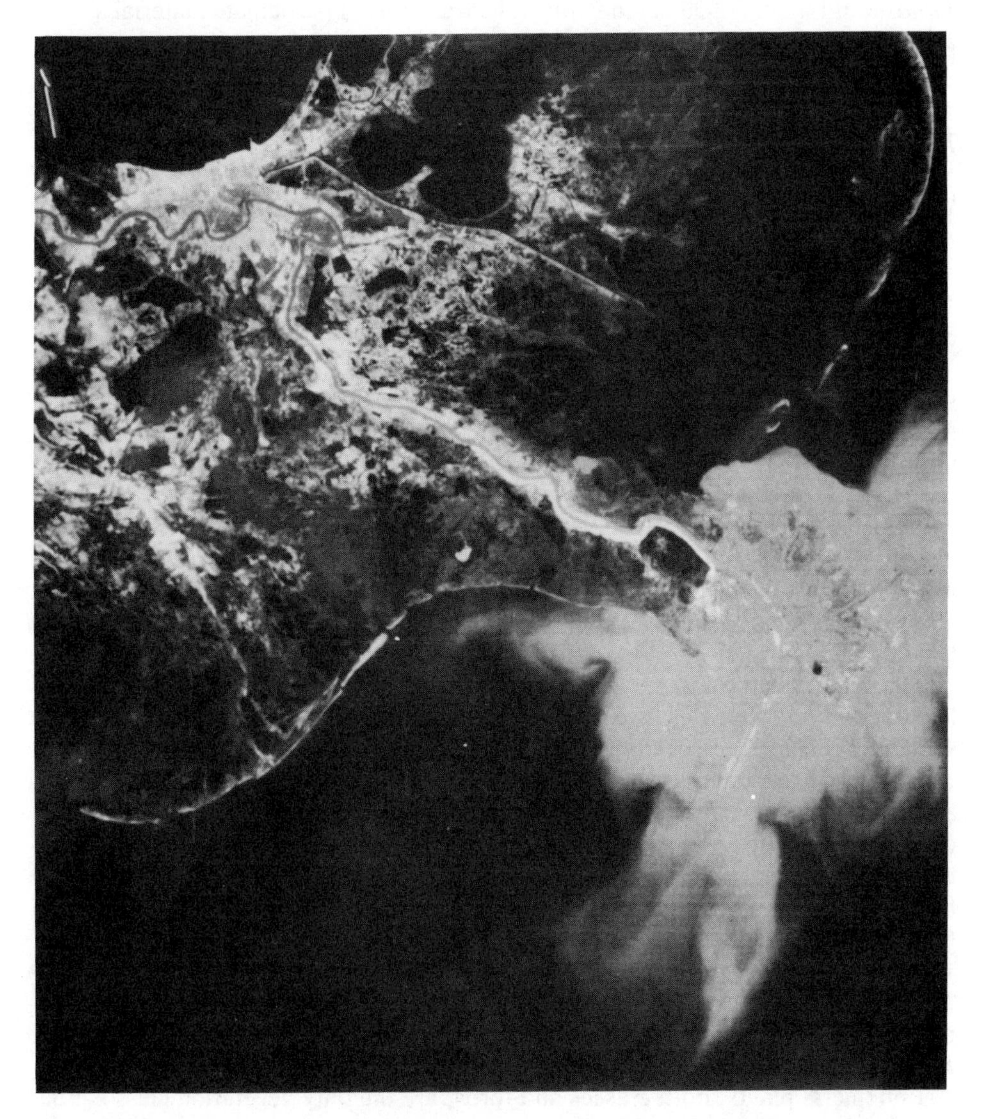

Figure 7-2-29. Discharge area of the Mississippi viewed with *Landsat* MSS in January 1973. Courtesy of NASA.

uncertain; it may be part of the Changjiang discharge presumably through the north passage, or it may partially originate from river discharge occurring to the north along the Chinese coast.

The size and shape of discharge plumes from rivers are observed to be a function not only of the amount of freshwater discharged, but also of the speed and direction of the wind. Rouse and Coleman (1976) showed that, for the Mississippi, the plume enlarges with rising river stage, but the areal extent of the turbid water can be related to discharge only in a general way because of the concentration and dispersal effects produced by the wind. The plumes of suspended sediment are readily detected during favorable wind conditions at distances greater than 75 km seaward of the various river mouths. The discharge area viewed with *Landsat* in Figure 7-2-29 demonstrates the transport of particulate material as well as the deposit of sediments in the estuary.

7-3 VEGETATION

Vegetation is a very dynamic feature at the land surface. At an early stage of the *Landsat* system, the application of satellite remote sensing became quite useful in monitoring the changes in vegetation due to natural or human-made activities. This has become more apparent in agriculture, in the shifting and substitution of crops, in forest exploitation, and in alterations and defoliation due to insects.

For investigations of forest areas, factors to be taken into account include the slope of the soil, soil background, and morphology of the plant canopy, especially the crown projections (see Fig. 7-3-1). Shadows are almost uniformly dark, and their characteristics are independent of the component species that cast them or of the ground coverage, which can be bare soil, grass litter, or all components combined. Graetz and Gentle (1982) showed that a dense stand of saltbush of 20% cover can obscure 8% shadow in mid-summer and some 35% in mid-winter. The relationship between the cover of vegetation and the amount of shadow on the intervening soil surface is weakly nonlinear only over certain ranges of reflectance levels.

The spectral properties of vegetation can be changed due to stress situations and defoliation caused by insects. Forest insects generally attack a particular tree species or species association that has a fairly well-defined and documented geographic distribution. Thus, the delineation of tree species via *Landsat* multispectral signature extraction and classification is of minor importance for forest canopy damage detection. Development of techniques for monitoring insect defoliation has therefore been slanted toward isolating and assessing changes in the forest canopy, due to fluctuations in the amount of leaf material per unit area (Williams and Stauffer, 1979).

The interaction between soil background and rangeland-type of vegetation has to be taken into account for image interpretation of shrubs. Table 7-3-1 summarizes the different reflectance values as recorded in different wavebands of the *Landsat* MSS for various rangeland landscape components.

Ephemeral plants, both grasses and forbs, are the only vegetation class similar to agriculture crops. The high IR reflectance of ephemerals is itself feeting for water-stress or aging results in a substantial reduction in reflectance in these

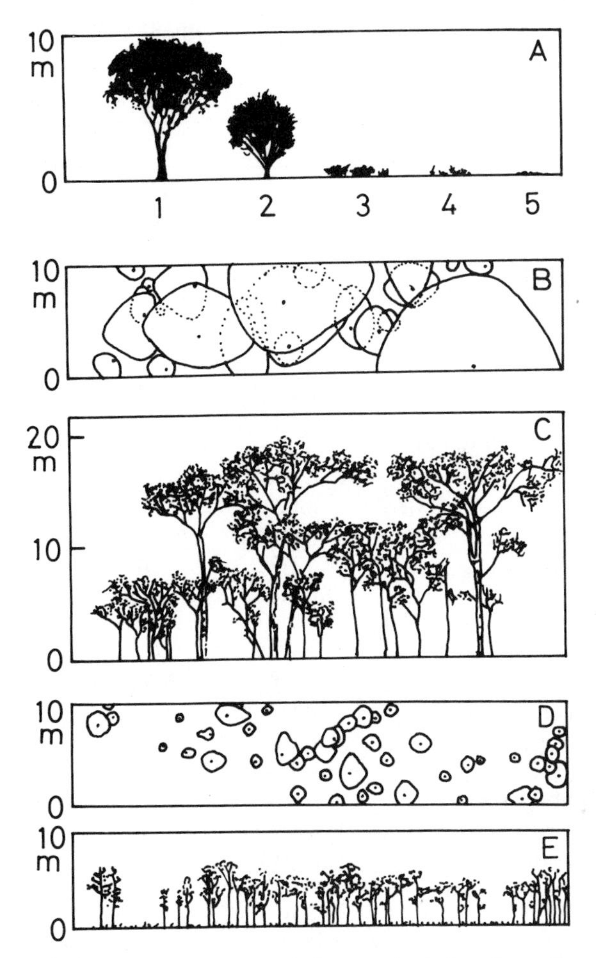

Figure 7-3-1. (*a*) Sketch illustrating the morphology and relative sizes of woody perennial and intershrub vegetation of semi- and Australian shrubland at Koonamore: 1. trees–eucalyptus; 2. arborescent shrubs–acacia; 3. low shrubs–atriplex; 4. grasses and forbs; 5. soil lichen crust and litter (Green, 1986). (*b*) *and* (*c*) Crown projection and profile diagram of tree (dbh ≥ 4.5 cm) of dry dipterocarp forest. (*d*) *and* (*e*) Crown projection and profile diagram of tree (dbh ≥ 4.5 cm of dry dipterocarp forest. Modified after Wacharakitti et al. (1985).

bands. Perennial shrubs do not exhibit high IR reflectance most of the time, which is largely determined by the structure and state of the leaf mesophyll cells, but may also be, in part, the result of leaf surface characteristics such as hairiness. Saltbush leaves, particularly when drought stressed, are covered with collapsed epidermal hairs that contain crystalline sodium chloride, which substantially increases the reflectance in the visible wavelengths. Dry grass (or standing-dead plant material) and litter (fallen-dead) plant material have very similar four-band reflectance values, and the yellow-gray litter has much the same values regardless of its origin, that is, whether the plant material is derived from dead and desiccated ephemerals, from grasses, or from shrubs (Graetz and Gentle, 1982).

In investigations of shrublands in the interior of the Australian continent (which support a large grazing industry), distinguishing the woody perennial vegetation

TABLE 7-3-1. Mean Reflectance Values in the *Landsat* Wavebands for Various Rangeland Landscape Components[a]

Component		*Landsat* MSS Wavebands 4	5	6	7
Bare soil	Undisturbed (53)	0.12 (0.009)	0.24 (0.009)	0.36 (0.012)	0.42 (0.012)
	Eroded (22)	0.15 (0.005)	0.35 (0.008)	0.54 (0.011)	0.54 (0.013)
	Calcareous (17)	0.10 (0.002)	0.21 (0.005)	0.34 (0.009)	0.38 (0.024)
	Lichen crusted (23)	0.06 (0.004)	0.12 (0.006)	0.22 (0.011)	0.25 (0.012)
Litter (29)		0.04 (0.001)	0.07 (0.003)	0.12 (0.004)	0.14 (0.005)
Perennial shrubs	Vigorous (28)	0.10 (0.005)	0.10 (0.006)	0.40 (0.012)	0.50 (0.016)
	Nonvigorous (95)	0.07 (0.002)	0.07 (0.002)	0.19 (0.004)	0.20 (0.004)
Grass	Vigorous (3)	0.09 (0.01)	0.11 (0.014)	0.43 (0.035)	0.50 (0.029)
	Nonvigorous (24)	0.06 (0.004)	0.07 (0.004)	0.16 (0.010)	0.15 (0.009)
Ephemerals	Vigorous (31)	0.04 (0.005)	0.04 (0.010)	0.42 (0.010)	0.53 (0.010)
Shadow (10)		0.01 (0.001)	0.01 (0.003)	0.01 (0.003)	0.02 (0.003)

[a] Standard errors are given in parentheses. The number of samples for each component are given in parentheses following the target name.

After Graetz and Gentle (1982). Reproduced with permission from *Photogrammetric Engineering & Remote Sensing*, copyright by the American Society for Photogrammetry and Remote Sensing.

from the smaller herbaceous vegetation and soil-encrusting lichen found between the shrubs is critical for range management; however, to do this with *Landsat* data alone is difficult. Green (1986) combined radar data from SIR-A and *Landsat* MSS data and concluded that radar return should be primarily determined by the percentage of the area occupied by shrubs, due to low topography and fine-textured soils. During periods when most of the vegetation was nonvigorous and spectrally homogeneous, SIR-A data, as a surrogate measure of shrub cover, allowed the reflectance due to shrubs in *Landsat* data to be separated from the reflectance due to the intervening ground. This method allows estimation of the intershrub reflectance properties that are related to herbaceous vegetation, lichen, and bare soil exposures.

7-3-1 Forests

The typical appearance of forest as given in satellite false-color images can be demonstrated with a *Landsat* color composite for the San Francisco area. As shown in Figure 7-3-2, the Santa Cruz Mountains, south of San Francisco, are covered with redwoods, indicated by the deep reds along the California coast. The hills to the north and east of the Bay area and southward support California oaks and evergreens mixed with chaparral, and the eastern slopes of the coast ranges are covered with grasses that turn a distinctive golden brown in summer. Tule marshes dominate the delta country to the north near the Sacramento River. The Sacramento Valley to the east is characterized by land use and farming for rice, cotton, barley, sunflower, sugar beets, beans, and tomatoes (Short et al., 1976).

For a quantitative approach of specie identification and for percentage of surface coverage, computer image analysis and classification techniques are being

Figure 7-3-2. *Landsat* color composite for the San Francisco area. Courtesy of NASA.

used, although the classical photo interpretation of *Landsat* images is still used even for operational application.

The various techniques of classification for *Landsat* MSS data in the application for tropical forests can be summarized (Vibulsresth et al., 1985) as follows:

1. By using the *band ratioing* method, one can replace band 5 in the false-color composite of bands 4, 5, and 7 with the vegetation index (VI) image to produce an enhanced image that reveals the forest pattern relating to its density. On the other hand, by replacing band 7 with the VI image, one can remove the shadow effect appearing on the image.
2. *Preliminary classification* of the dry season image in the form of a color map was checked in the field to relate forest density classes to actual forest types. It was found that the high-density forest is a dry evergreen forest, the medium-density forest is dry dipterocarp and disturbed forests, and the low-density forest is a mixed land use of disturbed forest, scrub, old clearing, and field crops.
3. In the *supervised classification,* seven classes of forest land use have been identified, based on the differences in leaf condition, stand density, and vegetation type. These classes are

    1. Dry evergreen forest (full leaf all year round)
    2. Dry dipterocarp forest class 1 (all leaves shedded in dry season)
    3. Dry dipterocarp forest class 2 (some trees are still green in dry season)
    4. Disturbed forest (consisting of secondary growth and some big trees)
    5. Old clearing (and grassland)
    6. Scrub forest (bushes and scrubs and young plantation)
    7. Agricultural land (field crops)

A dry season image has an advantage over the wet season image in that it can differentiate a dry dipterocarp forest into two levels according to their leaf conditions. Furthermore, each class can clearly be defined in dry season when taking into account the distribution of mean and standard deviations. The two-band crossing plot between band 7 and VI ratio illustrates the separation between different classes. The classification results of the composite of the dry and wet season images were found to be analogous to those obtained from dry season single data. The results of such supervised classification after undergoing a filtering process are summarized in Table 7-3-2.

Although tropical regions have frequent cloud coverage, Umali and Amaro (1985) completed a forest inventory for the Philippines, using *Landsat* multispectral data in the digital mode and classifying the following types of forest-related cover: (1) full-canopy closure dipterocarps, (2) partial-canopy closure dipterocarps—reproduction areas and brush, (3) mangrove forest, (4) mossy or high elevation forest, and (5) non-forest-land wetlands, marshy areas, and small water bodies in the proximity of forested areas. The measurements for the major forest types were compiled for each *Landsat* scene and then appropriately combined for the following geographical areas: Mindanao, Luzon, Mindoro, Palawan, and Visayas. The final results of the *Landsat* forest inventory are presented in Table

TABLE 7-3-2. Summary of the Supervised Classification of Tropical Forest[a]

Input Data Sets (Classes)	Dry Evergreen Forest		Dry Dipterocarp Forest Class 1		Dry Dipterocarp Forest Class 2		Disturbed Forest		Old Clearing	
SINGLE DATE										
1. Dry season										
4 bands	35.1	43.6	10.0	12.4	3.2	4.0	7.4	9.2	16.1	20.0
3 bands + VI (replace band 5)	36.3	45.1	10.2	12.7	3.3	4.1	6.8	8.5	15.6	19.4
2. Wet season										
4 bands	33.1	41.2	5.4	6.7	7.7	9.6	7.5	9.3	12.8	15.9
3 bands + VI (replace band 5)	32.9	40.9	5.3	6.6	8.4	10.5	6.7	8.3	12.8	15.9
TWO DATE										
1. 4 bands	33.3	41.4	5.0	6.2	10.3	12.8	8.5	10.6	15.1	18.8
2. 3 bands + VI (replace band 5)	33.9	42.2	5.3	6.6	10.4	12.9	7.9	9.8	14.9	18.5

[a] The first column under each heading is given in km^2; the second in %.
After Vibulsresth et al. (1985).

7-3-3; the area totals for each forest type are listed by region and for the entire nation and show that 38% of the nation is covered by forest.

Apart from forest species, the distribution of trees as monitored by remote sensing techniques is influenced by age, density, and exposure of the slope. In spite of a certain amount of confusion, it is possible to distinguish in the northern latitudes mature and overmature balsam fir stands, black spruce stands, level of regeneration after cutovers or fires, and even damage caused by insect epidemic. By using unsupervised digital classification, Beaubien (1979) found that it is gener-

TABLE 7-3-3. Forest Mangrove and Wetlands Inventory. Area Measurements Summary by Major Islands (in Hectares). IMAGE-100 Results

Island or Region	Total Land Area	Forest Full Closure	Forest Partial Closure	Forest Obscured by Clouds	Mangrove	Mossy	Wetlands	IMAGE-100[b] Total Forest	IMAGE-100 Forest Cover (%)
Mindanao[a]	10,199,840	2,411,869	1,135,765	721,685	39,810	2,622	36,408	4,311,751	42.27
Palawan	1,489,626	675,681	312,324	87,030	34,853	30	92	1,109,918	74.51
Luzon	11,625,410	2,177,671	1,406,442	578,837	10,924	52	213,160	4,113,926	35.39
Mindoro	1,024,457	272,190	187,412	30,877	6,701	4,516	35,418	501,696	48.97
Visayas	5,660,622	474,356	775,698	101,474	13,845	—	67,806	1,365,373	24.12
								11,402,664	38.00
IMAGE-100									
Total[b]	11,402,664	5,951,767	3,817,641	1,519,903	106,133	7,220	352,884		
Percentage	38.00	19.84	12.73	5.07	0.35	0.02	1.18		

[a] Area measurements include Basilan and Sulu Archipelago.
[b] IMAGE-100 total forest does not include wetlands.
After Umali and Amaro (1985).

ally possible to map hardwood, mixed wood, and two or three types of softwood stands, depending on the area. On the other hand, it is impossible to separate regeneration and mature stands within mixed or hardwood forest types. The most important issue is that the accuracy of a computer classification depends on factors specific to each study, but particularly on the type of terrain to be mapped.

In consequence, different sampling techniques for ground checking are required and have to be adapted according to the inventory strategy. For instance, Talbot and Markon (1986) described sample areas based on a combination of helicopter-ground survey, aerial photo interpretation, and digital *Landsat* data. Major steps in the *Landsat* analysis involved preprocessing, geometric correction, derivation of statistical parameters for spectral classes, spectral class labeling of sample areas, preliminary classification of the entire study area using a maximum-likelihood algorithm, and final classification utilizing ancillary information, such as digital evaluation data.

Problems arise in the refinement of a vegetation canopy if more than one category is contained in the spectral classes. Most frequently, these occur where variation in slope and aspect cause inconsistent reflectance properties or different vegetation types have similar reflectance properties because of terrain influence. Site moisture is another factor altering the reflectance characteristics of a vegetation type. Therefore, as pointed out by Talbot and Markon (1986), dwarf shrub wetlands, for instance, can be consistently mistaken for open and closed conifer forests because all have low reflectance in the IR.

Although certain limitations have to be taken into account in developing forest inventories, the real advantage of satellite acquired data sets is time sequences. For instance, Table 7-3-4 gives the forest area of Thailand by region as obtained from *Landsat* studies and shows the changes within nine years. It may be seen that there is an average annual decrease of 3.26% of the forest area. These studies of forest area using satellite remote sensing were quite elementary in nature, but they enabled the monitoring of the rate of deforestation as an input to policy-level

TABLE 7-3-4. Forested Area by Region in Thailand and Area Changed from 1973–1982

Region	Total Area (km^2)	Forested Area[a] 1973		Forested Area[b] 1982		Decrease in Forested Area over 9-Year Period		Average Annual Decrease (%)
		km^2	%	km^2	%	km^2	%	
Northern	169,644.29	113,595	66.96	87,756	51.73	25,839	22.75	2.53
Northeastern	168,854.34	50,671	30.01	25,886	15.33	24,785	48.91	5.43
Eastern	36,502.50	15,036	41.19	8,000	21.92	7,036	46.79	5.20
Central	67,398.70	23,970	35.56	18,516	27.47	5,454	22.75	2.53
Southern	70,715.19	18,435	26.07	16,442	23.25	1,993	10.81	1.20
Total	513,115.02	221,707	43.21	156,600	30.52	65,107	29.37	3.26

[a] Source: *Landsat* 1 imagery.
[b] Source: *Landsat* 2 imagery.
After Chaudhury (1985).

decision making to prevent further devastation of the forest ecosystem of Thailand (Chaudhury, 1985).

Investigations using SAR are of particular interest for tropical forest monitoring because active microwave systems can penetrate clouds and record the amount of energy returned from surface features. In cloud-covered zones, where sensors that operate in visible and IR wavelengths have limited utility, microwave sensors may provide information about forest area and structure that would be otherwise unobtainable (Sader, 1987).

Recent investigations have reported qualitative comparisons of SAR signals and forest targets with data collected from *Seasat* and SIR-A (Wu, 1981, 1984), SIR-B (Mueller et al., 1985), and aircraft systems (Hoffer et al., 1985; Ford and Wickland, 1985; Paris, 1985; Evans et al., 1986). Physical models of radar backscatter from vegetation components have been derived from experimentation with agricultural crops (Attema and Ulaby, 1978; Fung, 1979; Hoekmann et al., 1982), but knowledge concerning the physical relationship between radar backscatter and forest vegetation is still lacking (Hoekmann, 1985).

Investigations of vegetation structure are important for understanding the potential role of microwave sensors for deriving estimates of forest biomass and vegetation structure. Crown shape, leaf size, leaf mass per volume, and age class were related to X-band (3-cm) backscatter differences in polarization (Hoekmann, 1985). In deciduous forests, the leaves were believed to be important as scatterers of X-band waves; however, L-band waves (24 cm) with HH polarization penetrated through leaves and were scattered back from the branches of trees (Sieber, 1985). Plant height, density, geometry, and presence or absence of standing water at the surface influence the SAR response, but incidence angle and polarization can also have a significant effect (Ormsby et al., 1985; Ulaby and Wilson, 1985; Hoekmann, 1985; Pampaloni and Paloscia, 1985).

7-3-2 Wetlands and Mangroves

Coastal wetlands are some of the most valuable and biologically highly productive areas along the coast, since they are an essential breeding ground for many ocean fish. For example, at least 70% of the U.S. East Coast fish, including 30% of the U.S. commercial catch, depend sometime during their life cycle on the shallow waters bordered by the coastal wetlands. Tidal cycles regulate nutrients and detritus, which support an ever-expanding biostructure that includes man (McEwen et al., 1976).

Wetlands are generally composed of coastal vegetation and are under the influence of the tide. The wetlands are the transition zone between the mainland and the sea and are influenced by storm surges and human activities. As agricultural activities, urban development and other activities compete in the wetlands (see Fig. 7-3-3), one can no longer find an undisturbed environment in the wetlands. Recognition of marshes has been a difficult task for inventories. The difficulty is due to the greater spectral variation within marshes than in other categories. Since marshes are often associated with subirrigated meadows and since the line defining the division between these two categories is not easily determined, even at ground level, limited ground resolution in imagery further complicates delineation. An additional complicating factor in the interpretation of both marshes and

Figure 7-3-3. Wetland area with various activities including agriculture and urban development. Reproduced with permission from the Environmental Research Institute of Michigan (ERIM).

subirrigated meadows in satellite data is the presence of nonliving vegetative material that may dominate portions of an image. For instance, the nonliving vegetative material masks the response of living vegetative material at various wavelengths. This can be shown with "mottled" data in *Landsat* MSS channel 6 due to the presence of the nonliving vegetation as well as to the infinite variation in mixtures of marsh vegetation, surficial water, and nonmarsh vegetation.

In tropical regions, mangrove forest stands are difficult to recognize apart from the land cover types in visual examination of *Landsat* imageries. This makes supervised classification quite difficult, too. In addition, as a result of economic exploitation, most of the mangrove areas in tropical countries are now in varying secondary and reproductive growth stages, with the original mangrove species having been replaced, in several cases, by nonmangrove types. In some areas of the Philippines, for instance, the majority of the vegetation stands, though some portions of which still show remnants of the primary mangrove forest, are second-growth mangroves with varying ages, canopy cover, and density. Interspersed with the stands are fishponds and logged-over areas overgrown by brush vegetation (Biña et al., 1978a).

An area of coastal environment with partial cover of mangroves is shown in Figure 7-3-4 for the region in Guinea Bissau.

Carter and Schubert (1974) showed that classification, delineation, and evaluation of coastal wetlands can be made on the basis of major vegetative associations and their spectral signatures (see Table 7-3-5). For the production of wetland maps, vegetation analysis look-up tables have been developed for use of *Landsat* MSS digital values for bands 4, 5, and 7, using seasonal spectral reflectance measurements from field observations. Based on these look-up tables, computer-generated maps at an approximate scale of 1 : 20,000 can be produced with a high degree of accuracy in the identification of wetland features and plant associations.

In development of look-up tables, the following basic classes typical of Central Atlantic coastal saline and near-saline wetlands were established: water (open), soil and/or bare sand, sandy mud flat, marsh vegetation, and upland vegetation.

TABLE 7-3-5. *Landsat* MSS Signature Analysis for Coastal Wetlands (October)

MSS 7 Digital Values	MSS 4 Digital Values		
	<23	23–26	>27
<5	Water (.)		
5–6	Organic mud flat (/)	Sandy mud flat (-)	
7–10	*Spartina alterniflora* (A)		
11–14	*Iva frutescens* (I)	Spoil (_)	
15–26	*Spartina patens*[a] (P)		Upland Vegetation (D)
		MSS 5	Digital Values
>27		<20 Grass (G)	>20 Sand (S)

[a] MSS 5 also used for separation.

After Carter and Schubert (1974). Reproduced with permission from the Environmental Research Institute of Michigan (ERIM).

Figure 7-3-4. Coastal environment with partial cover of mangroves in Guinea Bissau. Courtesy of SPOT-Image.

The marsh vegetation subdivided into five major plant associations: *Spartina alterniflora* (saltmarsh cordgrass) association; *Spartina patens* (saltmeadow cordgrass) association; *Iva frutescens* (marsh elder) association; organic mud flat association (low, sparse vegetation or detritus with occasional bare spots); and *Juncus roemerianus* (needlerush) association (Carter and Schubert, 1974).

7-3-3 Agricultural Crops and Land Use

The identification of agricultural crops is based on the spectral reflectance characteristics of the different crop types, which vary as a function of different factors, such as phenology, growing conditions, soil fertility, management practices, and weather conditions. The existence of two or more crops with similar reflectance values in a particular area can lead to misclassification, which means that a careful selection of the acquisition date, which makes use of different phenological dynamics of the crops in question, can often overcome this problem, provided the crop calendar of area is known (Reichert, 1984). Simultaneous evaluation of data taken at different times of a growing season, however, increases significantly the accuracy of crop identification.

All four bands of *Landsat* MSS are useful for land use inventories; however, the bands covering the ranges 0.6–0.7 μm and 0.7–1.1 μm provide most of the information for the interpreter. Using computerized methods for digital processing of CCTs, one can categorize up to 18 levels of land use with a high precision. So far, land use inventories have been carried out at various levels of classification for use along with other types of information for land and resource management: identifying and preserving land suited for agriculture, forest location and management, prediction of water usage and stream management, and so on.

The different agricultural practices and resulting spectral responses can be recognized in images without special processing of the original data tapes. An example of this is given for the Sacramento Valley in Figure 7-3-5, where rice growing areas and other land use classes can be identified.

Similar observations have been made of the border between the Great Plains of northcentral Montana (U.S.) and the province of Alberta (Canada) (Fig. 7-3-6), where the different land use patterns result in different spectral responses. Through the higher resolution and specific spectral bands of SPOT data, land use patterns and related structures can be distinguished more easily than with the *Landsat* MSS and TM data (Fig. 7-3-7). This is particularly important for applications in areas where the field is much smaller than in northern latitudes, for instance.

An example of agricultural activities observed from space is given in Figure 7-3-8, which shows El Gezira situated between the white Nile to the north and the blue Nile going south. The fertile soil, which was deposited from the blue Nile about 12,000 years ago, is responsible for the agricultural development of the region. Based on a resolution of about 10 m, it is possible to identify the irrigation and cultivation structures of farms together with the respective rotation of crops.

The strong contrast between bare surface deposit and vegetation is illustrated with a *Landsat* image over the Nile Delta region in Egypt (see Fig. 7-3-9). The major crop in the delta is cotton, which accounts for about 25% of the total crop area. Since rainfall in the area is below 25 cm per year, farming in this region is

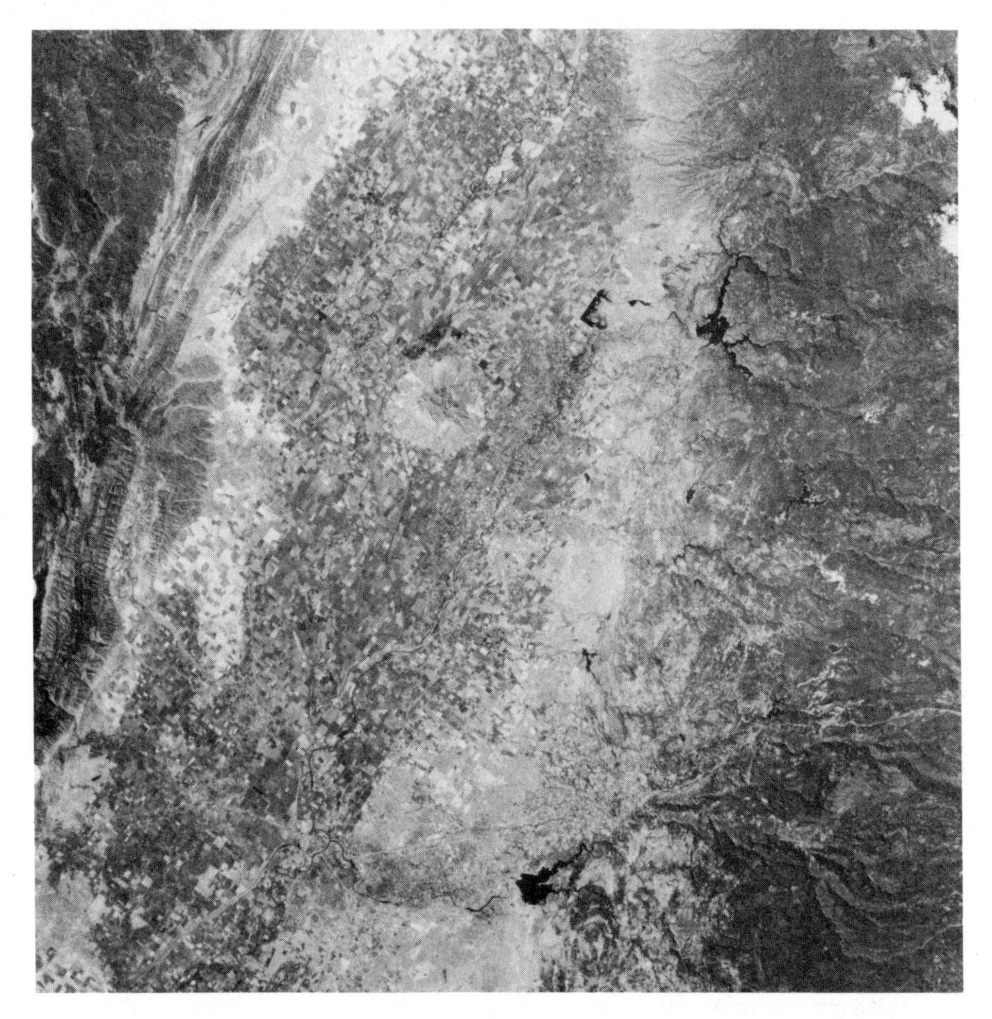

Figure 7-3-5. Major rice growing areas and other agricultural land use classes in the Sacramento Valley. Courtesy of NASA, Lyndon B. Johnson Space Center.

entirely dependent on irrigation through canals (Short et al., 1976). In comparison with Figure 7-3-9, the data of portions shown in Figure 7-3-10 give the advantages of high resolution, since crop size in this particular region is smaller than, for instance, in the United States or in Europe.

Tucker et al. (1984) used data from the AVHRR onboard *NOAA-6* and *NOAA-7* in order to study the large-scale vegetation dynamics in the Nile River Valley and Nile Delta of Egypt. Definite trends with respect to time were observed, which correlated with growing conditions and agricultural practices. Results of the study indicated that these data are useful for monitoring large areas of vegetation, demonstrated that atmospheric effects must be understood quantitatively for specific inventory purposes, and showed that the addition of a thermal channel is valuable for detecting areas of visible and subvisible clouds. It was also shown

Figure 7-3-6. Land use brings out the sharp definition of the border separating the Great Plains of northcentral Montana and the province of Alberta, Canada.

that data with a resolution of only about 1 km can still provide valuable information with respect to land use or cycles of vegetated surfaces.

One has to keep in mind that satellite images are not specifically recorded for themes; rather, they contain multidisciplinary information that either can be extracted for deriving thematic maps or are used for broader maps on land use, for example. An analysis of TM data for land use mapping is given in Figure 7-3-11.

Types of vegetables have been the focus of *Landsat* MSS studies by Ryerson et al. (1979, 1981) and by Zhu et al. (1983), and several *Landsat* MSS investigators have included vegetables among other crops of interest (e.g., Morse and Card, 1983). Although spectral separability of vegetables was generally achievable with four MSS bands, crop inventory with MSS data can be hindered by the small, irregularly shaped fields and the lack of continuous crop canopy.

Figure 7-3-7. SPOT data showing land use pattern and related structures. Courtesy of SPOT-Image.

Figure 7-3-8. Agricultural activities in the El Gezira region in Sudan. Photo was taken with the metric camera on the Space Shuttle. Courtesy of Zeiss, Federal Republic of Germany.

Computer-processed *Landsat* data have shown that crop species can be determined to a high degree. Bauer and Cipra (1973), for instance, used a 2000 mi^2 area as a test site and yielded improved results through the use of temporal and spatial data. One area was classified first, using a training set with 12 corn fields, 12 soybean fields, and 2 to 3 fields for each of the "other" classes. An overall test performance of 83% was achieved for the three classes of corn, soybeans, and "other." Quantitative results from this classification are shown in Table 7-3-6.

Safir and Myers (1973) pointed out that recognition percentages are high for those vegetation classes that have mature and uniform canopies at the time the data are collected. Corn, soybeans, and trees (forest) were classified accurately because they met this criterion. The class of senescent or senescing vegetation included observations from field beans, wheat stubble, and grass. These canopies were characterized by nonuniform distributions of dead and dying vegetation

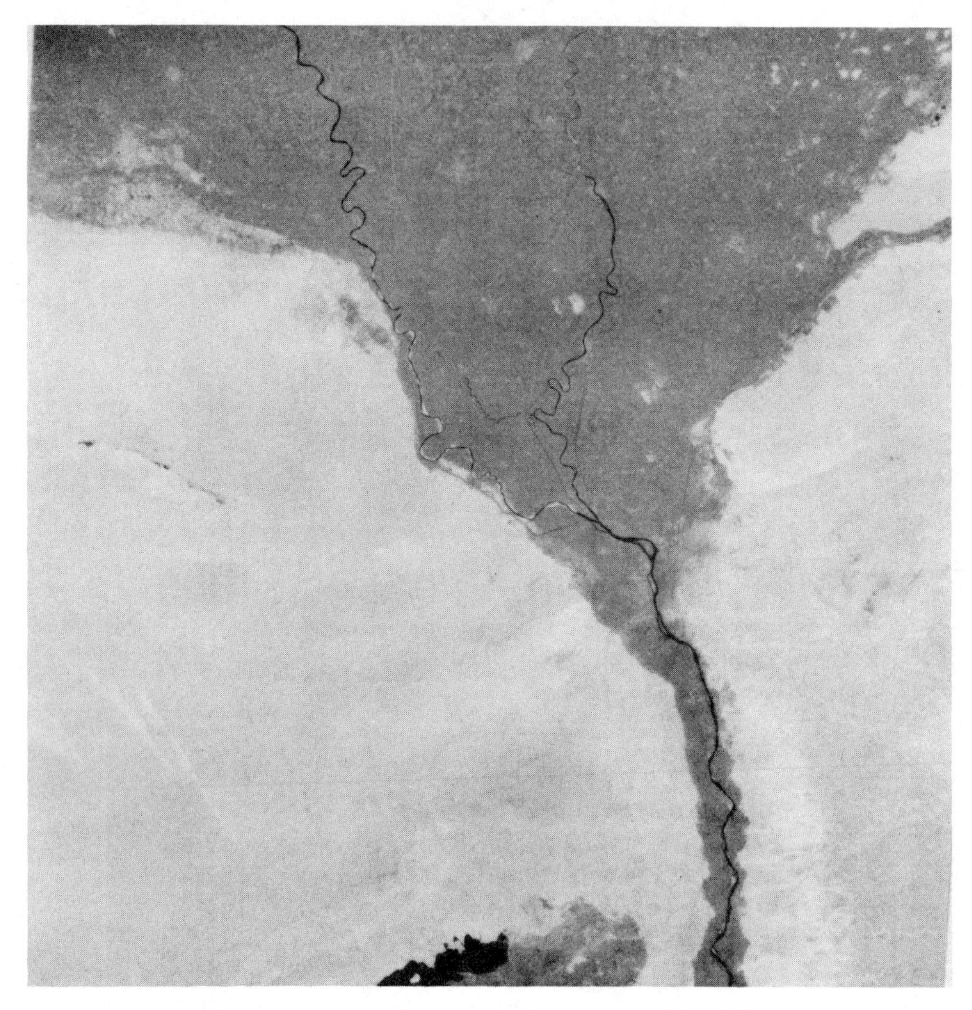

Figure 7-3-9. *Landsat* image over the Nile Delta region in Egypt.

TABLE 7-3-6. Classification of Corn, Soybean, and "Other" Test Fields by Computer Analysis of MSS Data

Crop	No. Points	No. Points Classified as Corn	Soybeans	"Other"	Percent Correctly Classified
Corn	3968	3367	357	244	85
Soybeans	1113	115	855	133	77
"Other"	295	16	50	234	79
Total	5376	3498	1262	611	
Overall Performance: 83%					

After Bauer and Cipra (1973).

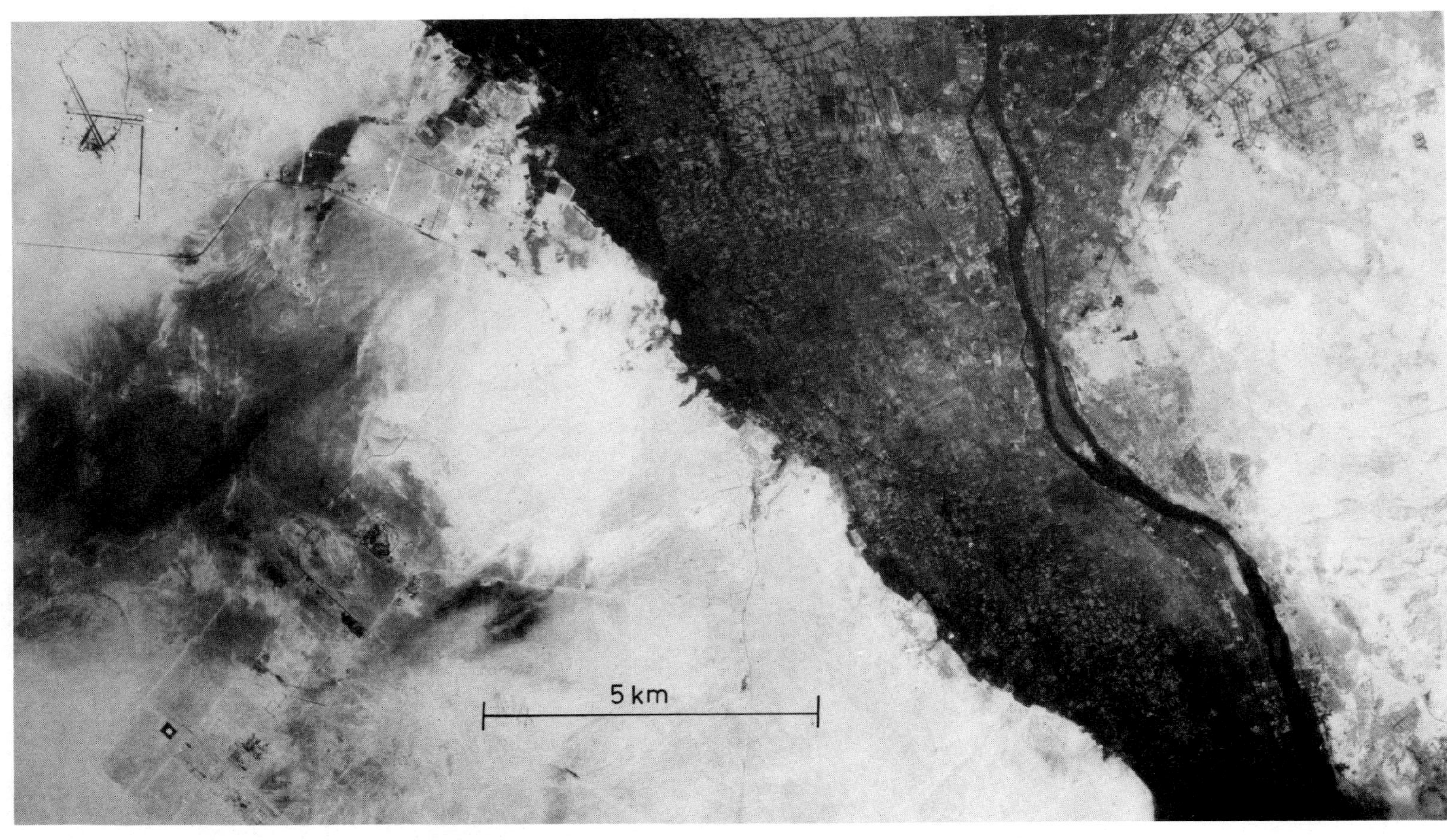

Figure 7-3-10. Nile Delta region taken with higher resolution than image in *Figure 7-3-9*. Courtesy of SPOT-Image.

(a)

Figure 7-3-11(a). Land use based on Landsat TM data in the People's Republic of China. See insert for color representation.

Figure 7-3-11(b). Example of land use pattern in Tangshan area, People's Republic of China. Rectangular area refers to the land use map given in *Figure 7-3-11(a)*. Courtesy of Prof. Yang Shiren, Academia Sinica, Beijing, People's Republic of China. See insert for color representation.

along with patches of healthier vegetation. For example, field beans had matured and had begun senescing, whereas soybeans and corn were more vigorous. Also, wheat stubble fields were dry and brown except for some that had been seeded to alfalfa or red clover.

Bizzell et al. (1973) used known fields for rice, barley, stubble, and so forth, and defined them to an automatic pattern recognition routine. The statistics for the required classes were computed, then each data point in the entire set was classified according to its spectral "signature" as defined by the four MSS bands and the generated statistical parameters. The results are shown in Table 7-3-7.

In addition to those mentioned in Table 7-3-7 other major crop types in the study area have been identified (e.g., safflower, alfalfa, cotton) as a result of the computer processing operations. Also, the utilization of image interpretation analysis of the computer-generated images yielded additional information related to crop condition or farming practices; for example, areas depicted as water in rectangular shapes were deduced and later confirmed to be flooded rice fields.

Based on analysis of one season's data on vegetable crops, *Landsat* TM data appear capable of providing reliable identification of vegetable crops. Testing of a single-date, supervised maximum likelihood classification showed accuracies of at least 90% for vegetables, with low errors of commission. Similarly, subjective testing found that visual image analysis of the digitally displayed image could easily identify most of the crops (Williams et al., 1987). Digital analysis provides a more rapid approach, but visual analysis provides higher accuracies. Where digital classifications were limited by varying growth stages and narrow fields, these fields could be recognized through visual interpretation. For single-date analysis, highest accuracies were obtained late in the season, when the crops were mature; however, for some crops, an additional scene, acquired midway through the growing season, may also be needed. Successful identification of vegetable crops would have to include crop calendars, the availability of TM data on the best dates for crop identification, and data on representative fields for training and testing. Crop reflectance measurements in the representative fields as an input to ground checks provide an important aid in analyzing the TM data, even when there is variation in planting date for the crops. When used together with a crop calendar, the field data can be used to determine the best dates for crop identification and to assess the regional crop separability with TM data.

The *Landsat* TM type of data fills a gap in many countries where *Landsat* MSS data have not been suitable for agricultural inventories. With the increasing avail-

TABLE 7-3-7. Classification Summary Report for Known Fields in Delta Mendota

Classes	Rice	Stubble	Burned	Turned	Other	Overall
No. pixels	13,528	6,483	4,155	22,820	153,915	200,901
% Total	6.7	3.2	2.1	11.4	78.6	100
Equivalent acreages	14,703	7,046	4,516	24,802	167,285	218,352
Training field accuracy (%)	92	97	90	98	93	94
Test field accuracy (%)	97	98	76	94	95	93

After Bizzell et al. (1973).

ability of higher-resolution data from *Landsat* TM and also SPOT, the visual image interpretation of TM imagery, as evaluated by Murai and Yanagida (1984) for color composites and black-and-white images of six bands, is helpful in the use of multispectral images (see Table 7-3-8).

One of the main constraints of crop identification with *Landsat* data is the relatively low spatial resolution. Whereas this proved to be sufficient for regions where large field sizes are predominant, such as the United States, Canada, or the Soviet Union, most European countries and the developing countries, have much smaller-structured agricultural systems.

Inventories of agricultural land use with *Landsat* showed that the field size also influences the overall correct automatic classification. This has been shown by Richardson et al. (1977); see Table 7-3-9. Since 10 *Landsat* pixels, corresponding to a field size of 7.5 ha, are required for the digital classification of a single field or a cluster of fields, all containing the same class, a major breakthrough has been done with SPOT. Corresponding figures are as follows: for *Landsat* TM (30 m × 30 m) a minimum field size of 0.9 ha, for the French SPOT or the German MOMS (20 m × 20 m) a minimum field size of 0.4 ha, and for SPOT in the panchromatic mode a field size of 0.1 ha (10 m × 10 m spatial resolution) (Reichert, 1984).

Although radar is not operationally flown yet and the interpretation of radar

TABLE 7-3-8. Interpretation of TM Imagery

Categories	Color Band 2 Band 3 Band 5	Band 1 (0.45–0.52 μm)	Band 2 (0.52–0.60 μm)	Band 3 (0.63–0.69 μm)	Band 4 (0.76–0.90 μm)	Band 5 (1.55–1.75 μm)	Band 7 (2.08–2.35 μm)
Forest	Dark red ++	Black +	Black −	Black −	Gray −−	Very dark gray ++	Dark gray +
Dense green	Red ++	Dark gray −	Dark gray +	Very dark gray −	White +	Dark gray −−	Gray +
Green	Pink +	Dark gray −	Gray +	Gray −	White +	Gray −−	Light gray −−
Dry soil	Light brown +	Light gray +	Light gray −	Light gray ++	Gray −	Light gray +	Light gray −
Normal soil	Brown −	Gray −	Light gray −	Gray +	Gray −	Gray −−	Light gray −−
Wet soil	Bluish purple −	Gray −	Gray −−	Gray −−	Dark gray +	Dark gray −−	Light gray −−
Bare soil	White ++	White +	White +	Light gray +	Gray −	Light gray −	Light gray −
Water	Light blue ++	Light gray −−	Light gray −	Light gray +	Black ++	Black ++	Black ++

Key: ++ clearly identified; + identified; −− not identified; − poorly identified.
After Murai and Yanagida (1984).

TABLE 7-3-9. The Effect of Field Size (1290 Fields), Plant Cover, and Plant Height (588 Fields) on Classification Results (Per Field Basis) for *Landsat* Data Collected January 21, 1973, Using MSS Channels 4, 5, and 7

Field Size Stratification			Crop Cover Stratification			Crop Height Stratification		
Field Size (ha)	Overall Correct Classification	Accumulative Total Fields	Crop Cover (%)	Overall Correct Classification	Accumulative Total Fields	Crop Height (cm)	Overall Correct Classification	Accumulative Total Fields
0	54.0	1290	0	55.4	588	0	55.4	588
2	57.0	1127	5	56.3	566	10	57.4	533
4	61.3	828	10	56.8	540	20	61.6	451
6	64.2	649	15	58.0	517	30	64.8	378
8	67.6	479	20	59.1	495	40	64.8	353
16	74.4	188	25	59.7	485	100	66.6	303
20	79.1	120	40	61.6	415	200	66.6	276
40	84.3	51	60	63.5	239	300	70.1	164
—	—	—	80	69.4	131	400	62.5	16

After Richardson et al. (1977). Reproduced with permission from *Photogrammetric Engineering & Remote Sensing*, copyright by the American Society for Photogrammetry and Remote Sensing.

data for land use inventories is not fully understood, the application of radar data has a promising future. The high ground resolution and cloud penetration capability make it highly desirable in countries with frequent cloud coverage or small field size. An example of SIR-A image over part of Brazil is shown in Figure 7-3-12, and an image of Quebec recorded by the SIR-B is shown in Figure 7-3-13.

7-3-4 Large-Scale Mapping of Vegetation

Monitoring of green vegetation cover on a larger scale has been of interest for a long time, because it is important in assessing the phenological development of the crops for an input into agrometeorological yield models. For this purpose, a vegetation index (VI) has been developed, which has been found useful for assessing forest conditions with respect to defoliation, harvesting, and clear cutting.

For mapping vegetation, in principle, several methods have been proposed, which have been documented by Logan and Strahler (1983). Two types of data compression procedures have tended to be useful largely because of the ease with which they can be formed: ratioing and principal components analysis. The most common ratios for vegetation monitoring have been summarized by Logan and Strahler (1983):

$$\frac{\text{MSS 5}}{\text{MSS 6}}$$

or vice versa, or MSS 7 in lieu of MSS 6. The ratio

$$\frac{\text{MSS 5}}{\text{MSS 4}}$$

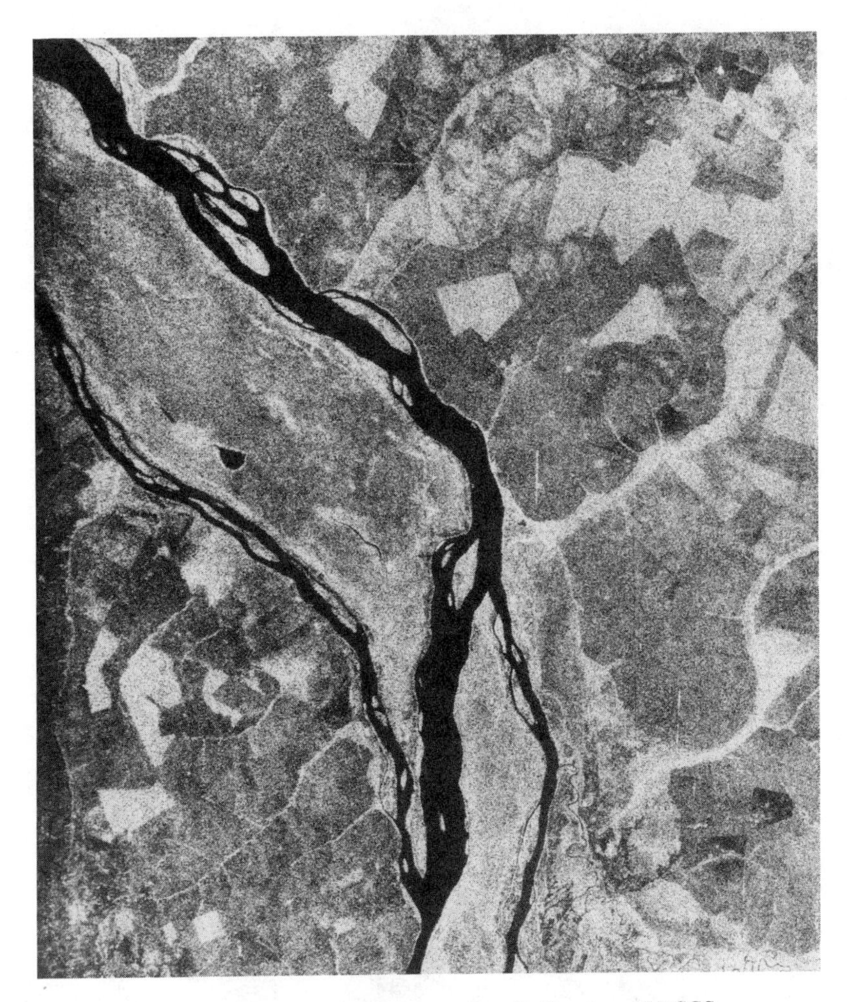

Figure 7-3-12. SIR-A image over Brazil. Courtesy of USGS.

has also been used to emphasize the key vegetation channels within *Landsat* data. A slight variation of this theme, which retains the use of all four bands, is the ratio of averages,

$$\frac{(\mathrm{MSS4} + \mathrm{MSS5})/2}{(\mathrm{MSS6} + \mathrm{MSS7})/2}$$

and the ratio of sums divided by the differences,

$$\frac{(\mathrm{MSS4} + \mathrm{MSS5})/(\mathrm{MSS6} + \mathrm{MSS7})}{(\mathrm{MSS6} + \mathrm{MSS7}) - (\mathrm{MSS4} + \mathrm{MSS5})}$$

The most useful for estimating relative greenness, the transform VI with MSS6 (TVI6), is given as

$$\sqrt{\frac{\mathrm{MSS6} - \mathrm{MSS5}}{\mathrm{MSS6} + \mathrm{MSS5}} + 0.5}$$

Figure 7-3-13. Image of Montreal, Quebec, as acquired by the SIR-B on October 7, 1984 with NASA's Space Shuttle *Challenger*. The area seen is approximately 15 km wide × 50 km long (10 mi × 30 mi). The center of the photo is the St. Lawrence River. Islands in the river are clearly visible as are at least five major bridges that cross the waterway. Cultivated fields at right display long, striplike patterns typical of French subdivisions of land. Courtesy of NASA, Jet Propulsion Laboratory.

The perpendicular vegetation index (PVI6) distinguishes the spectral response of green vegetation from the response contributed by background soils:

$$\sqrt{(\text{SOIL5} - \text{MSS5})^2 + (\text{SOIL6} - \text{MSS6})^2}$$

where

$$\text{SOIL5} = -0.498 + 0.543 \times \text{MSS5} + 0.498 \times \text{MSS6}$$

$$\text{SOIL6} = 2.734 + 0.498 \times \text{MSS5} + 0.498 \times \text{MSS6}$$

Ratioing techniques are popular because they can often be easily performed digitally as well as photographically, and they also tend to reduce the effects of shadowing and atmospheric degradation.

Large area crop investigation experiment (LACIE) was designed to develop and apply a technology to monitor wheat production in regions throughout the world. It utilized quantitative multispectral data collected by *Landsat* in concert with current weather data and historical information and exploited high-speed digital computer processing of data and mathematical models for extracting information in a timely and objective manner. The results from three crop years of focused experimentation indicated that the current technology can successfully monitor wheat production in regions having similar characteristics to those of the wheat areas in the Soviet Union and the U.S. hard red winter wheat area, and that with additional applied research significant improvements in capabilities to monitor wheat in important production regions can be expected.

Utilization of *Landsat* full-frame imagery allowed samples to be drawn only from agricultural areas and required only 2% of the area to be analyzed, with the contribution of sampling error to the area estimate being less than 2%. The digital, computer-aided, statistical pattern recognition techniques employed under the LACIE program were designed to take advantage of the changing spectral response of crop types over time in order to maximize the accuracy of the area measurement. In addition, environmental satellites have been used to refine the precipitation analyses based upon cloud patterns. Yield models were developed in order to make estimates early in the season, throughout the growing season, and at harvest. For winter wheat in the northern hemisphere, these estimates began in December and were updated until harvest in June or July. Spring wheat yield estimates began as early as March and were revised monthly through August or September. This allowed assessments of potential yield to be made almost at the time the plant emerged from the ground.

Global indices for vegetation have been monitored since 1982, with the *NOAA-7* satellite using the data from the VHRR. The products indicate a tremendous potential for estimating fluctuations in vegetation, which on a large scale are important as a tool for drought or environmental assessments.

The use of red and near IR spectral data for estimating total dry matter accumulation is based upon the ability to spectrally estimate the photosynthetically active biomass (i.e., green leaf area or green leaf biomass) and the relationship of the intensity and duration of the photosynthetically active biomass through time to the total dry matter production (Markham et al., 1981; Tucker et al., 1981). Ratio and linear combinations of red and photographic IR spectral data have been used in a variety of different vegetational situations, including tropical forests, grass lands and agricultural crops (Tucker, 1979, 1980a).

The VI used for global monitoring is produced by mapping mosaics computed from the ratio

$$VI = \frac{Ch_2 - Ch_1}{Ch_2 + Ch_1}$$

where Ch_1 and Ch_2 are the count values from measurements in the spectral range 0.6 to 0.7 μm and 0.7 to 1.1 μm, respectively (Yates and Tarpley, 1982). The VI is generated daily from daytime *NOAA-7* data for the northern and southern hemispheres. A seven-day composite is produced, which consists of the maximum VI found in the daily maps.

To reduce cloud contamination, a seven-day maximum composite is produced from the daily arrays. For each composite period, only the largest VI is retained at each array location. This eliminates clouds from the composite, except for areas with persistent cloud coverage during the time of investigations. Maximum value compositing greatly improves the product because all atmospheric and scan angle effects act to reduce the local time. A single day's map consists of a mosaic of the daytime portion of 14 orbital swaths. The maps are on a $\frac{1}{16}$ submesh grid of the standard 65 $\times$ 65 polar stereographic projection used by the National Meteorological Center and the Air Force Global Weather Center in the United States. Resolution in the mapped data ranges from 15 km at the equator to 30 km at the poles (Tarpley et al., 1982).

Clouds, water, and snow have larger reflectance in the visible than in the near IR, which results in a negative VI. Rock and bare soil have similar reflectances in these two bands, so their VIs are near zero. In scenes with vegetation, the index ranges from 0.1 to 0.6, the higher values indicating greater density and greenness of the plant canopy. Atmospheric effects such as scattering by dust and aerosols, Rayleigh scattering, and subpixel-sized clouds all act to increase Ch_1 with respect to Ch_2 and to reduce the computed VIs.

Tucker et al. (1983) cautioned that with the current interest in measuring "green leaf area" from satellites, it must be realized that satellite "green leaf area" measurements take into account plant canopy, structural and optical differences of all the species present in the field of view, whereas destructive ground-based green leaf area or green leaf biomass measurements usually do not. Tarpley et al. (1982) pointed out that the disadvantage of the described method is that under completely clear conditions, it selects the greenest area within an array location and causes the map to be less representative of typical vegetation conditions.

The identification of vegetation and the determination of its acreage is not sufficient for the estimation of the yield expected at harvest time or for forecasting the growing cycle. Amount and distribution of precipitation, soil types, and soil conditions have to be known. In addition, the vigor of vegetation, which is often affected by parasites or diseases, has to be monitored.

Moisture availability is critical to the normal development of plant biomass, and it has been demonstrated that crop yield and pasture production are directly related to it. It is therefore possible to assess the likelihood of crop shortfall or reduced biomass production in rangelands by monitoring the soil moisture conditions throughout a growing season. Relatively simple water-balance models, which use standard meteorological data, can be used to estimate moisture availability for plant growth. Henricksen and Durkin (1986) found a strong correlation ($r = 0.99$) between rates of change of the normalized difference vegetation index (NDVI) derived from the AVHRR data and threshold values of a soil moisture index at the beginnings and ends of growing periods. The moisture index

$$\frac{P + S}{ETp}$$

relates precipitation (P), stored soil moisture (S), and potential evapotranspiration (ETp) in a simple moisture balance model that requires inputs of standard monthly meteorological data. Trends have been detected in values of the NDVI during vegetation growth cycles and suggest that useful minima exist at the begin-

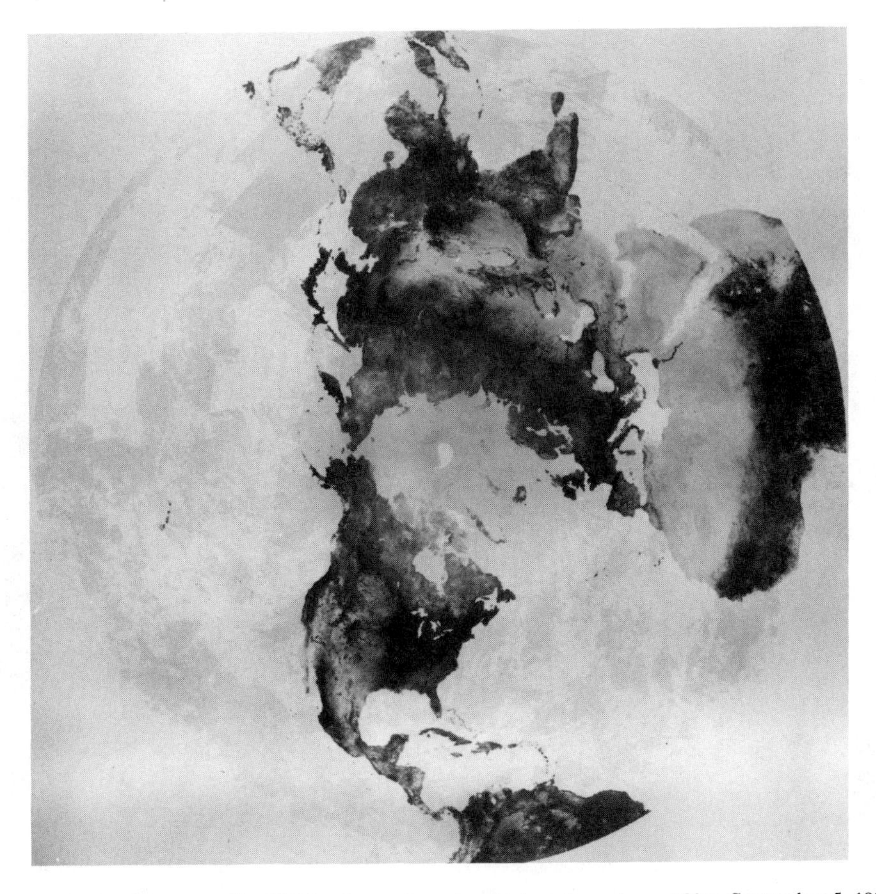

Figure 7-3-14. The normalized VI for the northern hemisphere from August 29 to September 5, 1982. Courtesy of NOAA.

nings and ends of growing periods. Below minima, growing periods are unlikely to have been initiated or to continue during a declining growth stage.

Based on data over Ethiopia, correlation analysis indicated a relation between moisture index and NDVI, with NDVI lagging in time, in most cases, by five or less weeks during the initial growth stage and six or more weeks during declining growth (Henricksen and Durkin, 1986).

Figure 7-3-14 shows the northern hemispheric normalized vegetation index (NVI) composite maps for August 29 through September 5, 1982. This map is a photographic display produced from 1024 × 1024 arrays of digital NVI values. The darker the gray shade is on the image, the greener the area is on the ground. Bodies of water, clouds, snow, and ice appear as white or light gray; deserts, as light gray; forests and cultivated areas, as medium to dark gray, depending on the greenness and density of the vegetation (Tarpley et al., 1982).

Images of the southwestern United States were produced from AVHRR sensor data aboard the *NOAA-7* satellite. The data shown in Figure 7-3-15 have been geometrically corrected using approximately 85 control points to produce images covering an area of 2,000,000 km^2, whereby "warm" colors indicate dense vegeta-

Figure 7-3-15. *NOAA 7* geometrically corrected false-color composite of the western United States. The VI is superimposed in red, while channel 1 (0.58–0.68 μm) is displayed in blue, and channel 2 (0.725–1.1 μm) is displayed in green. Reproduced with permission from the Environmental Research Institute of Michigan (ERIM). See insert for color representation.

tion, the cooler greens and blues show sparse vegetation and arid or snow-covered areas.

The VI produced from NOAA/AVHRR data is especially useful for its rapid repeat cycle, which allows time-critical events to be monitored frequently. The broad field of view (2250-km swath) also covers large regions to be examined at one instant in time. Thus, NOAA/AVHRR data are ideally suited to large area inventory and monitoring programs where great details are not required.

In an area such as the Sahel, it is possible to monitor changing conditions on a large scale. Such information can be used as a management tool and to monitor and predict droughts. Monitoring vegetation, for instance, can be applied to drought forecasting, and the data can be useful for disaster-related situations (see Fig. 7-3-16).

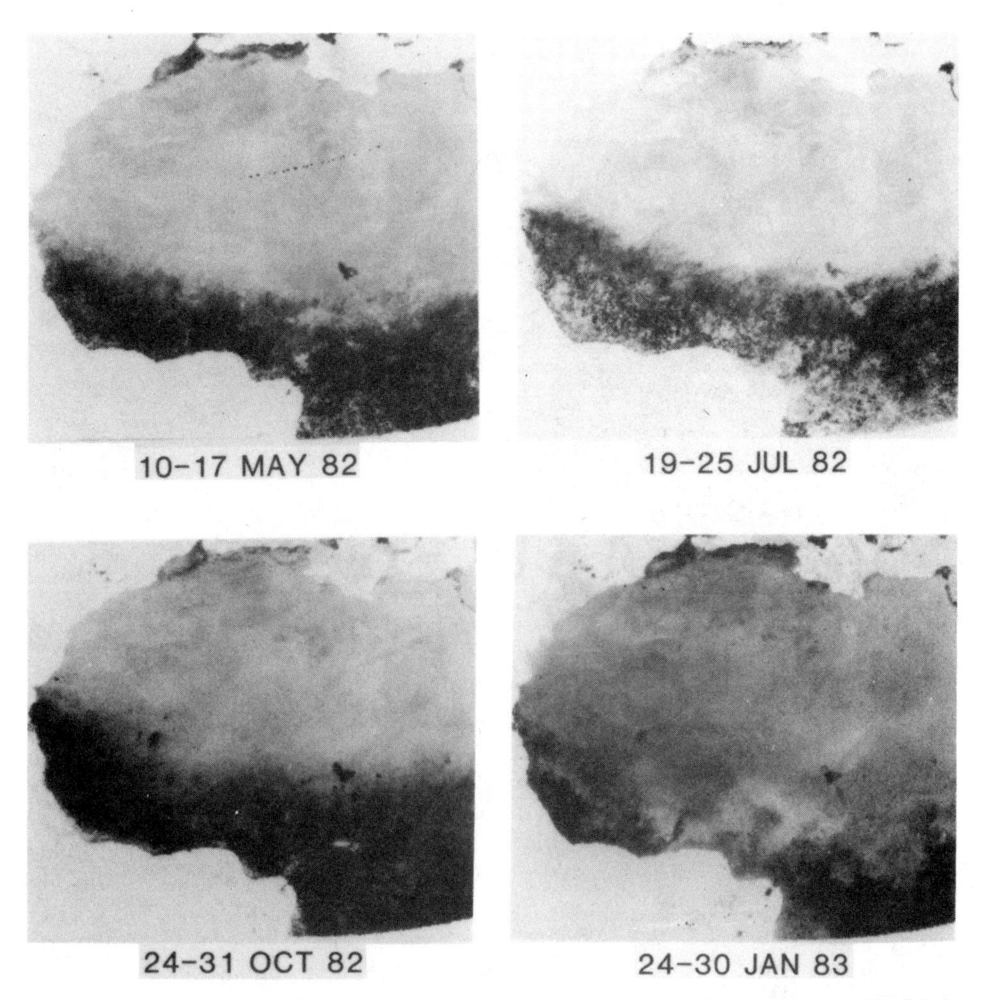

Figure 7-3-16. The vegetation index as management tool in drought forecasting over the Sahel region. Courtesy of NOAA. *Continued on next page.*

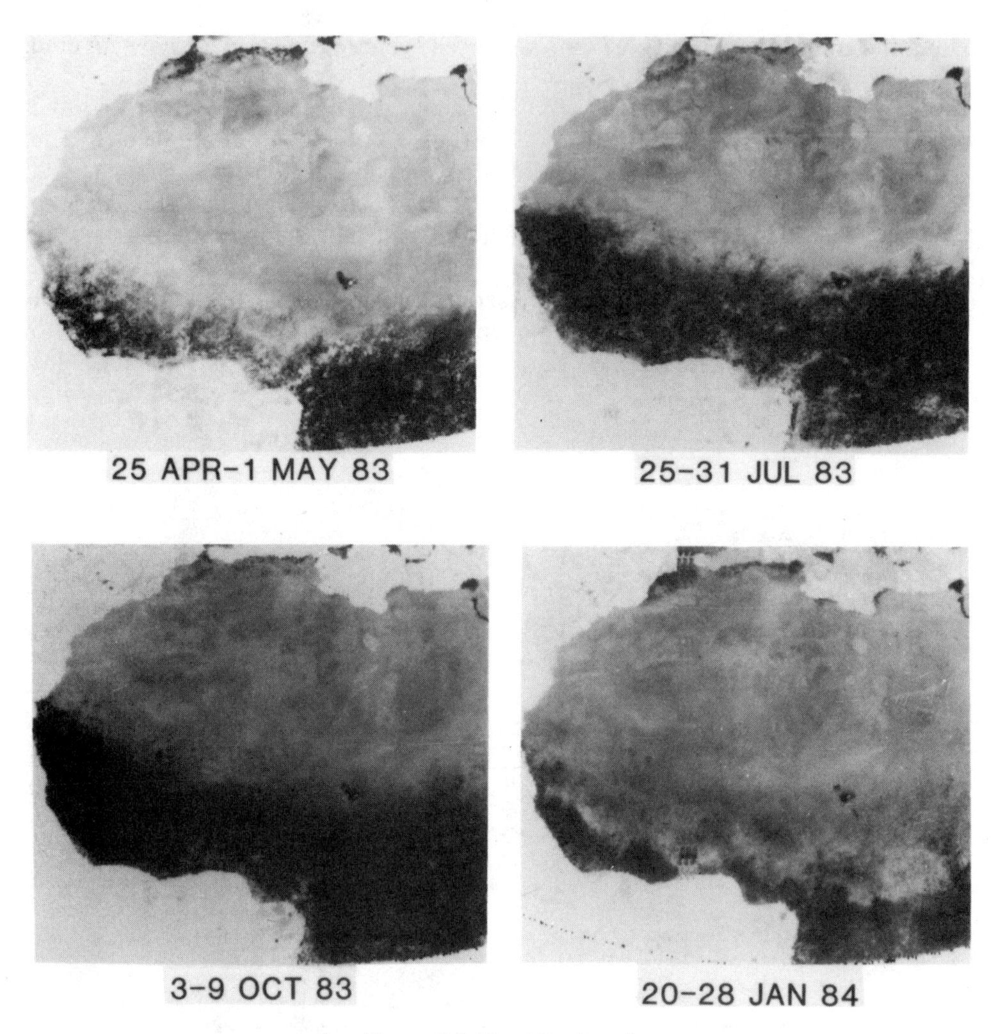

Figure 7-3-16. (*Continued*)

In monitoring the drought areas in the Sahel region with the vegetation index (Hock, 1984), Hock showed that the mean VIs for 1984 suggested more stress in Niger, northern Burkina Faso, and southern Mali than occurred in the preceding year. By June 1984, it was possible to detect the severity of the drought using the VIs and the satellite derived precipitation estimates.

Tucker et al. (1985) also integrated the spectral VI in order to quantify the yearly vegetation response for Africa. This integrated spectral index produced a zonation of Africa in which deserts and semideserts, dry savannas and dry grasslands, and forests and wet savannas were differentiated from each other, and tropical forest and tropical savanna were found to have similar integrated index values.

In LACIE the green index number was developed in order to detect moisture stress using *Landsat* data with the objective to apply the procedure over a large

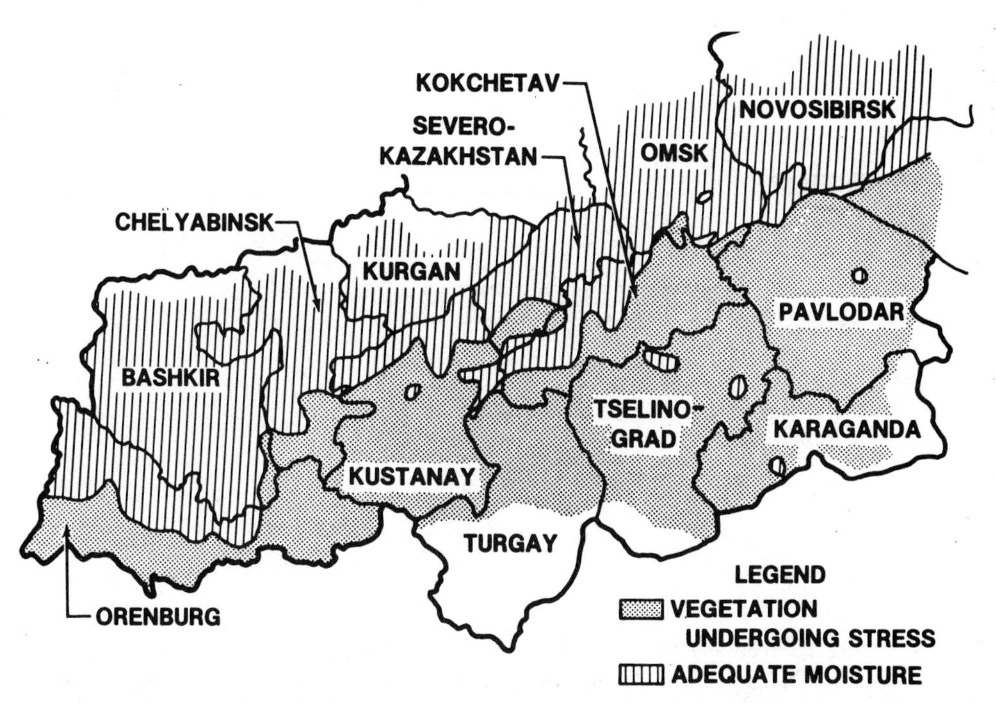

Figure 7-3-17. Area of stressed vegetation in the USSR delineated for July 1977. Courtesy of Dr. D. R. Thompson, NASA Lyndon B. Johnson Space Center.

area. The basic idea is that a particular weighted difference of the infrared and visible channels of *Landsat* data represents growing vegetation. The monitoring program showed that parts of the USSR spring wheat region were undergoing moisture stress during the *Landsat* passes of July 1977 (Thompson and Wehman, 1979). The analysis shown in Figure 7-3-17 indicates that *Landsat* is a valuable tool to monitor drought conditions as well as vegetation stress even in areas where ground information might not be available.

REFERENCES

Part of the referenced material in this book is based on actual documentation published by NASA, NOAA, USGS, among others. Some of the original material can be retrieved through two major libraries: the Library of the Goddard Space Flight Center (Greenbelt, Maryland) and the Technical Information Service of the American Institute of Aeronautics and Astronautics (New York, New York).

The library at GSFC provides a collection of essential scientific and technical literature, access to materials not held by the library, technical information support, including computer-assisted literature searches, and copies of publications and other materials by means of circulation, acquisition, interlibrary loan, photocopy, and selective dissemination. While most document searches are made by direct line to a computer data base, one can refer to the printed version of the same information and the abstract/index periodicals available in the main reading room.

The best sources for NASA-sponsored documents are:

Scientific and Technical Aerospace Reports (STAR) issued by NASA, and
International Aerospace Abstracts (IAA) issued by the American Institute of Aeronautics and Astronautics under contract to NASA.

One can request to be trained to use either of the two terminals for direct access to the NASA computer data base, RECON (retrieval by remote console).

The library at GSFC maintains a collection of documents prepared for internal use by the civil service staff and other qualified users. Under the revised NASA publications system, most GSFC documents are now published as technical papers and memoranda.

Literature searches draw on the following online, real-time resources:

1. The NASA data base, RECON. Two direct lines and terminals link the Library and the NASA computerized records at NASA's Scientific and

Technical Information Facility (STIF). A thesaurus of terms is available at each terminal. STIF makes weekly additions from its own collection of aerospace-related materials, published as STAR, and also from IAA.

2. The NASA library network data base, NASA Library Network (NALNET).
3. Data bases of all government technical agencies, including those of the Department of Defense, the National Oceanic and Atmospheric Administration, the Department of Energy, the National Technical Information Service, and others.
4. Commercial data bases, including those of the professional societies, major scientific publishing houses, and news and legislative sources such as the Lockheed *DIALOG,* the System Development Corporation *ORBIT,* and the *New York Times Information Bank.*
5. The new numerical data bases for financial, physical, chemical and engineering data.

The Technical Information Service (TIS) of the AIAA acquires, processes and disseminates technical and engineering information. Using extensive international resources, trained staff produce and provide information services (e.g., *International Aerospace Abstracts* and the Aerospace Database) designed to facilitate multiple applications of the comprehensive information base.

Coverage in the library spans the full spectrum of scientific and technical aerospace disciplines: atmospheric and space sciences, aircraft and aerospace systems, information systems, propulsion and energy conversion, structures and materials, design testing and management.

The AIAA library provides long-distance or on-site access to:

1. More than 1600 current journals, as well as significant holdings in recent and historical journal issues;
2. More than 26,000 books;
3. Over 750,000 NASA and AIAA microfiche, including substantial collections of older and rare reports;
4. Extensive foreign coverage (50% of the total from outside the United States);
5. Conference and meeting proceedings throughout the world;
6. A comprehensive reference collection.

REFERENCES CITED IN TEXT

Acuna, M. H., C. S. Scearce, J. B. Seek, and J. Scheifele. The Magsat vector magnetometer—A precision fluxgate magnetometer for the measurement of the geomagnetic field. *NASA Tech. Memo.,* **NASA TM-79656,** 1978.

Albert, N. R., and P. S. Chavez. Computer-enhanced Landsat imagery as a tool for mineral exploration in Alaska. *Proc. 1st Annu. W. Pecora Mem. Symp.,* Sioux Falls, 1975.

Allen, W. A., H. W. Gausman, A. J. Richardson, and J. R. Tomas. Interaction of isotropic lights with a transparent plate. *J. Opt. Soc. Am.,* p. 59, 1969.

Allison, L. J. An analysis of Tiros II radiation data recorded over New Zealand at night. *NASA Tech. Note,* **NASA TN-D-1910,** 1964.

Allison, L. J. *Geological Applications of Nimbus Radiation Data in the Middle East,* Doc. X-901-76-164. NASA, GSFC, Greenbelt, MD, 1976.

Allison, L. J., and G. Warnecke. The interpretation of Tiros radiation data for practical use in synoptic weather analysis. *NASA Tech. Note,* **NASA TN-D-2851,** 1965.

Allison, L. J., T. J. Schmugge, and G. Byrne. A hydrological analysis of east Australian floods using Nimbus-5 Electrically Scanning Microwave Radiometer data. *NASA Tech. Memo.,* **NASA TM-79689,** 1979.

Anderson, R. K., J. P. Ashman, F. Bittner, G. R. Farr, E. W. Ferguson, V. J. Oliver, and A. H. Smith. *Application of Meteorological Satellite Data in Analysis and Forecasting,* Tech. Rep. 212. Air Weather Service (MAC), U.S. Air Force, 1969.

Attema, E. P. W., and F. T. Ulaby. Vegetation modelled as a water cloud. *Radio Sci.,* **13**(2), 357–364, 1978.

Barton, I. J. Dual-channel satellite measurements of sea surface temperature. *Q. J. R. Meteorol. Soc.,* **109,** 356–378, 1983.

Basharinov, A. E., A. S. Gurvitch, and S. T. Igorov. Features of microwave passive remote sensing. *Proc. Int. Symp. Remote Sens. Environ., 7th, 1971,* Vol. 1, 1971, pp. 119–131.

Bauer, M. E., and J. E. Cipra. Identification of agricultural crops by computer processing of ERTS MSS data. In *Symposium on Significant Results Obtained from the Earth Resources Technology Satellite—1,* Vol. 1, Sect. A. NASA, Washington, DC, 1973, pp. 205–212.

Beaubien, J. Forest type mapping from Landsat digital data. *Photogramm. Eng. Remote Sens.,* **45**(8), 1135–1141, 1979.

Beer, T. Microwave sensing from satellite. *Remote Sens. Environ.,* **9,** 65–85, 1980.

Bell, E. E., and I. L. Eisner. Infrared radiation from the White Sands at White Sands National Monument, New Mexico. *J. Opt. Soc. Am.,* **46,** 303–304, 1956.

Bell, E. E., I. L. Eisner, J. B. Young, A. Abolins, and R. Z. Oetjen. *Infrared Techniques and Measurements,* Final Eng. Rep. Contract AF 33 (616)-3312. Ohio State Res. Found., 1957.

Benigno, J. A. Fish detection through aerial surveillance. Paper presented during the Technical Conference on Fish Finding Purse Seining and Aimed Trawling, Reykjavik, 1970, pp. 24–30.

Beran, E., E. Merrit, and D. Chang. *Interpretation of Baroclinic Systems and Wind Fields as Observed by the Nimbus 2 MRIR,* Final Report, Contract NAS 5-10334. Allied Research Associates, Concord MA, 1968.

Bernstein, R. L. Sea surface temperature estimation using the NOAA-6 satellite advanced very high resolution radiometer. *J. Geophys. Res.,* **87**(C12), 9455–9456, 1982.

Biña, R. T., R. S. Jara, B. R. de Jesus, Jr., and E. N. Lorenzo. *Mangrove Inventory of the Philippines Using Landsat Multispectral Data and the Image 100 System,* Res. Monogr. No. 2. Natural Resources Management Center, Philippines, 1978a.

Biña, R. T., K. Carpenter, W. Zacher, R. S. Jara, and J. B. R. Lim. *Coral Reef Mapping Using Landsat Data: Follow-Up Studies,* Res. Monogr. No. 3. Natural Resources Management Center, Philippines, 1978b.

Bizzell, R. M., L. C. Wade, H. L. Prior, and B. Spiers. The results of an agricultural analysis of the ERTS-1 MSS data at the Johnson Space Center. In *Symposium on Significant Results Obtained from the Earth Resources Technology Satellite—1,* Vol. 1, Sect. A. NASA, Washington, DC, 1973, pp. 189–196.

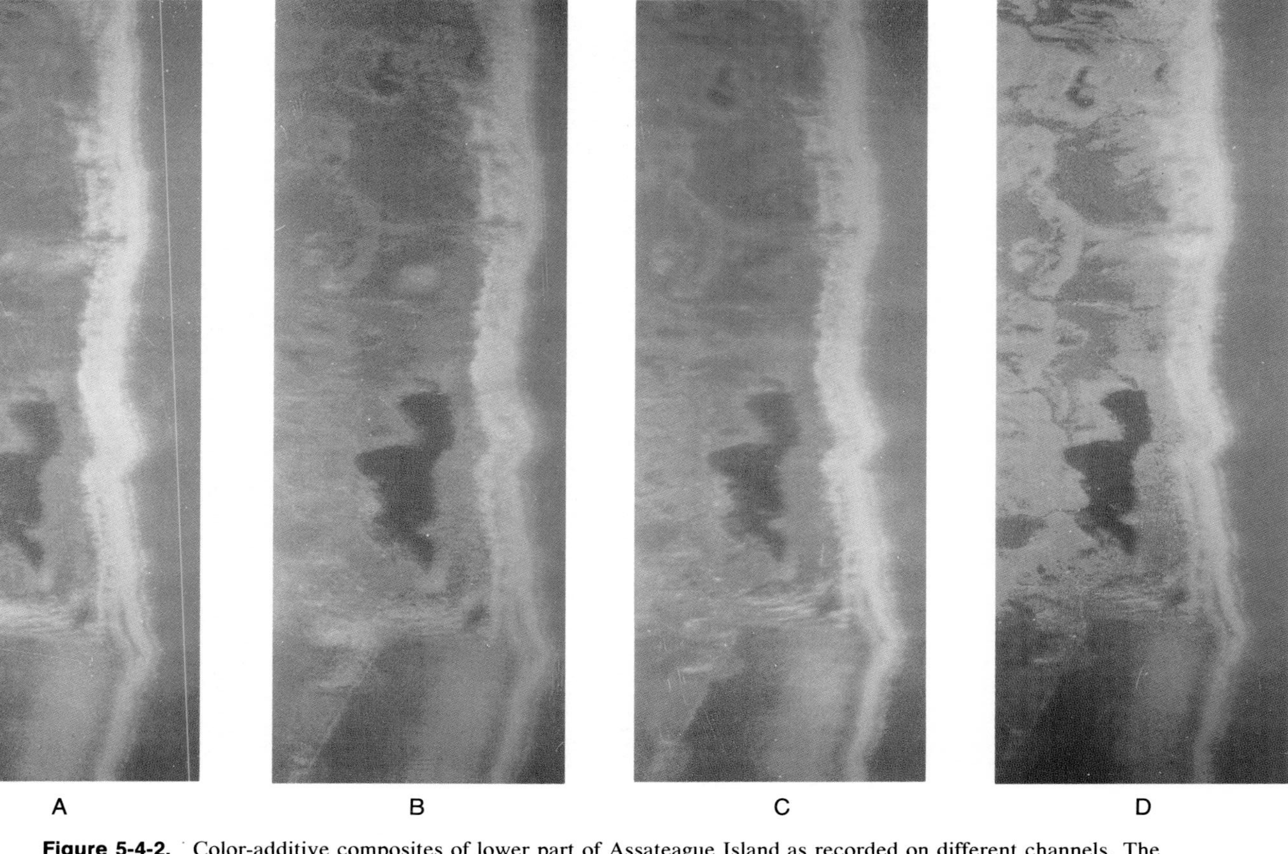

Figure 5-4-2. Color-additive composites of lower part of Assateague Island as recorded on different channels. The images were projected through viewer filters given in Table 5-4-1. After Short and MacLeod (1972).

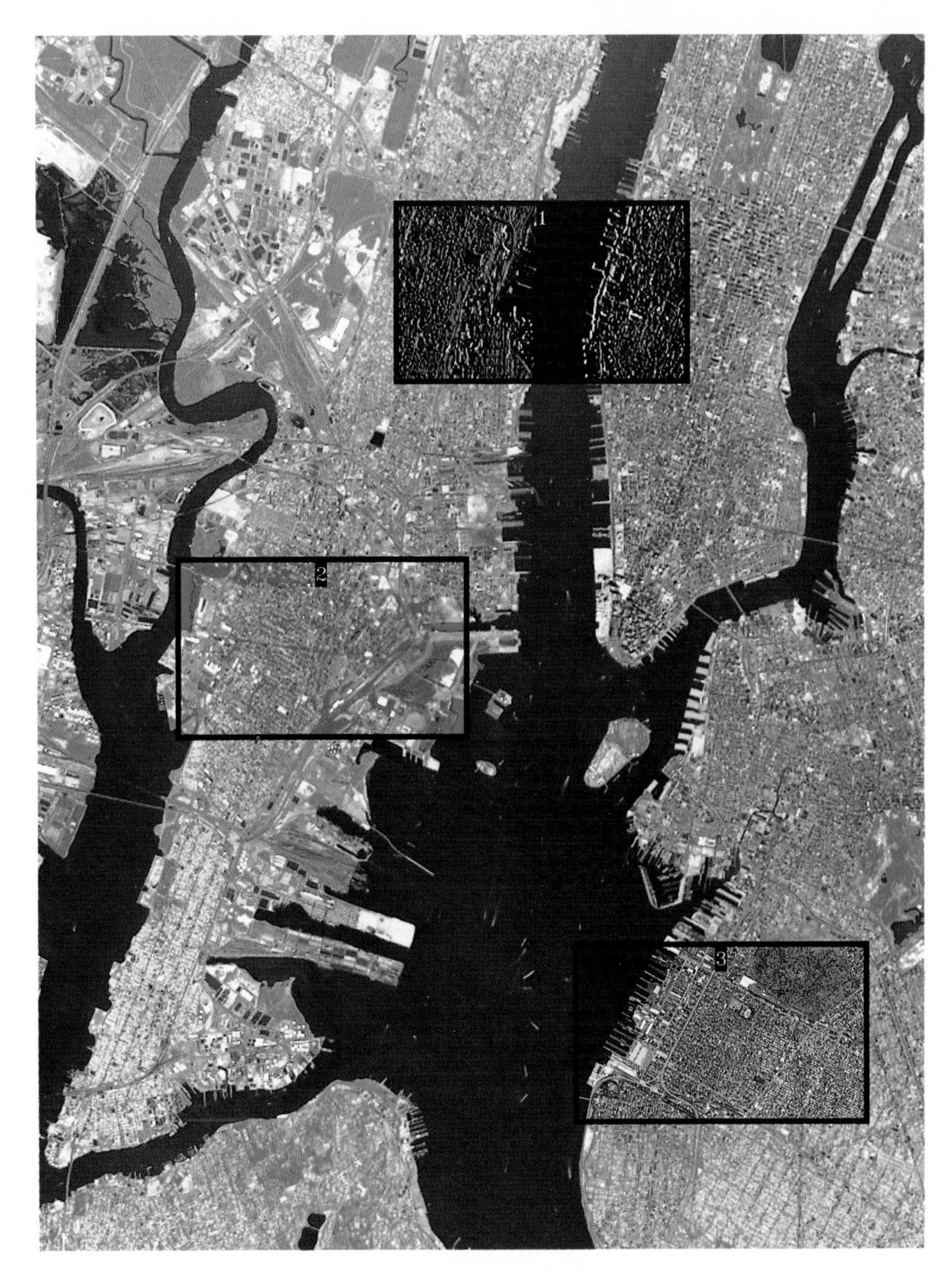

Figure 5-5-6. Digital processing and enhancement techniques as applied on multispectral image of Manhattan. See text for explanation of insets. Courtesy of ERDAS.

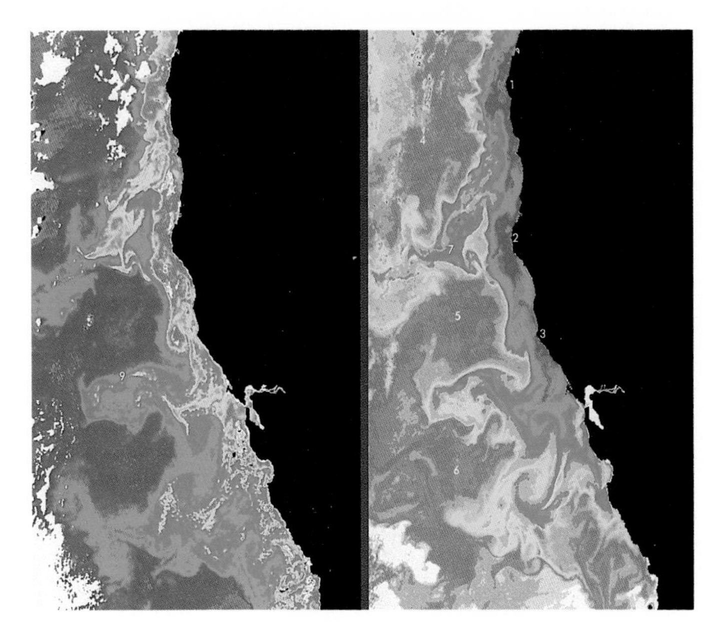

Figure 6-2-23. Temperature and pigment distributions in the coastal upwelling along the California coast, July 7, 1981. Courtesy of NASA.

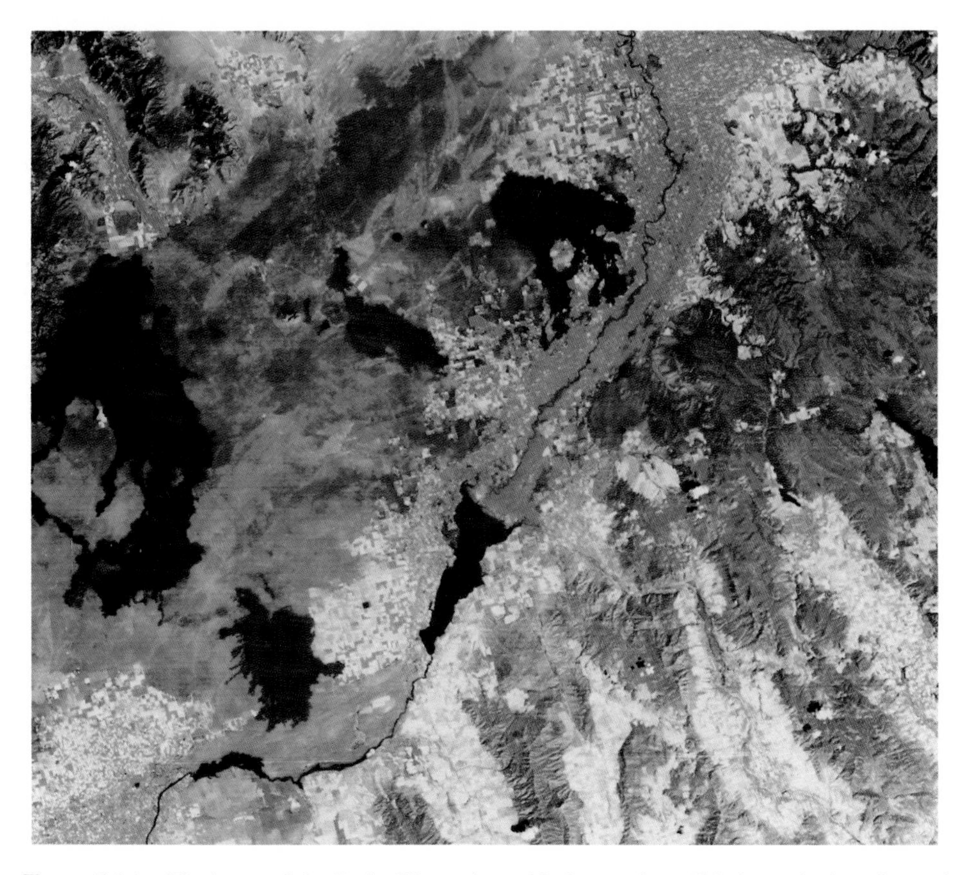

Figure 7-1-1. The image of the Snake River taken with the *Landsat* MSS shows the lava flows of Tertiary to Quaternary age which appear in varying shades of dark blue, representing basalt lavas. Courtesy of NASA.

Figure 7-1-3. (A) Geological map of Wadi Wassat area. (B) Contrast-enhanced ratio color-composite image of the Wadi Wassat area (digital scanner data from *Landsat* scene 1226-07011). Courtesy of Dr. H. Blodget, NASA GSFC.

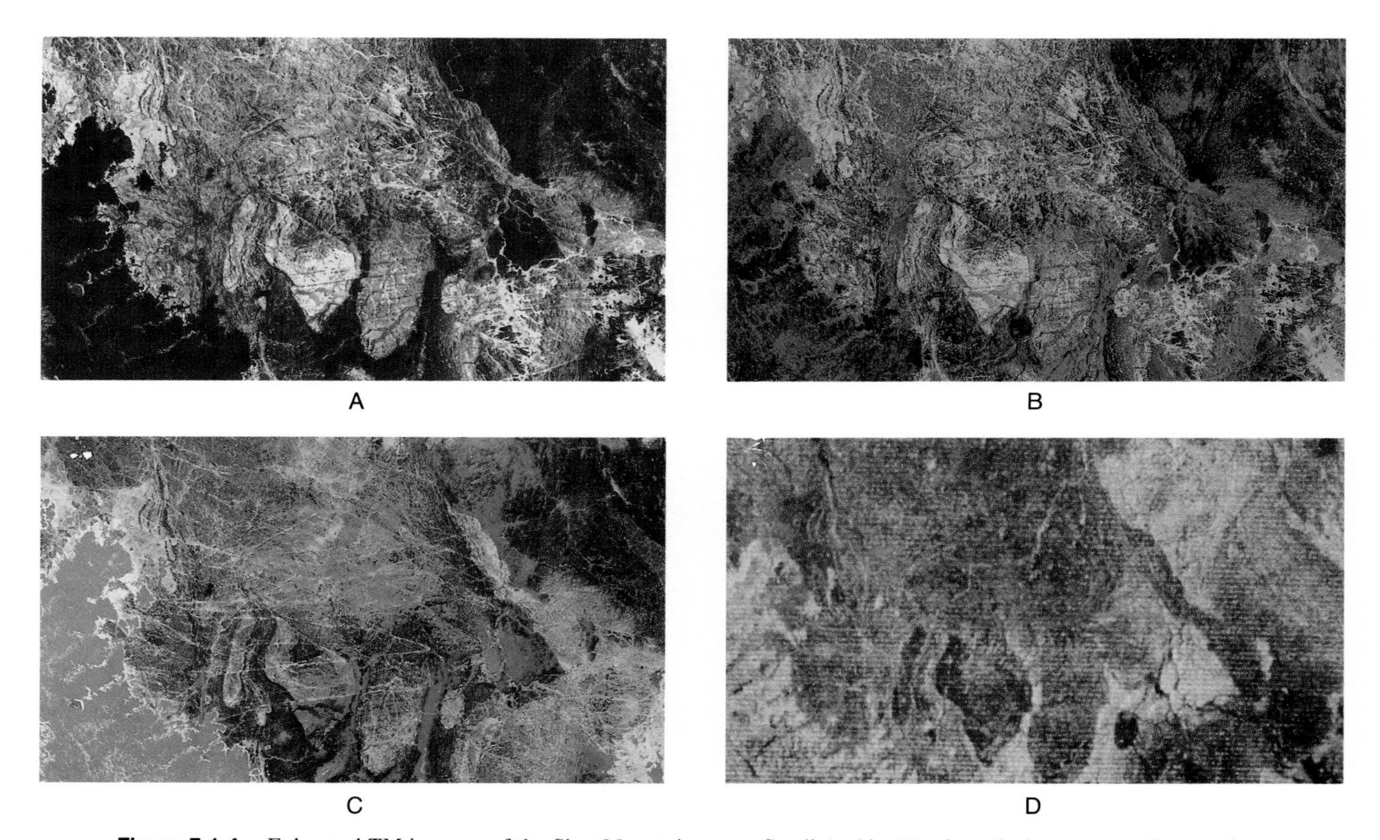

Figure 7-1-4. Enhanced TM imagery of the Sirat Mountains area, Saudi Arabia. Numbers designate areas discussed in text. (A) TM bands 2, 3, and 4 displayed as blue, green, and red, respectively; linear contrast stretch. (B) TM bands 1, 2, and 7 displayed as red, green, and blue, respectively; decorrelation stretch. (C) TM ratios 3/2, 3/7, and 5/7 displayed as red, blue, and green; histogram equalization stretch. (D) MSS ratios 4/5, 5/6 and 6/7 displayed as blue, green, and red; histogram equalization stretch. Courtesy of Dr. H. Blodget, NASA GSFC.

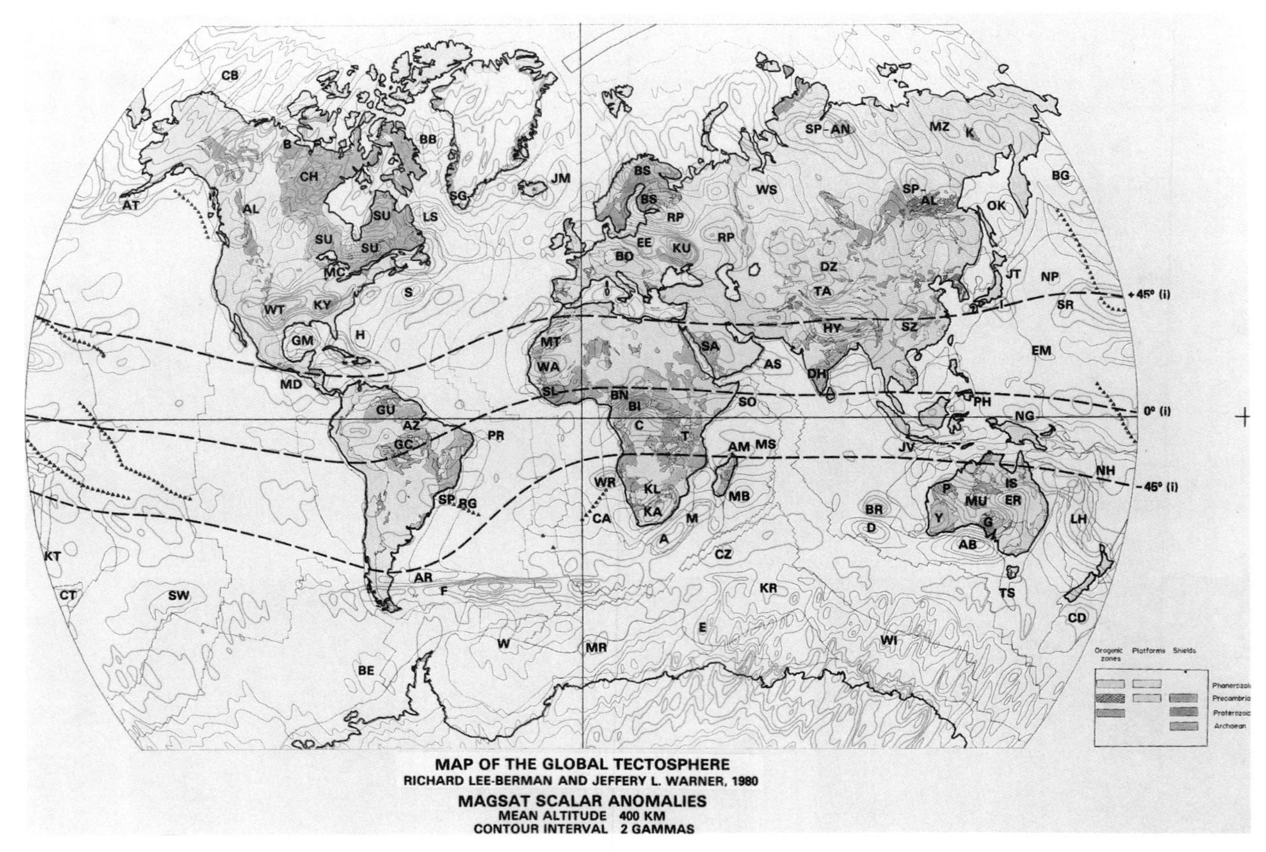

Figure 7-1-25. *Magsat* scalar anomalies superimposed on a simplified tectonic province map. Positive anomalies in red, negative anomalies in blue. Contour interval is 2 nT. Courtesy of Dr. R. A. Langel, NASA GSFC.

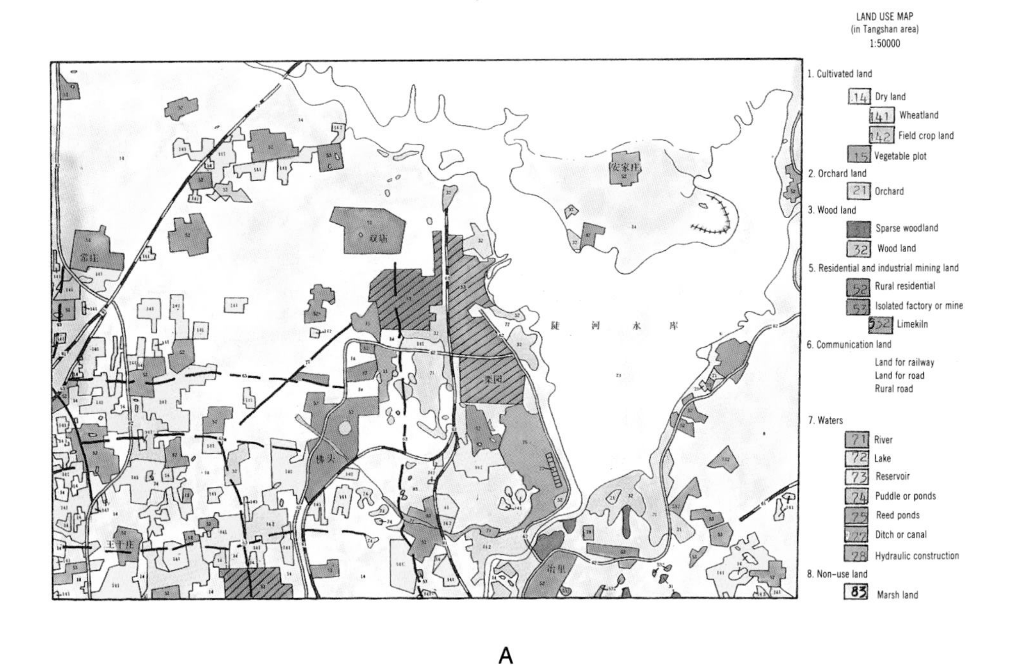

A

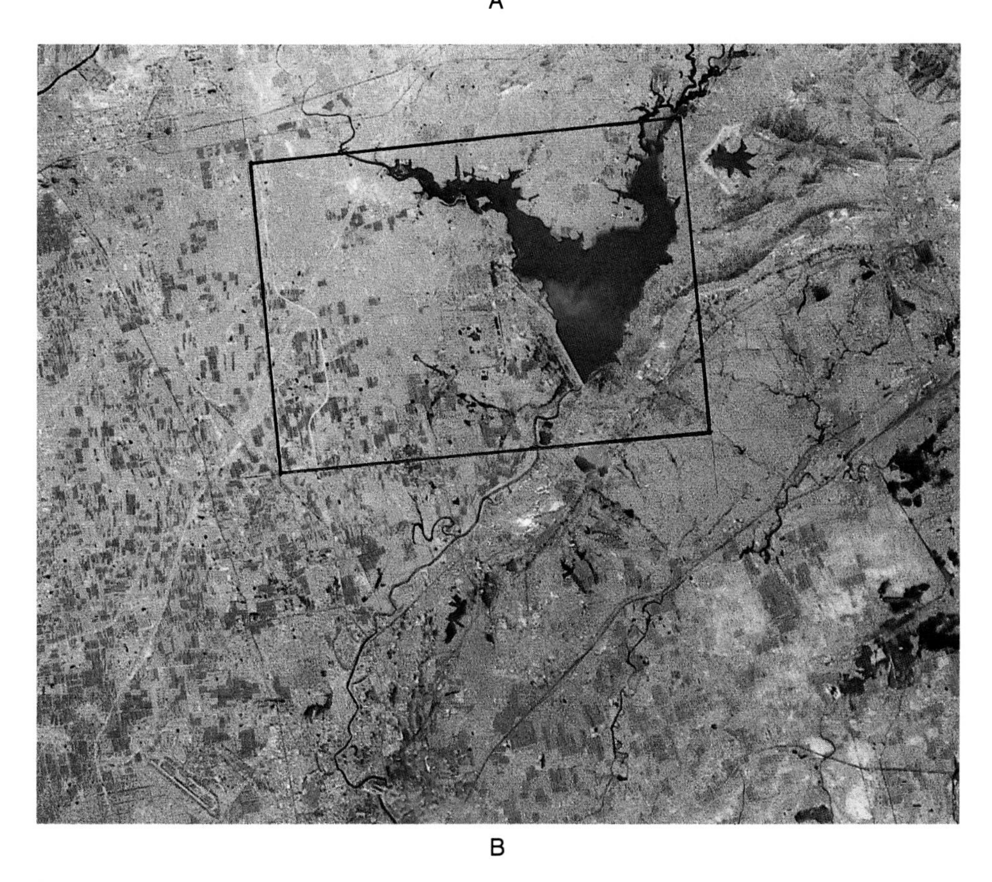

B

Figure 7-3-11. (A) Land use based on Landsat TM data in the People's Republic of China. (B) Example of land use pattern in Tangshan area, People's Republic of China. Rectangular area refers to the land use map given in (A). Courtesy of Prof. Yang Shiren, Academia Sinica, Beijing, People's Republic of China.

Figure 7-3-15. *NOAA 7* geometrically corrected false-color composite of the western United States. The VI is superimposed in red, while channel 1 (0.58–0.68 μm) is displayed in blue, and channel 2 (0.725–1.1 μm) is displayed in green. Reproduced with permission from the Environmental Research Institute of Michigan (ERIM).

Blodget, H. W., F. J. Gunther, and M. H. Podwysocki. Discrimination of rock classes and alteration products in southwestern Saudi Arabia with computer-enhanced Landsat data. *NASA Tech. Pap.,* **1327,** 1978.

Blodget, H. W., C. G. Andre, and R. F. Marcell. Enhanced rock discrimination using Landsat-5 Thematic Mapper (TM) data. *Racing into Tomorrow 1985 ACSM-ASPRS Fall Conv.,* 1985, pp. 912–921.

Blom, R. G., and M. Daily. Radar image processing for rock-type discrimination. *IEEE Trans. Geosci. Remote Sens.,* **GE-20**(3), 343–351, 1982.

Blom, R. G., and T. H. Dixon. Interpretation and application of spaceborne imaging radar data to geological problems. In K.-H. Szekielda (Ed.), *Satellite Remote Sensing for Resources Development.* Graham & Trotman, London, 1986, pp. 185–215.

Blom, R. G., R. E. Crippen, and C. Elachi. Detection of subsurface features in Seasat radar images of Means Valley, Mojave Desert, California. *Geology,* **12,** 346–349, 1984.

Bodechtel, H. Thematic mapping of natural resources with the modular optoelectronic multispectral scanner (MOMS). In K.-H. Szekielda (Ed.), *Satellite Remote Sensing for Resources Development.* Graham & Trotman, London, 1986, pp. 121–136.

Brachet, G. SPOT: The first operational remote sensing satellite. In K.-H. Szekielda (Ed.), *Satellite Remote Sensing for Resources Development.* Graham & Trotman, London, 1986, pp. 59–80.

Bramson, M. A., I. L. Zel'manovich, and G. I. Kuleshova. *Nasa Tech. Rep.,* **NASA TR-R-F-319,** 1965.

Braun, H. M., and G. Rausch. Present status of microwave remote sensing from space with respect to natural resources monitoring. In K.-H. Szekielda (Ed.), *Satellite Remote Sensing for Resources Development.* Graham & Trotman, London, 1986, pp. 23–58.

Brown, O. B., R. H. Evans, J. W. Brown, H. R. Gordon, R. C. Smith, and K. S. Baker. Phytoplankton blooming of the U.S. East Coast: A satellite description. *Science,* **229,** 163–167, 1985.

Burke, H. K., K. R. Hardy, and N. K. Tripp. Detection of rainfall rates utilizing spaceborne microwave radiometers. *Remote Sens. Environ.* **12,** 169–180, 1982.

Campbell, W. J. Ice lead and polynya dynamics. In R. S. Williams and W. D. Carter (Eds.), *ERTS-1: A New Window on Our Planet.* U.S. Government Printing Office, Washington, DC, 1976a, pp. 340–342.

Campbell, W. J. Seasonal metamorphosis of sea ice. In R. S. Williams and W. D. Carter (Eds.), *ERTS-1: A New Window on Our Planet.* U.S. Government Printing Office, Washington, DC, 1976b, pp. 343–345.

Campbell, W. J., R. O. Ramseir, W. F. Weeks, and P. Gloersen. An integrated approach to the remote sensing of floating ice. *Proc. Can. Symp. Remote Sens., 3rd,* 1975a, pp. 39–72.

Campbell, W. J., W. F. Weeks, R. O. Ramseier, and P. Gloersen. Geophysical studies of floating ice by remote sensing. *J. Glaciol.,* **15**(73), 305–328, 1975b.

Capehart, B. L., J. J. Ewel, B. R. Sedlik, and R. L. Myers. Remote sensing survey of *Melaleuca. Photogramm. Eng. Remote Sens.,* **43**(2), 197–206, 1977.

Carter, V., and J. Schubert. Coastal wetlands analysis from ERTS MSS digital data and field spectral measurements. *Proc. Int. Symp. Remote Sens. Environ., 9th, 1974,* 1974, pp. 1241–1260.

Charnell, R. L., and G. A. Maul. An oceanographic observation of the New York Bight from ERTS-1. *NOAA Tech. Rep.,* **NOAA TR ERL 262**(9), 1973.

Chaudhury, M. U. Satellite remote sensing applications to study forest ecosystems. In

Regional Training on Digital Image Interpretation for Forestry Application. ESCAP/NRMC, 1985, pp. 67–79.

Chavez, P. S., Jr., A. Berlin, and W. Mitchell. Computer enhancement techniques of Landsat MSS digital images for land use/land cover assessments. In *Proceedings of the Sixth Annual Remote Sens. of Earth Resources Conference*. Tennessee Space Institute, University of Tennessee, Tullahoma, 1977.

Cheney, R. E., and J. G. Marsh. Seasat altimeter observations of dynamic topography in the western North Atlantic. *J. Geophys. Res.*, **86,** 473–483, 1981.

Cheney, R. E., W. H. Gemmell, and M. K. Shank. Tracking a Gulf Stream ring with SOFAR floats. *J. Phys. Oceanogr.* **6,** 741–749, 1976.

Chiu, H.-Y., and W. Collins. A spectroradiometer for airborne remote sensing. *Photogramm. Eng. Remote Sens.*, **44,** 507–517, 1978.

Clark, R. N. *The Spectral Reflectance of Water-Mineral Mixtures at Low Temperatures,* PSD Publ. No. 252. MIT, Cambridge, MA, 1981.

Clarke, G. L., G. C. Ewing, and C. J. Lorenzen. Spectra of backscattered light from the sea obtained from aircraft as a measure of chlorophyll concentration. *Science,* **167,** 1119–1121, 1970.

Collins, W. Remote sensing of crop type and maturity. *Photogramm. Eng. Remote Sens.*, **44,** 43–55, 1978.

Conaway, J. Dependence of microwave emission upon sea state. In *Significant Accomplishments in Science*. NASA, GSFC, Washington, DC, 1969, pp. 16–19.

Corbell, R. P., C. J. Callahan, and W. J. Kotsch (Eds.). *The GOES/SMS User's Guide*. NOAA, NESS, and NASA, Washington, DC, 1980.

Curran, P. J., and H. D. Williamson. Sample size for ground and remotely sensed data. *Remote Sens. Environ.* **20,** 31–41, 1986.

Curtis, J. D., V. S. Frost, and L. F. Dellwig. Geological mapping potential of computer enhancement images from the Shuttle Imaging Radar: Lisbon Valley Anticline, Utah. *Photogramm. Eng. Remote Sens.*, **52**(4), 525–532, 1986.

Daily, M. I., T. Farr, C. Elachi, and G. Schaber. Geologic interpretation from composited radar and Landsat imagery. *Photogramm. Eng. Remote Sens.*, **45**(8), 1109–1116, 1979.

Darch, J. P. A study of moisture conditions in the pantanal of Brazil using satellite imagery. *Remote Sens. Environ.*, **8,** 331–348, 1979.

Dennis, R. E., F. G. Everdale, D. F. Johnson, I. C. Sheifer, R. P. Stump, and K. W. Turgeon. *Application of Satellite Imagery to West African Fisheries*. NESDIS/NOAA, U.S. Dept. of Commerce, Washington, DC, 1986.

De Rycke, R. J., and P. K. Rao. Eddies along a Gulf Stream boundary viewed from a very high resolution radiometer. *J. Phys. Oceanogr.*, **3,** 490–492, 1973.

Deutsch, M. Optical processing of ERTS data for determining extent of the 1973 Mississippi River Flood. In R. S. Williams, Jr. and W. D. Carter (Eds.), *ERTS 1: A New Window on Our Planet*. U.S. Government Printing Office, Washington, DC, 1976, pp. 209–213.

Dietrich, G. *Allgemeine Meereskunde*. Borntraeger, Berlin, 1957.

Drennan, K. L. Some potential applications of remote sensing in fisheries. In *Proceedings of the Symposium on Remote Sensing in Marine Biology and Fishery Resources*. Texas A & M University, 1971, pp. 66–74.

Duncan, C. P. An eddy in the subtropical convergence southwest of South Africa. *J. Geophys. Res.* **73,** 531–534, 1968.

Edgerton, A. T., and D. T. Trexler. *Radiometric Detection of Oil Slicks,* Contract No. DOT-C-G-93,288A. U.S. Coast Guard Applied Technology Division, 1970.

Elachi, C. *Physics and Techniques of Remote Sensing*. Wiley, New York, 1987.

Elachi, C., W. E. Brown, J. B. Cimino, T. Dixon, D. L. Evans, J. P. Ford, R. S. Saunders, C. Breed, H. Masursky, J. F. McCauley, G. Schaber, L. Dellwig, A. England, H. MacDonald, P. Martin-Kaye and F. F. Sabins. Shuttle imaging radar. *Science,* 996–1003, 1982.

Elliott, B. A. Anticyclonic rings and the energetics of the circulation of the Gulf of Mexico. Ph.D. Thesis, Dept. of Oceanography, Texas A & M University, 1979.

Evans, D. L., T. G. Farr, J. P. Ford, T. W. Thompson, and C. L. Werner. Multipolarization radar images for geologic mapping and vegetation discrimination. *IEEE Trans. Geosci. Remote Sens.,* **GE-24**(2), 246–256, 1986.

Evans, R. H., K. S. Baker, O. B. Brown, and R. C. Smith. Chronology of warm-core ring 82B. *J. Geophys. Res.,* **90,** 8803–8811, 1985.

Ewing, G. C. Remote spectrography of ocean color as an index of biological activity. In *Proceedings of the Symposium on Remote Sensing in Marine Biology and Fishery Resources*. Texas A & M University, 1971, pp. 66–74.

Eyton, J. R. Landsat multitemporal color composites. *Photogramm. Eng. Remote Sens.,* **49,** 231–235, 1983.

Feldman, G. C. Satellite color observations of the phytoplankton distribution in the eastern equatorial Pacific during the 1982–1983 El Niño. *Science,* **224**(4678), 1069–1071, 1984.

Feldman, G. C. Variability of the productive habitat in the eastern equatorial Pacific. *Eos,* **67**(9), 106–108, 1986.

Ferns, D. C., S. J. Zara, and J. Barber. Application of high resolution spectroradiometry to vegetation. *Photogramm. Eng. Remote Sens.,* **50**(12), 1725–1735, 1984.

Ford, J. P., and D. E. Wickland. Forest discrimination with multipolarization imaging radar. *Int. Geosci. Remote Sens. Symp. Proc.* (*IGARSS 1985*), Vol. 1, pp. 462–465, 1985.

Ford, J. P., R. G. Blom, M. L. Bryan, M. I. Daily, T. H. Dixon, E. Elachi, and E. C. Xenos. Seasat views North America, the Caribbean, and Western Europe with imaging radar. *JPL Publ.,* **80-67,** 1–141, 1980.

Foster, J. L., D. K. Hall, A. T. C. Chang, A. Rango, L. J. Allison, and B. C. Diesen, III. The influence of snow depth and surface air temperature on satellite derived microwave brightness temperature. *NASA Tech. Memo.,* **NASA TM-80695,** 1980.

Frey, H. Magsat scalar anomaly distribution: The global perspective. *Geophys. Res. Lett.,* **9**(4), 277–280, 1982.

Fuglister, F. C. Cyclonic rings formed by the Gulf Stream, 1965–66. In A. Gordon (Ed.), *Studies in Physical Oceanography*. Gordon & Breach, New York, 1971, pp. 137–168.

Fung, A. K. Scattering from a vegetation layer. *IEEE Trans. Geosci. Electron.,* **17,** 1, 1979.

Gausman, H. W., W. A. Allen, R. Cardenas, and A. J. Richardson. Relation of light reflectance to cotton leaf maturity (*Gossypium hirsutum L.*). *Proc. Int. Symp. Remote Sens. Environ., 6th, 1969,* Vol. 2, 1969, pp. 1123–1141.

Gausman, H. W., W. A. Allen, R. Cardenas, and A. J. Richardson. Relation of light reflectance to histological and physical evaluations of cotton leaf maturity (*Gossypium hisutum L.*). *Appl. Opt.,* **9,** 545–552, 1970.

Gloersen, P., W. Nordberg, T. J. Schmugge, T. T. Wilheit, and W. J. Campbell. Microwave signatures of first-year and multiyear sea ice. *J. Geophys. Res.,* **78**(18), 3564–3572, 1973.

Gloersen, P., T. T. Wilheit, T. C. Chang, and W. Nordberg. Microwave maps of the polar ice of the earth. *Bull. Am. Meteorol. Soc.,* **55,** 1442–1448, 1974.

Gonzalez, F. I., R. C. Beal, W. E. Brown, J. F. R. Gower, D. Lichy, D. B. Ross, C. L.

Rufenach, and R. A. Shuchman. Seasat synthetic aperture radar ocean wave detection capabilities. *Science,* **204,** 1418–1421, 1979.

Gordon, H. R., D. K. Clark, J. L. Mueller, and W. A. Hovis. Phytoplankton pigments from the Nimbus-7 Coastal Zone Color Scanner: Comparisons with surface measurements. *Science,* **210,** 63–66, 1980.

Gordon, H. R., D. K. Clark, J. W. Brown, O. B. Brown, R. H. Evans, and W. W. Broenkow. Phytoplankton pigment concentrations in the Middle Atlantic Bight: Comparison of ship determination and CZCS estimates. *Appl. Opt.,* **22,** 20–36, 1983.

Graetz, R. D., and M. R. Gentle. The relationships between reflectance in the Landsat wavebands and the composition of an Australian semi-arid shrub rangeland. *Photogramm. Eng. Remote Sens.,* **48**(11), 1721–1730, 1982.

Green, G. M. Use of SIR A and Landsat MSS data in mapping shrub and intershrub vegetation at Koonamore, South Australia. *Photogramm. Eng. Remote Sens.,* **52**(5), 659–670, 1986.

Grew, G. Remote detection of water pollution with MOCS: An imaging multispectral scanner. In *Proceedings of the Second Conference on Environmental Quality Sensors.* National Environment Research Center, Las Vegas, NV, 1973, pp. 18–38.

Guha, P. K., and S. B. Mallick. Digital MSS and spectral reflectance data in lithologic discrimination. *ITC J.,* pp. 42–46, 1985.

Hallikainen, M. T., and P. A. Jolma. Retrieval of the water equivalent of snow cover in Finland by satellite microwave radiometry. *IEEE Trans. Geosci. Remote Sens.* **GE-24**(6), 855–862, 1986.

Harris, T. F. W., R. Legeckis, and D. Van Forest. Satellite infrared images in the Agulhas Current System. *Deep-Sea Res.,* **25,** 543–548, 1978.

Hay, C. M. Remote sensing measurement technique for use in crop inventories. In C. Johansen and J. Sanders (Eds.), *Remote Sensing for Resource Management.* Soil Conservation Society of America, Ankeny, IA, 1982, pp. 420–433.

Heath, D. F., A. J. Krueger, H. R. Roeder, and B. D. Henderson. The solar backscatter ultraviolet and total ozone mapping spectrometer (SBUV/TOMS) for Nimbus G. *Opt. Eng.,* **14**(4), 323–331, 1983.

Henricksen, B. L., and J. W. Durkin. Growing period and drought early warning in Africa using satellite data. Unpublished manuscript, 1986.

Hock, J. C. Monitoring environmental resources through NOAA's polar orbiting satellites. *ITC J.,* pp. 263–268, 1984.

Hodarev, U. K., B. S. Dunaev, B. N. Rodionov, A. L. Serebryakyan, Y. M. Tchesnokov, and V. S. Etkin. Some possible uses of optical and radio-physical remote measurements for earth investigations. *Proc. Int. Symp. Remote Sens. Environ., 6th, 1971,* Vol. 1, 1971, pp. 99–118.

Hodgson, R. A. Regional linear analysis as a guide to mineral resource exploration using Landsat (ERTS) data. *Proc. 1st Annu. W. Pecora Mem. Symp.,* Sioux Falls, 1975, pp. 155–171.

Hoekmann, D. H. Radar backscattering of forest stands. *Int. J. Remote Sens.,* **6**(2), 325–343, 1985.

Hoekmann, D. H., Krul, and E. P. W. Attema. A multilayer model for radar backscattering from vegetation canopies. *Proc. Int. Geosci. Remote Sens. Symp.,* Munich, Vol. 2, No. TA-1, 1982, pp. 41–47.

Hoffer, R. M., S. E. Davidson, P. W. Mueller, and D. F. Lozano-Garcia. A comparison of X- and L-band radar data for discriminating forest cover types. In *Proceedings of*

Pecora 10: Remote Sensing in Forest and Range Resource Management. Colorado State University, Fort Collins, CO, 1985, pp. 578–592.

Hovis, W. A. The surface composition mapping radiometer (SCMR) experiment. In R. R. Sabatini (Ed.), *Nimbus 5 User's Guide.* NASA, GSFC, Washington, DC, 1972, pp. 49–58.

Hovis, W. A., D. K. Clark, F. Anderson, R. W. Austin, W. H. Wilson, E. T. Baker, D. Ball, H. R. Gordon, J. L. Mueller, S. Z. El-Sayed, B. Sturm, R. C. Wrigley, and C. S. Yentsch. Nimbus 7 Coastal Zone Color Scanner: system description and initial imagery. *Science,* **210,** 60–63, 1980.

Huerte, A. R., R. D. Jackson, and D. F. Post. Spectral response of a plant canopy with different soil backgrounds. *Remote Sens. Environ.,* **19,** 37–53, 1985.

Jackson, R. J., T. J. Schmugge, and P. O'Neill. Passive microwave remote sensing of soil moisture from an aircraft platform. *Remote Sens. Environ.,* **14,** 135–151, 1984.

Jones, W. L., L. C. Schroeder, and J. L. Mitchell. Aircraft measurements of the microwave scattering signature of the ocean. *IEEE Trans. Antennas Propag.,* **AP-25,** 1, 1977.

Jordan, R. L. The Seasat—A synthetic aperture radar system. *IEEE J. Oceanic Eng.,* **OE-5**(2), 154–164, 1980.

Justus, C. G., M. V. Paris, and J. D. Tarpley. Satellite-measured insolation in the United States, Mexico and South America. *Remote Sens. Environ.,* **20,** 57–83, 1986.

Kachhwaha, T. S. Spectral signatures obtained from Landsat digital data for forest vegetation and land-use mapping in India. *Photogramm. Eng. Remote Sens.,* **49,** 685–689, 1983.

Knepper, D. H., and G. L. Raines. Determining stretch parameters for lithologic discrimination on Landsat MSS band-ratio images. *Photogramm. Eng. Remote Sens.,* **51**(1), 63–70, 1985.

Knipling, E. B. Physical and physiological basis for the reflectance of visible and near infrared radiation from vegetation. *Remote Sens. Environ.,* **1,** 155–159, 1970.

Koopmans, B. N. Spaceborne imaging radars, present and future. *ITC J.,* pp. 223–232, 1983.

Krinsley, D. B. Lake fluctuations in the Shiraz and Nerez Playas of Iran. In R. S. Williams, Jr. and W. D. Carter (Eds.), *ERTS 1: A New Window on Our Planet.* U.S. Government Printing Office, Washington, DC, 1976, pp. 143–148.

Lancaster, E. R., T. Jennings, M. Morrissey, and R. A. Langel. Magsat vector magnetometer calibration using Magsat geomagnetic field measurements. *NASA Tech. Memo.,* **NASA TM-82046,** 1980.

Langel, R. A. The magnetic earth as seen from Magsat, initial results. *Geophys. Res. Lett.,* **9**(4), 239–242, 1982.

Langel, R. A., J. Berbert, T. Jennings, and R. Horner. Magsat data processing: A report for investigators. *NASA Tech. Memo.,* **NASA TM-82160,** 1981.

Langel, R. A., J. D. Phillips, and R. J. Horner. Initial scalar anomaly map from MAGSAT. *Geophys. Res. Lett.,* **9**(4), 1982a.

Langel, R. A., G. Ousley, J. Berbert, J. Murphy, and M. Settle. The Magsat mission. *Geophys. Res. Lett.,* **9,** 243–245, 1982b.

Lathrop, R. G., and T. M. Lillesand. Use of thematic mapper data to assess water quality in Green Bay and Central Lake Michigan. *Photogramm. Eng. Remote Sens.,* **52**(5), 671–680, 1986.

La Violette, P. E., and J. M. Hubertz. Surface circulation patterns off the east coast of Greenland as deduced from satellite photographs of ice floes. *Geophys. Res. Lett.,* **2**(9), 400–402, 1975.

Lawrence, R. D., and J. H. Herzog. Geology and forestry classification from ERTS-1 digital data. *Photogramm. Eng. Remote Sens.,* **41,** 1241–1251, 1975.

Lee, K., D. H. Knepper, and D. L. Sawatzky. Geologic information from satellite images. In F. Shahrokhi (Ed.), *Remote Sensing of Earth Resources,* Vol. III. Tennessee Space Institute, University of Tennessee, Tullahoma, 1974, pp. 411–447.

Leese, J., W. Pichel, B. Goddard, and R. Brower. An experimental model for automated detection, measurement and quality control of sea surface temperatures from ITOS-SR data. *Proc. Int. Symp. Remote Sens. Environ., 7th, 1971,* 1971, pp. 625–646.

Legeckis, R. V. Application of synchronous meteorological satellite data to the study of time dependent sea surface temperature changes along the boundary of the Gulf Stream. *Geophys. Res. Lett.,* **2**(10), 435–438, 1975.

Legeckis, R. V. Satellite observations of the influence of bottom topography on the seaward deflection of the Gulf Stream off Charleston, South Carolina. *J. Phys. Oceanogr.,* **9**(3), 1979.

Logan, T. L., and A. H. Strahler. Optimal Landsat forest transforms. *Machine Processing of Remotely Sensed Data Symposium,* 1983.

Lyon, R. J. P., and E. Z. Burns. Infrared analysis of the lunar surface from an orbiting spacecraft. *Proc. Int. Symp. Remote Sens. Environ., 2nd, 1962,* 1963, pp. 309–328.

Lyon, R. J. P., and J. W. Patterson. Infrared spectral signatures—a field geological tool. *Proc. Int. Symp. Remote Sens. Environ., 4th, 1966,* 1966, pp. 215–330.

Malan, O. G. How to use transparent diazo colour film for interpretation of Landsat images. *COSPAR Tech. Manual Ser.,* Manual No. 6, 1976.

Markham, B. L., D. S. Kimes, and C. J. Tucker. Temporal spectral response of a corn canopy. *Photogramm. Eng. Remote Sens.,* **28,** 1599–1605, 1981.

Marsh, J. G., A. C. Brenner, B. D. Beckley, and T. V. Martin. Global mean sea surface based upon the Seasat Altimeter data. *J. Geophys. Res.,* **91**(No.B.3), 3501–3506, 1986.

Martin, C. D., and B. Liley. *ERTS Cloud Cover Study,* Final Report, Contract NAS 5-11231. Allied Research Associates, Concord, MA, 1971.

Martin, F. L., and V. V. Salomonson. Statistical characteristics of subtropical jet stream features in terms of MRIR observations from Nimbus 2. *J. Appl. Meteorol.,* **9,** 508–520, 1970.

Matson, M., and D. R. Wiesnet. New data base for climate studies. *Nature (London),* **289**(5797), 451–456, 1981.

Maul, G. A., and H. R. Gordon. On the use of the Earth Resources Technology Satellite in optical oceanography. *Remote Sens. Environ.,* **4,** 95–128, 1975.

Mayer, K. E. Identification of conifer species groupings from Landsat digital classifications. *Photogramm. Eng. Remote Sens.,* **48,** 1607–1614, 1981.

McClain, E. P., W. G. Pichel, and C. C. Walton. Comparative performance of AVHRR-based multichannel sea surface temperatures. *J. Geophys. Res.,* **90**(C6), 11,587–11,601, 1985.

McClellan, W. D., J. P. Meiners, and D. G. Orr. Spectral reflectance studies on plants. *Proc. Int. Symp. Remote Sens. Environ., 2nd, 1962,* 1963, pp. 403–413.

McCulloch, A. W. The temperature humidity infrared radiometer. In R. R. Sabatini (Ed.), *Nimbus 5 User's Guide.* NASA, GSFC, Washington, DC, 1972, pp. 11–47.

McEwen, R. B., W. L. Rosco, and V. Carter. Coastal wetland mapping. *Photogramm. Eng. Remote Sens.,* **42**(2), 221–232, 1976.

McFarland, M. J., and B. J. Blanchard. Temporal correlation of antecedent precipitation with Nimbus 5 ESMR brightness temperatures. In *Preprints of Second Hydrometeorol-*

ogy Conference, Toronto, Canada. Am. Meteorol. Soc., Boston, MA, 1977, pp. 311–315.

McGinnis, D. F., J. A. Pritchard, and D. R. Wiesnet. Snow depth and snow extent using VHRR data from the NOAA-2 satellite. *NOAA Tech. Memo.,* **NESS 63,** 1975.

McMillin, L. M., and D. S. Crosby. Theory and validation of the multiple window sea surface temperature technique. *J. Geophys. Res.,* **89**(C3), 3655–3661, 1984.

Meier, M. F. Monitoring the motion of surging glaciers in the Mount McKinley Massif. In *Symposium on Significant Results Obtained from the Earth Resources Technology Satellite—1,* Vol. 1, Sect. A. NASA, Washington, DC, 1973a, pp. 185–187.

Meier, M. F. Evaluation of ERTS imagery for mapping and detection of changes of snow-cover on land and on glaciers. In *Symposium on Significant Results Obtained from the Earth Resources Technology Satellite—1,* Vol. 1, Sect. A. NASA, Washington, DC, 1973b, pp. 863–875.

Menenti, M., A. Lorkeers, and M. Vissers. An application of thematic mapper data in Tunisia—estimation of daily amplitude in near surface soil temperature and discrimination of hypersaline soils. *ITC J.,* 1986.

Miller, L. D., and R. L. Pearson. Areal mapping program of the IBP grassland biome: Remote sensing of the productivity of the shortgrass prairie as input into biosystem models. *Proc. Int. Symp. Remote Sens. Environ., 7th, 1971,* Vol. 1, 1971, pp. 175–205.

Moore, G. K. Prospecting for ground water with Landsat images. Paper presented at the Technical Session on Hydrologic Applications of Remote Sensors, World Water Conference, Mar de Plata, 1977.

Moore, G. K. *Ground Water Applications of Remote Sensing,* Open File Rep. 82-240. U.S. Dept. of Interior, EROS Data Center, 1982.

Moore, G. K., and F. A. Waltz. Objective procedures for lineament enhancement and extraction. *Photogramm. Eng. Remote Sens.,* **49**(5), 641–647, 1983.

Morse, A., and D. H. Card. Benchmark data on the separability among crops in the southern San Joaquin Valley of California. *Proc. Int. Symp. Remote Sens. Environ., 17th, 1983,* 1983, pp. 907–914.

Morse, B. J. Spatial and temporal distribution of Northern Hemisphere snow cover. *NOAA Tech. Rep.,* **NESDIS 6,** 1983.

Mourad, A. G., S. Gopalapillai, and M. Kuhner. The significance of the Skylab Altimeter experiment results and potential applications. *NASA Tech. Memo.,* **NASA TM-X-58168,** 1887–1909, 1975.

Mueller, J. L. Remote measurement of chlorophyll concentration and secchi-depth using principal components of the ocean's color spectrum. In *Fourth Annual Earth Resources Program Review,* Vol. 4. NASA Manned Spacecraft Center, Houston, TX, 1972, pp. 105-1–105-13.

Mueller, P. W., R. M. Hoffer, and D. F. Lozano-Garcia. Interpretation of forest cover on microwave and optical satellite imagery. In *Proceedings of Pecora 10: Remote Sensing in Forest and Range Resource Management.* Colorado State University, Fort Collins, CO, 1985.

Mulders, M. A., and G. F. Epema. The Thematic Mapper: A new tool for soil mapping in arid areas. *ITC J.,* 1986.

Murai, S., and S. Yanagida. Spectral separability of Thematic Mapper imagery in agricultural area. In *Proceedings of the Third Asian Agricultural Remote Sensing Symposium.* Tokai University, ESCAP/Thailand Remote Sensing Centre, 1984, pp. 39–59.

Nace, R. L. Are we running out of water? *Geol. Surv. Circ.* (*U.S.*), **536,** 1967.

NASA Geodynamics Program. An overview. *NASA Tech. Pap.* **2147,** 1983.

Noble, V. E. Multispectral remote sensing for monitoring of marine pollution. In *Marine Pollution and Sea Life.* FAO, Rome, 1972, pp. 500–505.

Nordberg, W. A., W. McCulloch, L. L. Foshee, and W. R. Bandeen, Preliminary results from Nimbus 2. *Bull. Am. Meteorol. Soc.,* **11,** 857–872, 1966.

Nordberg, W. A., J. Conaway, D. Ross, and T. Wilheit. Measurements of microwave emission from foam covered, wind driven sea. *J. Atmos. Sci.,* **28,** 429, 1971.

O'Brien, H. W., and R. H. Munis. Red and near infrared spectral reflectance of snow. *Cold Reg. Res. Eng. Lab. Rep.,* **322,** 1975.

Ormsby, J. P., B. J. Blanchard, and A. J. Blanchard. Detection of lowland flooding using active microwave systems. *Photogramm. Eng. Remote Sens.,* **51**(3), 317–328, 1985.

Pampaloni, P., and S. Paloscia. Experimental relationships between microwave emission and vegetation features. *Int. J. Remote Sens.,* **6**(2), 315–323, 1985.

Paris, J. F. On the use of L-band multipolarization airborne SAR for survey of crops, vineyards and orchards in a California irrigated agricultural region. *JPL Publ.,* **85-39,** 63, 1985.

Parker, C. E. Gulf Stream rings in the Sargasso Sea. *Deep-Sea Res.,* **18,** 981–993, 1971.

Parks, W. L., and R. E. Bodenheimer. Delineation of major soil associations using ERTS-1 imagery. In *Symposium on Significant Results Obtained from the Earth Resources Technology Satellite—1,* Vol. 1, Sect. A. NASA, Washington, DC, 1973, pp. 121–125.

Pfeiffer, H. J., R. A. Lindhurst, and J. L. Gallagher. *American Society of Photogrammetry,* Part II. ASP, Lake Buena Vista, FL, 1973, pp. 1004–1016.

Piech, K. R., and J. E. Walker. Interpretation of soils. *Photogramm. Eng. Remote Sens.,* 87–94, 1974.

Podwysocki, M. H., F. J. Gunther, and H. W. Blodget. *Discrimination of Rock and Soil Types by Digital Analysis of Landsat Data,* Doc. X-923-77-17. NASA, GSFC, Greenbelt, MD, 1977.

Polcyn, F. C. Applications of multispectral sensing to marine resources surveys. In *Proceedings of the Symposium on Remote Sensing in Marine Biology and Fishery Resources.* Texas A & M University, 1971, pp. 194–217.

Polcyn, F. C. *NASA/Cousteau Ocean Bathymetry Experiment—Remote Bathymetry Using High Gain Landsat Data,* Tech. Rep. NAS 5-22597. NASA/GSFC, Washington, DC, 1976.

Post, A., M. F. Meier, and L. R. Mayo. Measuring the motion of the Lowell and Tweedsmuir surging glaciers of British Colombia, Canada. In R. S. Williams, Jr. and W. D. Carter (Eds.), *ERTS-1: A New Window on Our Planet.* U.S. Government Printing Office, Washington, DC, 1976, pp. 180–184.

Pouquet, J. Geopedological features derived from satellite measurements in the 3.4–4.2 μm and 0.7–1.3 μm spectral regions. *Proc. Int. Symp. Remote Sens. Environ., 6th, 1969,* Vol. 2, 1969, pp. 967–988.

Pratt, D. A., and C. D. Ellyett. The thermal inertia approach to mapping of soil moisture and geology. *Remote Sens. Environ.,* **8,** 151–168, 1979.

Pyle, J. A., and J. C. Farman. Antarctic chemistry to blame. *Nature (London),* **329,** 103–104, 1987.

Raney, R. K. Selected features of the Seasat satellite. In K.-H. Szekielda (Ed.), *Satellite Remote Sensing for Resources Development.* Graham & Trotman, London, 1986, pp. 99–114.

Rao, V. R., E. J. Brach, and A. R. Mack. Crop discriminability in the visible and near infrared regions. *Photogramm. Eng. Remote Sens.,* **44**(9), 1179–1184, 1978.

Raschke, E., and W. R. Bandeen. A quasi-global analysis of tropospheric water vapor content from Tiros 4 radiation data. *J. Appl. Meteorol.,* **6,** 468–481, 1967.

Reichert, P. Remote sensing for agricultural resources survey. In *Proceedings of the Third Asian Agricultural Remote Sensing Symposium.* Tokai University, ESCAP/Thailand Remote Sensing Centre, 1984, pp. 33–37.

Reichert, P. Digital image processing. In *Regional Training on Digital Image Interpretation for Forestry Application.* ESCAP/NRMC, 1985, pp. 121–152.

Richardson, A. J., C. L. Wiegand, R. L. Torline, and M. R. Gautreaux. Landsat agricultural land-use survey. *Photogramm. Eng. Remote Sens.,* **43**(2), 207–216, 1977.

Ritchie, J. C., J. R. McHenry, F. R. Schiebe, and R. B. Wilson. The relationship of reflected solar radiation and the concentration of sediment in the surface water of reservoirs. In F. Sharokhi (Ed.), *Remote Sensing of Earth Resources,* Vol. III. Tennessee Space Institute, University of Tennessee, Tullahoma, 1974.

Robinson, I. S., N. C. Wells, and H. Charnock. The sea surface thermal boundary layer and its relevance to the measurement of sea surface temperature by airborne and spaceborne radiometers. *Int. J. Remote Sens.,* **5,** 19–45, 1984.

Rohde, W. G., J. K. Lo, and R. A. Pohl. EROS data center Landsat digital enhancement techniques and imagery availability. In *Proceedings, Remote Sensing Science and Technology Symposium, 1977.* Canadian Journal of Remote Sensing, Ottawa, Canada, 1978.

Rouse, L. J., and J. M. Coleman. Circulation observations in the Louisiana Bight using Landsat imagery. *Remote Sens. Environ.,* **5,** 55–66, 1976.

Ryerson, R. A., P. Mosher, V. R. Wallen, and N. E. Stewart. Three tests of agricultural remote sensing for crop inventory in eastern Canada: Results, problems and prospects. *Can. J. Remote Sens.,* **5**(1), 53–66, 1979.

Ryerson, R. A., R. J. Brown, J. L. Tambay, L. A. Murphy, and B. McLaughlin. A timely and accurate potato estimate from Landsat: Results of a demonstration. *Proc. Int. Symp. Remote Sens. Environ., 15th,* 1981, pp. 587–597.

Sader, S. S. Forest biomass, canopy structure, and species composition relationships with multipolarization L-band Synthetic Aperture Radar data. *Photogramm. Eng. Remote Sens.,* **53**(2), 193–202, 1987.

Scherz, J. P. Lake water quality mapping from Landsat. Paper presented at the Eleventh International Symposium on Remote Sensing of the Environment, Ann Arbor, MI, 1977.

Schmugge, T. J. Microwave approaches in hydrology. *Photogramm. Eng. Remote Sens.,* **46**(4), 495–507, 1980.

Schmugge, T. J., T. Wilheit, W. Webster, and P. Gloersen. Remote sensing of soil mosture with microwave radiometers. *NASA Tech. Note,* **NASA TN-D-8321,** 34, 1976.

Schmugge, T. J., J. M. Meneely, A. Rango, and R. Neff. Satellite microwave observations of soil moisture variations. *Water Resour. Bull.,* **13**(2), 265–281, 1977.

Schneider, S. R., D. F. McGinnis, and J. A. Gatlin. Use of NOAA/AVHRR visible and near-infrared data for land remote sensing. *NOAA Tech. Rep.,* **NESS 84,** 1981.

Schroeder, M. Spacelab metric camera experiments. In K.-H. Szekielda (Ed.), *Satellite Remote Sensing for Resources Development.* Graham & Trotman, London, 1986, pp. 81–92.

Scofield, R. A., and V. J. Oliver. A scheme for estimating convective rainfall from satellite imagery. *NOAA Tech. Memo.* **NESS 86,** 1977.

Seliger, H. H., and W. D. McElroy. *Light: Physical and Biological Action.* Academic Press, New York, 1965.

Sevchenko, N. A., and V. A. Florinskaya. The reflection and transmisison spectra of various modifications of silicon dioxide in the wavelength region 7–24 microns. *Dokl. Akad. Nauk SSSR,* **109,** 1115–1118, 1956.

Shen, H., J. Li, and H. Zhu. Transport of the suspended sediments in the Changjiang Estuary. In *Sedimentation on the Continental Shelf with Special Reference to the East China Sea.* China Ocean Press, 1983, pp. 389–399.

Short, K., and D. Linder. West Coast ocean features. *Ocean Mon.,* p. 15, 1983.

Short, N. M. *The Landsat Tutorial Workbook, Basics of Satellite Remote Sensing.* 1982.

Short, N. M., and P. D. Lowman. *Earth Observations from Space: Outlook for the Geological Sciences,* Doc. X-650-73-316. NASA, GSFC, Greenbelt, MD, 1973.

Short, N. M., and N. H. MacLeod. *Analysis of Multispectral Images Simulating ERTS Observations,* Doc. X-430-72-118. NASA/GSFC, Greenbelt, MD, 1972.

Short, N. M., and L. M. Stuart, Jr. *The Heat Capacity Mapping Mission (HCMM) Anthology.* NASA, Washington, DC, 1982.

Short, N. M., P. D. Lowman, and S. C. Freden. *Mission to Earth: Landsat Views the World.* NASA [*Spec. Publ.*] *SP,* **NASA SP-360,** 1976.

Sieber, A. J. Statistical analysis of SAR images. *Int. J. Remote Sens.,* **6**(9), 1555–1572, 1985.

Singer, R. B. *Near-infrared Spectral Reflectance of Mineral Mixtures: Systematic Combinations of Pyroxenes, Olivine and Iron Oxides,* PSD Publ. No. 258. MIT, Cambridge, MA, 1980.

Smith, W. L., and R. Koffler. National Environmental Satellite Service, NOAA, U.S. Dept. of Commerce, Washington, DC, 1970 (private communication).

Smith, W. L., P. K. Rao, R. Koffler, and W. R. Curtis. The determination of sea surface temperature from satellite high resolution infrared window radiation measurements. *Mon. Weather Rev.,* **98,** 604–611, 1970.

Smith, W. L., W. P. Bishop, V. F. Dvorak, C. M. Hayden, J. H. McElroy, F. R. Mosher, V. J. Oliver, J. F. Purdom, and D. Q. Wark. The meteorological satellite: Overview of 25 years of operation. *Science,* **231,** 455–462, 1986.

Spooner, D. L. Spectral reflectance of aquarium plants and other natural underwater materials. *Proc. Int. Symp. Remote Sens. Environ., 6th, 1969,* Vol. 2, 1969, pp. 1003–1016.

Steiner, D., and T. Gutermann. *Russian Data on Spectral Reflectance of Vegetation, Soil and Rock Types,* Final Technical Report. U.S. Army European Research Office, 1966.

Steranka, J., L. J. Allison, and V. V. Salomonson. Application of Nimbus 4 THIR 6.7 μm observations to regional and global moisture and wind field analyses. *J. Appl. Meteorol.,* **12,** 386–395, 1973.

Stoerz, G. E., and W. D. Carter. Hydrogeology of closed basins and deserts of South America, ERTS-1 interpretations. In *Symposium on Significant Results Obtained from the Earth Resources Technology Satellite—1,* Vol. 1, Sect. A. NASA, Washington, DC, 1973, pp. 695–705.

Strogonov, B. P. *Physiological Basis of Salt Tolerance of Plants.* Translated by Israel Program for Scientific Translations, Daniel Davey & Co., New York, 1962.

Strong, A. E. Great Lakes temperature maps by satellite (IFYGL). *Proc.—Conf. Great Lakes Res.,* **17,** 321–333, 1974.

Strong, A. E. Chemical whitings in the Great Lakes as viewed by Landsat. Paper presented at the 11th Annual AWRA Meeting on Aerospace Technology Applications for Environmental Hydrology, Baton Rouge, 1975.

Strong, A. E., and R. J. De Rycke. Ocean current monitoring employing a new satellite sensing technique. *Science,* **182,** 482–484, 1973.

Strong, A. E., and E. P. McClain. Improved ocean surface temperatures from space—Comparisons with drifting buoys. *Bull. Am. Meteorol. Soc.,* **85,** 138–142, 1984.

Stumpf, H. G., and P. K. Rao. Evaluation of Gulf Stream eddies as seen in satellite infrared imagery. *J. Phys. Oceanogr.,* **5,** 388–393, 1975.

Sydor, M. Turbidity in Lake Superior. In R. S. Williams, Jr. and W. D. Carter (Eds.), *ERTS-1/A New Window on Our Planet.* U.S. Government Printing Office, Washington, DC, 1976, pp. 153–156.

Szekielda, K.-H. Satellite investigations on the dynamics of the subtropical and antarctic convergence zone. *Dtsch Hydrogr. Z.,* **36,** 25–43, 1983.

Szekielda, K.-H., and J. T. Duvall. Pattern recognition of suspended material. *J. Cons., Cons. Int. Explor. Mer,* **36,** 205–216, 1976.

Szekielda, K.-H., and D. McGinnis. Investigations on eutrophication of coastal regions. Part IV. The Changjiang River and the Huanghai Sea. *SCOPE/UNEP Sonderb.,* No. 58, pp. 49–84, 1985.

Szekielda, K.-H., D. Suszkowski, and P. Tabor. Skylab investigations of the upwelling off the northwest coast of Africa. *J. Cons., Cons. Int. Explor. Mer,* **37,** 205–213, 1977.

Szekielda, K.-H., D. McGinnis, and R. Gird. Investigations with satellites on eutrophication of coastal regions. Part II. *SCOPE/UNEP Sonderb.,* No. 55, pp. 55–84, 1983.

Talbot, S. S., and C. J. Markon. Vegetation mapping of Nowitna National Wildlife Refuge, Alaska, using Landsat MSS digital data. *Photogramm. Eng. Remote Sens.,* **52**(6), 791–799, 1986.

Tapley, B. D., G. H. Born, H. H. Hagar, J. Lorell, M. E. Parke, J. M. Diamante, B. C. Douglas, C. C. Good, R. Kolenkiewicz, J. G. Marsh, C. F. Martin, S. L. Smith, III, W. F. Townsend, J. A. Whitehead, H. M. Byrne, L. S. Fedor, D. C. Hammond, and N. M. Mognard. Seasat altimeter calibration: Initial results. *Science,* **204,** 1410–1412, 1979.

Taranik, J. V. Characteristics of the Landsat system multispectral data system. *Geol. Surv. Open-File Rep. (U.S.)* **78-187,** 1978.

Taranik, J. V., and C. M. Trautwein. Integration of geological remote-sensing techniques in subsurface analysis. *Geol. Surv. Open-File Rep. (U.S.)* **76-402,** 1976.

Taranik, J. V., G. K. Moore, and C. A. Sheehan. *A Workshop Exercise on Targeting Ground-water Exploration in South-Central Arizona Using Landsat Imagery.* U.S. Dept. of Interior Geological Survey, Washington, DC, 1976.

Tarpley, J. D., S. R. Schneider, and R. L. Money. *Global Vegetation Indices from the NOAA-7 Meteorological Satellite,* NOAA Publ. NOAA, Washington, DC, 1982.

Thompson, D. R., and O. A. Wehman. Using *Lansat* digital data to detect moisture stress. *Photogram. Eng. Remote Sens. Environ.* **45** 201–207, (1979).

Thompson, J. D., G. H. Born, and G. A. Maul. Collinear-track altimetry in the Gulf of Mexico from Seasat: Measurements, models, and surface truth. *J. Geophys. Res.,* **88**(No. C3), 1625–1636, 1983.

Thompson, T. W., and A. Laderman. *Seasat—A Synthetic Aperture Radar: Radar System Implementations.* Jet Propulsion Laboratory, California Institute of Technology, Pasadena, 1976.

Tinney, L. R., J. R. Jensen, and J. E. Estes. Urban land use: Remote sensing of ground basin permeability. *Symposium NASA Tech. Memo.,* **NASA TM-X-58168,** 2199–2232, 1975.

Tomiyesu, K. Remote sensing of the earth by microwaves. *Proc. IEEE,* **62,** 86–92, 1974.

Tucker, C. J. Asymptotic nature of grass canopy spectral reflectance. *Appl. Opt.,* **16,** 1151–1157, 1977a.

Tucker, C. J. Spectral estimation of grass canopy variables. *Remote Sens. Environ.,* **6,** 11–26, 1977b.

Tucker, C. J. An evaluation of the first four Landsat Thematic Mapper reflective sensors for monitoring vegetation: A comparison with other satellite sensor systems. *NASA Tech. Memo.,* **NASA TM-79617,** 1978.

Tucker, C. J. Red and photographic infrared linear combinations for monitoring vegetation. *Remote Sens. Environ.,* **8,** 127–150, 1979.

Tucker, C. J. Nondestructive estimation of green-leaf biomass. *Grass Forage Sci.,* **35,** 177–182, 1980a.

Tucker, C. J. Remote sensing of leaf water content in the near infrared. *Remote Sens. Environ.,* **10,** 23–32, 1980b.

Tucker, C. J., and E. L. Maxwell. Sensor design for monitoring vegetation canopies. *Photogramm. Eng. Remote Sens.,* **24,** 1399–1410, 1976.

Tucker, C. J., B. N. Holben, J. H. Elgin, Jr., and J. E. McMurtrey. Remote sensing of total dry matter accumulation in winter wheat. *Remote Sens. Environ.,* **11,** 171–189, 1981.

Tucker, C. J., C. Vanpraet, and E. Boerwinkel. Satellite remote sensing of total dry matter production in the Senegalese Sahel. *Remote Sens. Environ.,* **13,** 461–474, 1983.

Tucker, C. J., J. A. Gatlin, and S. R. Schneider. Monitoring vegetation in the Nile Delta with NOAA 6 and NOAA-7 AVHRR imagery. *Photogramm. Eng. Remote Sens.,* **50**(1), 53–61, 1984.

Tucker, C. J., J. R. G. Townshend, and T. E. Goff. African land-cover classification using satellite data. *Science,* **227,** 369–375, 1985.

Turgeon, K. W. Satellite remote sensing: Applications for marine fisheries. Summary paper for UN ESCAP Remote Sensing Centre, 1986.

Ulaby, F. T., and E. A. Wilson. Microwave attenuation properties of vegetation canopies. *IEEE Trans. Geosci. Remote Sens.,* **GE-23**(5), 746–753, 1985.

Umali, R. M., and M. Amaro. Forest inventory of the Phillippines using Landsat Multispectral Scanner digital data. In *Regional Training on Digital Image Interpretation for Forestry Application.* ESCAP/NRMC, 1985, pp. 67–79.

United Nations General Assembly Resolution. *Principals Relating to Remote Sensing of the Earth from Space,* A/RES/41/65. United Nations, New York, 1986.

Van Liere, W. J. Applications of multispectral imagery to water resources development planning in the Lower Mekong Basin (Khmer Republic, Laos, Thailand and Viet Nam). In *Symposium on Significant Results Obtained from the Earth Resources Technology Satellite—1,* Vol. 1, Sect. A. NASA, Washington, DC, 1973, pp. 713–741.

Venkataratman, L. Monitoring and managing soil and land resources in India using remotely sensed data. In *Proceedings of the Third Asian Agricultural Remote Sensing Symposium.* Tokai University/ESCAP/Thailand Remote Sensing Centre, 1984, pp. 287–311.

Vibulsresth, S., K. Jirapayoongchai, D. Srisaengthong, C. Silapathong, and S. Wongparn. On the use of Landsat MSS data to forest classification in Thailand. In *Regional Remote Sensing Programme Pilot Projects on Forestry in Thailand, the Philippines and Indonesia.* UNDP/ESCAP Regional Remote Sensing Programme, 1985, pp. 86–96.

Vukovich, F. M. An investigation of a cold eddy on the eastern side of the Gulf Stream using NOAA 2 and NOAA 3 satellite and ship data. *J. Phys. Oceanogr.,* **6,** 605–612, 1976.

Wacharakitti, S., P. Pradistapongs, and P. Saguantham. Application of colour and colour infrared aerial photographs for ecological research and landuse studies. In *Regional Remote Sensing Programme Pilot Projects on Forestry in Thailand, the Philippines and Indonesia*. UNDP/ESCAP Regional Remote Sensing Programme, 1985.

Wang, J. R., J. E. McMurtrey, III, E. T. Engman, T. J. Jackson, T. J. Schmugge, W. I. Gould, J. E. Fuchs, and W. S. Glazer. Radiometric measurements over bare and vegetated fields at 1.4 GHz and 5 GHz frequencies. *Remote Sens. Environ.*, **12,** 295–311, 1982a.

Wang, J. R., T. J. Schmugge, W. I. Gould, W. S. Glazer, J. E. Fuchs, and J. E. McMurtrey. A multi-frequency radiometric measurement of soil moisture content over bare and vegetated fields. *Geophys. Res. Lett.*, **9**(4), 416–419, 1982b.

Warnecke, G., L. L. J. Allison, M. McMillin, and K.-H. Szekielda. Remote sensing of ocean currents and sea surface temperature changes derived from the Nimbus 2 satellite. *J. Phys. Oceanogr.*, **1,** 45–60, 1971.

Watson, K. Geological properties of thermal infrared images. *Proc. IEEE,* **63,** 128–137, 1975.

Weisblatt, E. A. Satellite remote sensing of coastal and estuarine environments. In *First Panamerican and Third National Congress of Photogrammetry, Photointerpretation and Geodesy,* 1974.

Werbowetzki, A. Atmospheric sounding user's guide. *NOAA Tech. Rep.* **NESS 83,** 1981.

Wezernak, C. T. The use of remote sensing in limnological studies. *Proc. Int. Symp. Remote Sens. Environ., 9th, 1974,* 1974, pp. 963–980.

White, P. G. A technique for the reduction and analysis of ocean spectral data. In *Fourth Annual Earth Resources Program Review,* Vol. 4. NASA Manned Spacecraft Center, Houston, TX, 1972, pp. 103-1–103-10.

Wilheit, T. T. The electrically scanning microwave radiometer (ESMR) experiment. In R. R. Sabatini (Ed.), *Nimbus 5 User's Guide*. NASA, GSFC, Greenbelt, MD, 1972, pp. 55–105.

Wilheit, T. T., J. S. Theon, W. E. Shenk, L. J. Allison, and E. B. Rodgers. Meteorological interpretations of the images from the Nimbus 5 Electrically Scanned Microwave Radiometer. *J. Appl. Meteorol.,* **15,** 166–172, 1976.

Williams, D. L., and M. L. Stauffer. Forest insect defoliation in Pennsylvania. In D. H. Williams and L. D. Miller (Eds.), *Monitoring Forest Canopy Alteration Around the World with Digital Analysis of Landsat Imagery*. NASA, Washington, DC, 1979, pp. 23–24.

Williams, V. L., W. R. Philipson, and W. D. Philpot. Identifying vegetable crops with Landsat thematic mapper data. *Photogramm. Eng. Remote Sens.,* **53**(2), 187–191, 1987.

Woolley, J. T. Reflectance and transmittance of light by leaves. *Plant Physiol.,* **47,** 656–662, 1971.

Work, E. A., and D. S. Gilmer. Utilization of satellite data for inventorying prairie ponds and lakes. *Photogramm. Eng. Remote Sens.,* **42**(5), 685–694, 1976.

Wu, S. T. *Analysis of Results Obtained from Integration of Landsat Multispectral Scanner and Seasat Synthetic Aperture Radar Data,* Rep. No. 189. NASA-Earth Resources Laboratory, NSTL, MS, 1981.

Wu, S. T. *Analysis of Data Acquired by Shuttle Imaging Radar and Landsat-4 Thematic Mapper over Baldwin County, Alabama,* Rep. No. 228. NASA-Earth Resources Laboratory, NSTL, MS, 1984.

Yates, H. W., and J. D. Tarpley. Machine processing of remotely sensed data with special emphasis on crop inventory and monitoring. In *Proceedings of the Role of Meteorologi-*

cal Satellites in Agricultural Remote Sensing. Laboratory for Applications of Remote Sensing, Purdue University, West Lafayette, IN, 1982, pp. 23–32.

Yates, H. W., T. D. Tarpley, S. R. Schneider, D. F. McGinnis, and R. A. Scofield. The role of meteorological satellites in agricultural remote sensing. *Remote Sens. Environ.*, **14,** 219–233, 1984.

Yentsch, C. S. The influence of phytoplankton pigments on the colour of seawater. *Deep-Sea Res.*, **7,** 1–9, 1960.

Yost, E., and S. Wenderoth. The reflectance spectra of mineralized trees. *Proc. Int. Symp. Remote Sens. Environ., 7th, 1971,* Vol. 1, 1971, pp. 269–280.

Young, J. D., and R. K. Moore. *Active Microwave Measurement from Space of Sea Surface Winds*. University of Kansas Space Technology Center, Lawrence, 1976.

Zhu, M. H., S. Y. Yan, W. R. Philipson, C. C. Yen, and W. D. Philpot. Analysis of Landsat for monitoring vegetables in New York mucklands. In *Proceedings of the 49th Annual Meeting of the American Society of Photogrammetry*. ASP, Falls Church, VA, 1983, pp. 343–353.

Zwally, H. J., T. T. Wilheit, P. Gloersen, and J. L. Mueller. Characteristics of Antarctic sea ice as determined by satellite-borne microwave imagers. *COSPAR 19th Plenary Meet.,* Philadelphia, 1976.

SELECTED MONOGRAPHS AND BOOKS ON REMOTE SENSING

Barrett, E. C., and D. W. Martin. *The Use of Satellite Data in Rainfall Monitoring*. Academic Press, Orlando, FL, 1981.

Bullard, R. K., and R. W. Dixon-Gough. *Britain from Space: An Atlas of Landsat Images*. Taylor & Francis, London, 1985.

Chagas, C. and V. Canuto (Eds.). *Remote Sensing and Its Impact on Developing Countries*. Pontifica Academiac Scientiarum Scripta Varia, Rome, 1987.

Chen, H. S. *Space Remote Sensing Systems*. Academic Press, Orlando, FL, 1985.

Colwell, R. N. (Ed.-in-chief), and J. E. Estes, D. S. Simonett, G. Thorley, and F. T. Ulaby (Assoc. Eds.). *Manual of Remote Sensing,* 2nd ed., 2 vols. American Society for Photogrammetry and Remote Sensing, Falls Church, VA, 1987.

Cracknell, A., and L. Hayes (Eds.). *Remote Sensing Yearbook 1987*. Taylor & Francis, London, 1987.

Curran, P. J. *Principles of Remote Sensing,* Longman Scientific and Technical Publication, London, 1986.

Deepak, A. (Ed.). *Remote Sensing of Atmospheres and Oceans*. Academic Press, Orlando, FL, 1980.

Deutsch, M., D. R. Wiesnet, and A. Rango (Eds.). *Satellite Hydrology*. American Water Resources Association, 1981.

Eden, M. J., and J. T. Perry. *Remote Sensing and Tropical Land Management*. 1986.

Elachi, C. *Physics and Techniques of Remote Sensing*. Wiley, New York, 1987.

Henderson, F. B., III, and B. N. Rock. *Frontiers for Geological Remote Sensing,* 4th Geosat Workshop. American Society of Photogrammetry, Falls Church, VA, 1983.

Henderson-Sellers, A. (Ed.). *Satellite Sensing of a Cloudy Atmosphere*. Taylor & Frances, London, 1984.

Hopkins, P. F. (Ed.). *Extraction of Information from Remotely Sensed Images*. American Society for Photogrammetry and Remote Sensing, Falls Church, VA, 1983.

Hord, R. M. *Digital Image Processing of Remotely Sensed Data*. Academic Press, Orlando, FL, 1982.

Hord, R. M. *Remote Sensing Methods and Applications*. Wiley, New York, 1986.

Kennie, T. J. M., and M. C. Matthews. *Remote Sensing in Civil Engineering*. 1985.

Lo, C. P. *Applied Remote Sensing,* Longman Scientific and Technical Publication, London, 1986.

Proceedings of Pecora 10: Remote Sensing in Forest and Range Resource Management. American Society for Photogrammetry and Remote Sensing, Falls Church, VA, 1987.

Rabchevsky, G. A. (Ed.-in-chief), and P.-M. Adrien, U. P. Boegli, R. R. Sabatini, A. C. Correa, and F. X. Lopez (Eds.-translators). *Multilingual Dictionary of Remote Sensing and Photogrammetry*. American Society for Photogrammetry and Remote Sensing, Falls Church, VA, 1984.

Robinson, I. S. *Satellite Oceanography: An Introduction for Oceanographers and Remote Sensing Scientists*. 1985.

Sabins, F. F. *Remote Sensing Principles and Interpretation,* 2nd ed. Freeman, New York, 1986.

Saltzman, B. (Ed.). *Satellite Oceanic Remote Sensing*. Academic Press, Orlando, FL, 1985.

Schander, E. *Physical Fundamentals of Remote Sensing*. Springer-Verlag, Berlin and New York, 1986.

Schowengerdt, R. A. *Techniques for Image Processing and Classification in Remote Sensing*. Academic Press, Orlando, FL, 1983.

Scorer, R. S. *Cloud Investigation by Satellite*. 1986.

Short, N. M. and R. W. Blair, Jr. *Geomorphology from Space*. NASA, Washington, DC, 1986.

Short, N. M., P. D. Lowman, and S. C. Freden. *Mission to Earth: Landsat Views the World,* NASA SP-360. NASA, Washington, DC, 1976.

Spot Simulation Applications Handbook. Proceedings of the 1984 Spot Symposium. American Society for Photogrammetry and Remote Sensing, Falls Church, VA, 1985.

Swain, P. H., and S. M. Davis (Eds.). *Remote Sensing: The Qualitative Approach*. McGraw-Hill, New York, 1978.

Szekielda, K.-H. (Ed.). *Satellite Remote Sensing for Resources Development*. Graham & Trotman, London, 1986.

Tsang, L., J. A. Kong, and R. T. Shin. *Theory of Microwave Remote Sensing*. Wiley, New York, 1985.

Young, T. Y., and K-S. Fu. (Eds.). *Handbook of Pattern Recognition and Image Processing*. Academic Press, Orlando, FL, 1986.

INDEX